Strategic Management in Telecommunications

For a listing of recent titles in the *Artech House Telecommunications Library*,
turn to the back of this book.

Strategic Management in Telecommunications

James K. Shaw

Artech House
Boston • London

Library of Congress Cataloging-in-Publication Data
Shaw, James K.
Strategic management in telecommunications / James K. Shaw.
p. cm. — (Artech House telecommunications library)
Includes bibliographical references and index.
ISBN 1-58053-018-4 (alk. paper)
1. Telecommunication—United States—Management. 2. Telecommunication policy—United States. 3. Strategic planning—United States. I. Title. II. Series.

HE7775 .S52 2000 99-089510
384'.068'4—dc21 CIP

British Library Cataloguing in Publication Data
Shaw, James K.
Strategic management in telecommunications. — (Artech House telecommunications library)
1. United States. Telecommunications Act 1996 2. Telecommunication—United States—Management 3. Strategic planning
I. Title
384'.068

ISBN 1-58053-018-4

Cover design by Igor Valdman

685 Canton Street

Norwood, MA 02062

International Standard Book Number: 1-58053-018-4
Library of Congress Catalog Card Number: 99-089510

10 9 8 7 6 5 4 3 2 1

For Anne, Alexander, and Aaron

Contents

Acknowledgments

Several individuals were decisive in bringing this research to fruition. I am indebted to Mark Walsh at Artech House for suggesting this avenue of scholarly and professional inquiry. This book would not have materialized in its finalized form in the absence of Mr. Walsh's inspiration and cajoling. Barbara Lovenvirth eased the burden of organizing and assimilating the vast array of data embedded in this material.

I also wish to thank Dr. Edward Davis, Assistant Dean in the College of Professional Studies at the University of San Francisco, for his kindness and support in recognizing the need for state-of-the-art research in telecommunications strategic management. He is a colleague and friend whose continued interest in this area of research signified inestimable value to the author. Among colleagues never reluctant to discipline the author in his treatment of the empirical evidence was Professor Eugene Kosso, who has also served well in his role as father-in-law.

I am particularly grateful to my wife and children, Anne, Alexander, and Aaron, who proved to be predictable and satisfying constants amid the tedium and disarray that I introduced to their lives in recent months. As always, I am eternally grateful to John Shaw, Ruth Shaw, and Lee Kosso for their continuing support and love.

Preface

It may be argued that any new book can be examined from three perspectives. There are the viewpoints of the author and audience, of course. There is, too, the historical retrospective as speculatively adjudged by writer and reader at the time of publication. The former are immediate and palpable, the last ephemeral and intangible. I have been unable to escape a consideration of the last concern while writing this manuscript.

I have taken something of a journey in compiling and interpreting the data contained in these pages. This research began as a straightforward attempt to elaborate and apply the principles and tools of contemporary strategic management for telecommunications professionals. After spending many months collecting and digesting strategic data within the industry, it occurred to me that I was confronting a parallel dilemma: I was dissecting the intertwined enigmas of "strategic content" and "economic context." A book that aspired to understanding one dynamic without illuminating the other would do little to facilitate the work of strategic planners.

One can no longer effectively manage strategy without demystifying the peculiarities of the emerging domestic and global economies—and what they portend for consumers, producers, suppliers, and distributors. For reasons empirically confirmed in this volume, it is now fact that the world economy rewards and punishes firms in a way that is fundamentally transitional for the modern enterprise. The "New Economy" is more than a bromide. We can now affirm that certain traditional laws of economics are being rewritten in the digital age, that forces unleashed during the past decade have so accelerated competitive change that strategic planners can no longer afford to dismiss them. Strategic management in modern telecommunications is thus at a crossroads: practitioners of strategy must now design and adopt seminal models, tools, and concepts to accommodate a revolutionary reconfiguration of the business environment. These tools must be accessible, responsive to unanticipated contingencies, and pliable.

My motive for writing this treatment of strategic management for telecommunications professionals has its roots in a refrain that permeates the professional literature. In discussions with my own clients, colleagues, and editor, I hear repeatedly the common frustration of practitioners: What models of strategic management are appropriate for communications companies, and how do I apply them? Equally important, what methods of strategic management have been most effective in contending with today's competitive landscape, and why have firms employing them been successful? The question is deceptively simple, but cajoles. Empirical and conceptual confirmation asserts the recent chaotic pattern of strategic planning within the industry. Deregulation, shortened product life cycles, technological innovation, converging industrial affiliations, and globalization have conspired to infuse the planning process with baffling unpredictability. Dogmatic certitudes about standard planning models have given way to demands for simple, flexible, adaptable planning strategies that can be executed at a moment's notice.

The reader should be aware that I do not pretend or imply rectitude about the strategic options or methods contained herein. Instead, a modest range of alternative planning models is elaborated, together with an assessment of their strengths and limitations. A debate about the efficacy of traditional and emerging planning methods is enjoined, and a serious attempt is made to catalog projected strategic problems for the industry through 2010. This book concerns itself, however, with more than an effort to display, apply, and evaluate planning models for this vital sector of the nation's economy.

I have concluded that knowledge of strategic management methodology, in and of itself, is inadequate for today's practitioners. Strategic content must be complemented by an analysis of economic context. It has become imperative to understand the newly emerging economic contours in which the industry must necessarily design and implement its strategy. Accordingly, the reader will note the following items in the attached Appendix: an annotated, edited listing of those sections of the Telecommunications Act of 1996 likely to impinge on strategic design during the coming decade; primary government documentation on the financial future of the communications sector and its prospective opportunities for business growth; a complete listing of traditional and emerging subfields of industrial specialization guiding interaction of the telephony, broadcasting and computer industries; and macroeconomic, microeconomic, and industrial forecasts for the domestic and global sectors extending through 2010. Testimony from practitioners, industry observers, consultants, and government officials is synthesized to deduce patterns of emerging strategic management.

The intention here is to present a cutting-edge discussion of contemporary telecommunications strategic management practices. I have attempted to restrain in tone what is likely to be provocative in substance. Contentiousness regarding the selection and interpretation of economic data is conceded. Material aimed at surveying modern strategic planning practices, however, is largely indisputable and fully supported by empirical evidence. The reader is encouraged to regularly review those websites identified at the end of this book; the theory and practice of strategic modeling is changing rapidly, as is the business and economic environment in which professionals apply their craft.

Finally, I have suggested here, somewhat cryptically, that I could not escape my own conjecture about the value of this research as perceived by readers in historical retrospective. Will the judgements, inferences, and forecasts embedded in this material be supported or repudiated by the passage of time? This speculation may appear immaterial or self-indulgent to some, but I pose the question in something other than rhetorical terms. The work of the strategic planner impels him or her to make congruent the strategy of the firm with the unfolding business environment. In short, I have placed myself in the role of the strategist while writing this manuscript. I have attempted to anticipate, and respond to, the problems strategic planners are likely to encounter in the years ahead.

I earnestly hope that readers find clarification and meaning in the menu of strategic options and methods arrayed in this text. The orthodoxy of contemporary strategic management is giving way to something very different and as yet unformed for the industry. Those willing to suspend the comforting certitudes of conventional strategy in favor of an exploration of this callow transformation will profit most from the exercise.

1

Foundations of Strategic Management in Telecommunications

Strategic management, the procedure by which a firm builds, monitors, and evaluates its future direction, is now characterized as a chaotic practice by many in the telecommunications industry. In an era marked by profound and extraordinary turbulence, every industry now encounters the simultaneous problems of proliferating competition, global reconfiguration, customer-driven fastidiousness, technological change, and organizational stability [1]. For telecommunications firms, this unprecedented period, and the era that beckons, have superimposed upon them the added uncertainties of unsystematic deregulation [2].

In contemplating the profundity of the transformation associated with this era, telecommunications companies operate within a duality: they must formulate strategic plans to define their mission and objectives while simultaneously allowing for adaptation to contingencies beyond their influence, and frequently, by extension, circumstances beyond their full comprehension. Firms must strive for precision while ensuring prompt adaptation to market change. Those only casually familiar with this dilemma might conclude that melodrama pervades the ambiance of the communications strategist. Whether real or perceptual, whether concrete or imagined, we can empirically verify that the industry is groping for alternative models of strategic management; traditional methods of planning no longer suffice in an industry where the pace of change is dictated by leaps in technology which defy standard techniques of forecasting [3].

This book has as its purpose an examination of the problems confronting strategic planners in the telecommunications industry, the challenges associated with unsystematic deregulation, and the burdens and opportunities triggered by an unfolding global communications market. The text reviews and evaluates the strengths and limitations of classic and emerging planning models, and their prospective application to the communications industry. Included in this analysis is an array of alternative planning methods intended to identify and project strategic options for the industry, along with lessons culled from other industrial sectors which might enlighten this process.

The author concedes, however, that much of this elaboration represents merely an attempt to inform the strategic management process. Unfortunately, or perhaps opportunely, the industry now operates in a "paradigmatic vacuum": there exists no consensus about the direction in which the industry will unfold or the optimal procedure by which to maximize market success [4]. We are thus relegated to the uncertainty a patron at a new restaurant inevitably experiences—we have but a menu, not a certitude regarding the dish most palatable.

1.1 Notes on the Organization and Use of This Research

A knowledge of strategic content fused with economic context informs the modern strategic management process for telecommunications practitioners. A momentum is gathering which leads us in this direction, as discussed below. Accordingly, the author has divided his own survey of veteran and emerging strategic tools as follows:

- Chapters 1, 2, and 3 elaborate the theories, practices, and methods of contemporary planning models. Included in this material is an analysis of the Telecommunications Act of 1996 and what it portends for prospective competition vis-a-vis deregulation and industry restructuring. Also examined is an assessment of the economics of technological convergence and how this dynamic has influenced both the domestic and global economies relative to industries other than the communications sector. The reader is urged to examine, while surveying this material, the Appendix, Sections A, D, and E. The body of economic evidence is growing that the telecommunications industry, as defined in this volume, is the single most vital underpinning of the macroeconomy; the macroeconomic restructuring now underway will later influence the level of demand for future communications products and services.
- Chapters 4, 5, and 6 isolate the determinants of telecommunications demand, supply, and prospective production. Included is a review of networked economies (electronic commerce). The reader will note an

attempt to define the likely future path of competitive behavior among telecommunications providers, and how these data can be exploited to the advantage of strategists. An extensive listing of research websites in strategic planning, economic analysis, and market research is attached to the Appendix. The author has carefully designed this database to reflect those sources of primary information adopted by successful strategists.

- Chapters 7 and 8 discuss and evaluate the strategic management process promulgated by leading firms within the telephony, broadcasting, and computer sectors. Particular emphasis is placed on those techniques applied in negotiating and implementing strategic affiliations, alliances, partnerships, mergers, acquisitions, and venture capital formations. The reader is urged to carefully examine the data arrayed in the Appendix, Section B. Source data enumerated in Section B (abridged government documentation) provide an overview of the transcendent economic and strategic importance of information technology when united with venture capital investment. There are important tertiary ramifications to this industry for all practitioners of strategic management.
- Chapter 9 supplies data of rising importance to international strategic planners. Inserted in the Appendix, Section C (Parts 1 and 2) are global (World Trade Organization) agreements outlining the future of multinational corporate development for the industry. Encased in this material is source data concerning the volume of trade in each subcategory of communications trade. The reader is advised to link data contained within this chapter to the theory of strategic alliances discussed in the preceding chapter. Increasingly, it is apparent that American telecommunications firms represent the vanguard of strategic design to be recapitulated later in other regions of the globe.
- Chapter 10 is a speculative piece focusing on issues of pending importance: mass customization, particle marketing, strategic adaptability, networked intelligence, strategic learning, product-lifecycle management. Included, too, is an overview of expert forecasts for the coming decade, and conjecture about the nature of strategic problems projected a decade beyond.

The book, given this format, seeks an answer to three questions:

1. What is the current practice of strategic management in telecommunications and how are practitioners responding to competitive change?

2. What will individuals, households, and organizations demand in the way of future communications products?
3. How do telecommunications firms position themselves to maximize future market share, profitability, and business opportunities? Distilled to its fundamentals, the text appraises the present state of strategic management—and the direction in which we might reasonably infer it is headed.

1.2 Definition and Meaning of Strategic Management

Because strategic management is a term that is used interchangeably and in different contexts, confusion as to its objectives and real meaning often results. Strategic management is the method by which a firm clarifies its mission, pursues its goals, marshals its resources, gauges its competitive environment, and evaluates its performance in the attainment of objectives. The expression "strategic planning" is typically used in its place. Until the 1970s, many enterprises referred to the same procedure as "business policy." Strategic management today is thus the term of choice, suggesting that effective strategic plans or policies must now be measured with some precision on a continuing basis [4].

Strategic management takes on heightened significance in an era in which emerging competition is compounded by sudden, jarring advances in technological innovation. Many firms hesitate in assuming the task while nevertheless acknowledging its inevitable imposition [5]. Despite the anxiety surrounding this responsibility, we can discern the traits of those corporations which have successfully applied the methodology [6]. Successful practitioners of strategic management:

- Integrate executive, managerial, and operational staff to formulate plans;
- Link short-term objectives with broader, long-range planning horizons;
- Establish contingencies to improve on-going planning effectiveness;
- Accurately forecast, within parameters, consumer preferences;
- Account for competitor innovations, incremental and radical;
- Design "learning environments" for employees to exploit all forms of creativity;
- Anticipate and respond to "customer-driven" requirements;
- Create, rather than adapt to, product obsolescence;
- Stimulate the pursuit of market opportunities by all organizational personnel;

- Embrace, rather than avoid, risk-taking; and
- Relentlessly pursue customer satisfaction as the principal goal of planning.

1.3 Historical Overview of Strategic Management

The essence of strategic management lies in an organizational commitment to maximize current business opportunities while simultaneously pursuing long-term market prospects. It is the tension between short-term considerations and longer-term objectives which paradoxically drives the contemporary management of strategy. If there are contradictions in reconciling the rectitude of the present with the intangible demands of the future, there is certainly a symbiosis as well. Success in the present lays the foundation of preparation for the future; a credibility in forecasting the future harnesses a firm's focus and energy. Strategic management is thus an enduring requirement of every firm, and is a procedure that must be refined in response to market competition and innovation.

Classic, or traditional, strategic management focuses on the design and implementation of strategic plans. A strategic plan is characterized by the following six steps:

1. A statement of the firm's mission (or reason for existence);
2. Elaboration of goals specified to meet the company's mission;
3. An evaluation of the competitive environment of the firm;
4. An audit of the firm's internal resources (available capital, technology, and labor);
5. Identification of action steps, or tactics, intended to facilitate the company's goals;
6. Implementation of tactics to reach the firm's goals.

Essentially, the mission of the firm defines its image and purpose in the marketplace. The organization's mission does more than clarify a marketplace profile; it serves to motivate staff and focus resources on the firm's principal pursuits. The goals of a company support its mission, and thus must define a congruence between the enterprise and its environment. Monitoring the competitive environment remains the most elusive and critical role in shaping a firm's evolving strategy. It is instructive to recall that this decade has seen more competitive change—technological, global, social, political, economic, and cultural—than that of any preceding period in history. In the last ten years the world has added, for example, nearly two billion consumers and producers to its competitive marketplace [7]. Recent methods in valuing a firm's internal assets, including its intellectual capital, now undergird internal auditing procedures. Elaboration and implementation of action steps in support of company goals remain difficult tasks, but successful enterprises have overcome internal obstacles to the challenge.

As the 1970s unfolded, the legitimacy of the classic planning model—the simple six-step method illustrated above—came under increasing skepticism by certain critics [8]. While there is elegance in straightforward planning, marketplace dynamism precipitated the application of systems analysis to strategic design and management [9]. Systems analysis planning incorporates inputs, decision processes, outputs, and feedback to produce and refine strategic plans on a continuing basis. Together, these planning procedures—classic and systems—outline the core of contemporary strategic management. These procedures, in terms of their similarities and differences, can be seen in Table 1.1 below.

You will note that there exist many variations in, and labels applied to, these models by the firms which utilize them. Distilled to their fundamentals,

Table 1.1
Strategic Management Formulation: Classical and Systems Procedures

Classical Model	Step One: a statement of the firm's mission (or reason for existence).
	Step Two: Elaboration of goals specified to meet the company's mission.
	Step Three: An evaluation of the competitive environment of the firm.
	Step Four: An audit of the firm's internal resources (available capital, technology, and labor).
	Step Five: Identification of action steps, or tactics, intended to facilitate the company's goals.
	Step Six: Implementation of tactics to reach the firm's goals.
Systems Model	Step One: Define Mission and Objectives.
	Step Two: Evaluate Economic Environment.
	Step Three: Perform Competitive Analysis.
	Step Four: Perform Internal Audit.
	Step Five: Identify Strategic Choices.
	Step Six: Elaborate Policies, Plans, and Administrative Apparatus in Support of Strategy.
	Step Seven: Implement Strategic Plan.
	Step Eight: Evaluate Strategic Effectiveness.
	Step Nine: Monitor Competitive Environment.
	Step Ten: Reformulate Strategic Plans in Response to Market Changes.

however, corporations veer in one direction—classic or systems—or the other. Some firms sputter in employing either model, and it is in these inadequacies that we find the impetus for new models, new ideas, and new approaches to strategic management. Many communications firms have been unable to effectively sustain either method in the design of strategic goals. We do know, and can empirically confirm, that communications firms find that the time required to develop classic or systems strategic management often inhibits their response to market change and evolution [10]. In this regard, both methods treat time as a constant resource; in fact, the rapidity of change now renders time an enemy, not an ally for planners. For those firms able to expedite the planning function, who can speed the design, implementation, and evaluation of strategic management, a distinct market advantage obtains.

1.3.1 Classic Strategic Management

As noted, the traditional, or classical strategic management method involves six systematic steps, illustrated in Table 1.1. The first step specifies the goals of the company. The second provides for an assessment of the competitive environment. The third evaluates the internal resources of the firm. The fourth step defines the systematic tactics, or action steps, geared toward achieving the company's goals. The final functions involve implementation of the action steps intended to assure the attainment of goals.

1.3.1.1 Mission and Goals

Unless and until a firm effectively defines its mission, the empirical evidence suggests important lessons as to the specification of goals [11]. For example, if the mission of a firm is to develop state-of-the-art software for electronic commerce, goals intended to carry out this mission should be focused and concrete, but sufficiently broad to allow future flexibility. Early in this century, most railroad firms defined goals so narrowly that only the movement of goods via railroading conformed to their mission. Had they committed themselves to the broader objective of transportation, a set of goals aimed at transporting commerce through alternative methods would have mitigated the encroachment of trucking and aviation. A narrowly defined mission leads to narrowly defined goals, thus preventing adaptation to change. Thus a firm committed to developing advanced software for electronic commerce should not confine its goals to satisfying the requirements of producers, but the needs of consumers as well.

Broader statements of mission and goals allow for greater innovation and market sensitivity. Innovation is constrained by attenuated missions and supporting goals. One recalls that in the 1970s, the software firm Visicalc so narrowly defined its mission and goals that it could not effectively replace its principal product with an improved version. The Lotus Development Corporation effectively supplanted its rival by expanding the scope of its mission and goals to include a wider range of users, and thus prospective consumers. These are lessons which hold particular import for communications firms.

In publicly-traded corporations, a traditional method for bringing both focus and flexibility to mission and goals lies in a simple axiom: "To maximize shareholder value." There are both advantages and limitations to this precept, including the risk of diminishing employee morale in the face of such a one-dimensional reference point. However, in setting a mission-and-goals statement, it is useful to consider unfolding shareholder value as a device to assess the efficacy of strategic goals. If newly defined goals do not support the value of a firm, they may reflect limited, unimaginative strategic thinking [12].

While broadly stated goals provide for a greater range of strategic options and product innovation, this is not to assert that they should be vague and intangible. Goals must be clearly defined and susceptible to measurement. A goal that is unmeasurable is one which diverts rather than exploits a firm's resources. Measurable goals include growth measurement, trend extrapolation, consumer satisfaction, and percentage change in market share, sales, and profits. An attainable mission will dictate goals clearly understood and easily evaluated on a regular basis. Proponents of classic strategic management argue that it is on this basis that their approach to strategy holds greatest merit relative to other prescribed models.

1.3.1.2 Analysis of the Competitive Environment

Practitioner experience and empirical research confirms the presence of these determinants as most incisive in defining a firm's competitive environment:

(1) Economy (Domestic and Global): undergirding demand for a firm's product are current and prospective economic conditions. These include interest rates, prevailing employment, aggregate consumer demand, business investment, taxation, inflation, industrial production, fiscal policy, and monetary policy. In this category we are concerned with linking the prospects of the firm with the evolution of the business cycle, as noted below. We will later review a continuing debate among telecommunications market planners about the efficacy and value of forecasting the economic climate in relation to a firm's product array and life-cycle. One can certainly contest the methodology of economists in discerning the relevance of the economy in terms of launching a new product line. It can hardly be debated, however, that to the extent that economic conditions influence unfolding consumer demand, an enterprise runs great risk in choosing to remain oblivious to this interaction.

Increasingly, the global economy impinges on domestic consumer demand for goods and services. In data outlined in subsequent chapters, we will note that the post-1991 economic era (the year in which the United States experienced its last recession), economic activity elsewhere has both positively and negatively influenced American balance-of-trade (BOT) and balance-of-payments (BOP). The immense American telecommunications market, by some estimates approaching 15% of gross domestic product (if broadly defined to include the computer, telephony, and broadcasting industries), is not impervious to these phenomena.

(2) Market Conditions: In assessing market demand for a product, it is essential to identify a firm's place in the developing business cycle, as elaborated in Figure 1.1. In this context, we are concerned with linking consumer and commercial demand for a product or service with the degree to which consumption must be financed. In other words, prospective purchases necessitating financing (borrowing/lending/leveraging) can only be forecasted accurately in the aggregate when consumption is tied to that portion of the business cycle which makes credit available at desirable interest rates. While this factor does not affect the purchase of all communications products and services, it is decisive in some instances; in other cases, the availability of credit is indirectly influential in defining future marginal demand for such products. Business investment in (and consumption of) communications products and services is also strongly tied to evolving points on the business cycle [13].

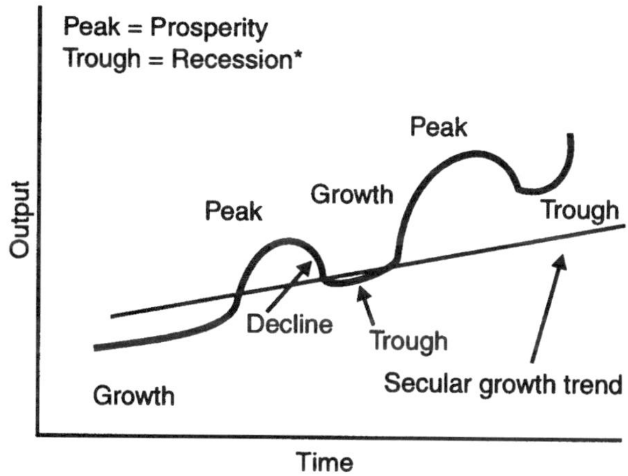

Figure 1.1 Transitions in the business cycle. Note that in the post-World War II era the average expansion has lasted approximately five years. The longest periods of expansion unpunctuated by recession are 1961–1970 and 1982–1990, which were marked by geometric growth in telecommunications infrastructure and services, as defined by convergence. The secular growth trend represents the average annual growth rate in the American economy (which has averaged between 3.0% and 3.5% since 1945).

(3) Competitive Landscape: This determinant involves successfully addressing a menu of questions: (a) Who are the firm's direct competitors? (b) Which firm or firms are likely to become competitors in both the short-term and long-term? (c) What are the probable new business formations likely to define the competitive marketplace (alliances, affiliations, partnerships, mergers, acquisitions)? (d) What are the emerging technological threats to the firm's viability, including prospective product substitutions? (e) What changes in consumer preference are likely to take hold in the short-run and long-run? (f) What is the current status of the supply-chain, and is it likely to change in the future (will the supplies the firm requires to conduct its business be available in perpetuity, and at what price will these supplies be provided)?

(4) Regulatory Climate: Regulation, in particular the continuity of regulation, is a critical determinant of traditional business planning. Regulation, the means by which government (federal, state, local) influences production, price, environmental, working, and safety conditions of industry, is subject to statutory control and bureaucratic discretion. The latter is far less predictable than the former, due to legislative deliberations. In the case of telecommunications, as outlined in Chapter 2, we find ourselves in the early stages of deregulation—a period marked by a relaxation, modification, or elimination of regulatory constraints as defined by the Telecommunications Act of 1996.

It is thus bureaucratic discretion and autonomy that now filters governmental intervention in this sector of the economy. We may surmise that those firms most adept at anticipating and preparing for likely bureaucratic interpretation of the Act are likely to profit most in this environment. In subsequent chapters, I pinpoint each section of the Act most likely to influence commercial activity in the communications sector, and suggest appropriate avenues for a firm's strategic response.

(5) Social Change: One can hardly dispute the significance of social change, that determinant which defines the future of civilization in terms of shared tastes, preferences, normative behavior, and living conditions. Embedded within this determinant is the crucial variable demographics. Demographics is the study of the statistical characteristics of the population, including family behavior, and trends in working conditions, political preferences, household formations, and related factors. Most decisively for the communications industry, as we explore in Chapter 3, is the relative size of each generation entering the economy. Generational size sets the parameters of future consumer demand and, by extension, the nature of product lines subsequently designed and reconstituted.

Market planners are preoccupied with deducing, estimating, and forecasting the fate of demographic variables. Such variables encompass income, occupation, regional origin, family structure, education, and, most significantly for communications usage, age. There is mounting empirical evidence that a comprehensive understanding of demographic determinants is most decisive to successful strategic planning, particularly with regard to firms which must plan in volatile sectors of the economy [14].

The growing importance of demographic analysis is outlined more fully in Chapter 3, but it is instructive to indicate the following: (a) Approximately eighty million Americans are approaching their peak years with regard to spending, debt accumulation, and prospective saving; (b) The Baby Boom generation, that group born between 1946 and 1964, is becoming increasingly fastidious with regard to the demand for customization of products and services, increasingly discerning with respect to quality and post-purchase satisfaction; (c) Generations X and Y, those groups born between the mid-1960s and 1980s, like their elders, are now cognizant about the strategic value of time (in short, they demand that technology be leveraged to replace mundane personal and professional tasks); (d) As a result of these preceding demographic phenomena, the appetite for information is accelerating, but the time necessary to assimilate knowledge is correspondingly contracting, imposing future threats and opportunities for communications providers; (e) As a result of these interrelated trends, the need to design strategic plans sensitive to the demands of customized product lines becomes more acute in the near term [15]. Economic and marketing specifics as to demographic trends are recapitulated in Chapters 4 and 5.

(6) Cultural Change: Allied with social change is cultural evolution. We may define cultural change as that body of knowledge which defines individual and shared human characteristics. Such characteristics include religion, education, customs, traditions, values, political attitudes, and a multitude of other variables. Organizations need to assert a knowledge of cultural factors for both internal and external reasons. Internally, they need to know how to recruit and manage staff of different backgrounds so as to thoroughly exploit the firm's creative powers. Externally, they require the most comprehensive understanding of those forces which influence consumer preferences, and must extend this knowledge to emerging global markets as well [16].

(7) Technological Change: The business world impels and compels change. It is in the field of technology that we find a unique paradox now reaffirmed: simultaneously, business facilitates profound social change through technological innovation; yet, business activity itself is also transformed by the same development. There was a time when technological innovation could be forecasted with comparative dexterity. Today, this has become an extraordinarily complex task for market forecasters, particularly in the telecommunications industry. There are several reasons for this emerging dilemma [17].

Formerly, most forecasters operated on the assumption that technological innovation was inevitable but largely incremental. Because innovation is defined as the improvement of existing product lines, one could safely presume that communication products would become more versatile, facile, cheaper, and faster. The essential forecasting task was thus confined to measuring product life-spans, with the emphasis on the calibration of developmental cycles [18]. Today, especially in the wake of deregulation, barriers to market entry between previously dichotomized communication sectors have fallen. The result is an unleashing of both creative and innovative forces. Creativity, in contrast to innovation, refers to the origination of fundamentally new products.

As regulatory barriers have become relaxed, the computer firm can now become a communications provider; the telephony firm can become a television broadcaster; the broadcasting company can intervene in the telephony market; and the cable television sector can launch new product lines on a cooperative basis with complementary software enterprises. In short, change is no longer limited to incremental developments by firms operating within a single sector. On the contrary, the contemporary and emerging marketplaces facilitate creative reconfigurations of the entire industry. Superimposed upon this realignment of enterprises is global dynamism of an unprecedented nature. Predicting technological change is no longer a straightforward task, but requires a reinterpretation of the method by which interacting firms create, innovate, and cooperate.

The integration of these qualitative and quantitative data—the economic, competitive, regulatory, social, cultural, and technological environments—defines market framework. This information and pertinent forecasts must be folded into the resulting strategic plan if the classic model is to be employed. To engineer the useful manipulation of this knowledge, it is necessary to perform a competitive audit. The schema outlined below provides a prototype of the procedure that traditional strategic planners seek to apply.

1.3.1.3 Evaluation of Internal Resources

Analysis of each component of the competitive environment leads inevitably to the third step in the traditional planning process, systematic evaluation of the firm's intrinsic resources. These resources include available capital, labor, technology, research and development capability, and productive capacity. Each of these resources is linked to the firm's prospective growth and profitability. In a veteran company, an evaluation of the enterprise's successful and unsuccessful product lines must be provided. A number of methodologies exist which allow a firm to audit the "effective success rate" of each product; among these are the growth-market share matrix, and variations on the theme of return-on-investment (ROI) [19].

While the accounting tools used to measure successful product lines are essential in established firms, they do not yield useful applications for emerging companies or planned product lines. In this context, it is essential to "benchmark" new product lines in the same industry sector. This approach presumes that analysis of projected demand must generate results consistent with industry standards, adjusted for comparable time horizons [20]. This method of analysis is elaborated in Chapters 4 and 5.

Internal resource audits, employed by traditional planners, are aimed at specifying the strengths and limitations of available capital, labor, technology, R&D, and productive capabilities relative to comparable resources which competitors bring to market. As noted, benchmarking—establishing industry and competitor standards—has become a preoccupation with planners. Those involved in strategic surveillance (sometimes referred to as "strategic intelligence") are committed to measuring the degree to which their own company can effectively meet the mobilization of identical competitor resources. Benchmarking methodology, discussed elsewhere in this text, has undergone substantial evolution in recent years, owing to dramatic technological, economic, and global transformations.

1.3.1.4 Company Goals and Action Steps

Having defined the firm's goals, and assessing its competitive environment, traditional planners move to identification and implementation action steps intended to achieve those objectives. Incorporating the results of the internal and external competitive audits, a firm positions itself to specify: (a) how to meet its goals; (b) mitigate competitive pressures; (c) develop its internal resources to the disadvantage of competitors; (d) exploit emerging consumer tastes; (e) anticipate future demand promulgated by the introduction of new products and services.

In Table 1.2 we isolate the contribution of each division of the firm—human resource, production, financial, sales, and marketing—to the resolutions of items (a) through (e). The hypothetical audit form provides a mechanism for identifying current and prospective contributions by each component of the firm to the management of strategy. Traditional planners contend that, while all personnel should contribute to the resolution of these questions, and should assist in developing the strategic plan, it is the executive who must finalize the adoption of strategy. It therefore becomes incumbent on executive officers to create a climate in which these activities are encouraged.

Common action steps, in response to newly defined plans, call for developing new products and reconfiguring veteran product lines; outlining new sales channels; identifying prospective alliances, affiliations, partnerships, and other agreements aimed at cooperative ventures; applying new sales tactics; adjusting capital structures; initiating new cost controls; refining advertising strategies; pursuing new suppliers; improving recruiting methods, developing

Table 1.2
Hypothetical Internal Audit Form: Measuring Contributions in Strategic Management

Function	(A) Goals	(B) Competitive Pressures	(C) Internal Resource Advantages	(D) Emerging Consumer Tastes	(E) Future Demand
Human Re-sources	0	+	+	+	+
Production	0	0	0	0	+
Finance	+	+	+	+	0
Sales	0	+	0	0	+
Marketing	+	0	+	+	+

+ = Contribution to strategic management.
0 = Detriment or non-contribution to strategic management.

outsourcing and independent contractor support; enhancing productivity measurement and management; reorganizing sales staff, territories, and compensation; strengthening customer relations; improving product quality; redefining brand and reputation.

1.3.1.5 Implementation of Action Steps

The final step in traditional strategic planning involves the planned implementation of the steps designed to achieve the firm's goals. It has often been noted that although virtually all firms design strategic plans, however nominally, few implement or monitor them upon adoption. There are a number of factors responsible for this phenomenon—among them, the stress of daily operational problems, unforseen contingencies, turnover of staff, changes in the competitive landscape—but this final step in the planning process has long been an organizational bane. It is obvious that the ultimate success of a strategic plan is contingent on the capacity of a firm to carry it out. Yet, many firms allow themselves to be diverted during the stage of implementation.

In studies conducted during the past two decades, it has become evident that incentives for employees to pursue the aims of strategic plans are often insufficient to ensure their success. Empirical evidence also suggests that internal communication between executives and midlevel managerial staff is also frequently inadequate. If the spate of continuing problems includes the pressures of day-to-day operations, personnel transitions, and competitive turbulence, these challenges, too, must be overcome. The body of evidence leads many traditional planners to suggest the following in implementing action plans:

- Link employee evaluations and incentive systems with the measurement of strategic objectives;

- Clearly outline, elaborate, and justify strategic objectives according to a time frame for measurement, thus enhancing the probability that the back-end of the planning process conforms to the front-end of the same;
- Reinforce the image, through executive leadership, that the firm's goals are consistent with the long-run professional objectives of managers and supporting staff;
- Reinforce the rationale that the strategic design is aimed at reducing time, cost, administrative minutia, and redundant meetings and procedures.

These interlocking benefits serve the interests of both organization and staff. In supporting case studies outlined in this volume, these threads in effective strategic management consistently surface.

It can be asserted without equivocation that there is little incentive for employees to adopt constructive action steps in the absence of such justification. Together, these five integrated steps form the procedure known as traditional strategic planning or strategic management.

1.4 Strategic Management and Systems Analysis

A modern departure from the classic mode of planning is the systems model. Systems analysis is a method of projecting strategic plans through the identification and application of inputs, process, outputs, feedback. Just as classic modeling constitutes a procedure to optimize strategic effectiveness, systems analysis may be described as an alternative process to formulate strategic plans. In both instances, strategic management is seen as process; it is not an end itself, but the means to strengthening the firm's image, reputation, brand, profitability and growth. Systems planning, first applied to business management during the 1960s, represented a radical departure from that concocted by traditionalists.

Systems planners are concerned principally with the measurement of outputs and feedback. In short, they have originated methods to specify in greater detail the comparative success and failure rates of strategic options. Second, by emphasizing the continuing measurement of feedback—the reaction of consumers, competitors, suppliers, and regulators—to newly designed strategic plans, they inevitably introduced the notion of continuing improvement in both the procedure and outcome of strategic plans. In other words, their view of strategic management holds that planning should be consistently reviewed and revised after the implementation stage. This approach to planning contrasts significantly with traditional planners, who develop contingencies in their plans, but do not systematically allow for expeditious change in a postimplementation procedure.

Viewed in strictly systems terms, the management of strategy is a "living organism," consistently subject to change and planned alteration. This method of analysis can be seen in the flow chart indicated in Figure 1.2. As noted below, systems planning would call for:

1. Defining mission and objectives (in this regard, the procedure is precisely the same as that outlined in traditional planning);
2. General environmental scanning (focusing on the identification of all key environmental variables impinging on the firm, including competitor behavior, demographic conditions, regulatory transitions, and so forth);
3. Industry and global assessment (sector and international market analysis);
4. Internal assessment (similar to the audit described earlier);
5. Generic strategy alternatives (comparative elaboration of every conceivable strategic scenario confronting the firm);
6. Strategic variations (isolation of the most desirable strategic options);
7. Strategic choice (identification of the optimal strategy);
8. Resources and structure (systematic evaluation of all internal resources that can be exploited to support the strategic plan);

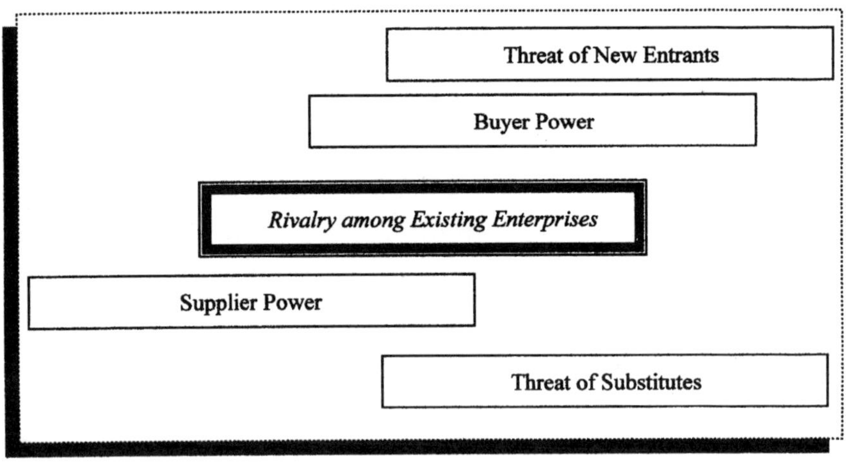

Figure 1.2 Competitive analysis (five-force) theory.

9. Implementation policies and plans (setting of administrative standards to tie employee performance to the requirements of the strategic plan);
10. Evaluation and control (quantitative and qualitative evaluation of the plan's effectiveness in achieving goals projected).

When viewed in their totality, these systems steps can be consolidated as follows: Steps 1 through 4 may be labeled as a diagnostic phase. Steps 5, 6 and 7 are committed to the identification of that choice among strategic options that maximizes the firm's long-term prospects. Steps 8, 9, and 10 may be collectively categorized as the implementation phase. In short, systems planners are procedurally committed to evaluating all strategic alternatives, selecting the most desirable strategic vehicle, then measuring its efficacy in the "feedback loop." In incorporating the advanced policies associated with this last step, systems analysts are dictating constant fine-tuning of strategic planning and management.

We may surmise that systems modeling is significantly more elaborate and demanding than the traditional alternative. It is precisely this comprehensiveness that systems advocates so proudly promote. Certainly thoroughness, particularly with respect to monitoring of feedback, distinguishes this methodology from that proposed by traditionalists. However, implicit in this model is a commitment, indeed the imperative, to invest enormous quantities of time in strategic development. While variations on systems modeling are commonly employed at many corporations, and taught by the leading business schools in the country, mounting dissatisfaction with the model has impelled alternatives to both this and classic modeling. Figure 1.2 specifies the thrust of Porter's analysis.

The telecommunications industry is experiencing a tumultuous era. A reasonable argument can be made that it is this sector of the economy, particularly in the aftermath of deregulation, that is the most dynamic and fluid of all industries. If the classic model of planning contains flaws in precision and postimplementation measurement, and if its systems cousin cannot be adapted in a timely and expeditious manner, it is inevitable that pragmatic alternatives will be pursued.

1.5 Recent Advances in Strategic Management

Classic and systems strategic management have undergone intense scrutiny by practitioners during the past decade. Seeking a practical alternative that synthesizes strengths while minimizing the dilemmas associated with both models, theorists have advanced two options in recent years [21]: Michael Porter's Five-Forces Competitive Design and the Co-opetition Model. These models are not intended to fundamentally replace either the straightforward five-step model embedded in the classic approach or the extensive flowchart method associated with systems analysis. Instead, these advances are geared toward augmenting the value of their predecessors by speeding strategic design and assuring their congruence with market forces. Both models stress the need to anticipate competitor behavior, and fuse new strategies with prospective business policy. In short, these schema imply that cooperating with competition can be as essential to survival as meeting new threats. In this, these approaches to strategic design constitute a revolt to accepted paradigms of planning.

1.5.1 Five-Force Model

Michael Porter, a professor of strategic planning at the Harvard Business School, has advanced a theory of strategic management which stresses interlocking strategies of cost reduction, product differentiation, and organizational focus. These themes of "competitive positioning" compel firms to monitor and assess competitor behavior (what some have termed strategic surveillance), and to benchmark that performance to evaluate company performance. In short, Porter's insights have given impetus to the use of benchmarking to determine the degree to which a firm has defined optimal strategic plans. It is estimated that in excess of one million managers have adopted this method in constructing strategic plans.

Porter's work includes three seminal volumes: *Competitive Strategy, Competitive Advantage,* and *Competitive Advantage of Nations.* Porter's essential thesis, now a ubiquitous influence in corporate strategic planning departments, holds that five horizontal and vertical dimensions clarify industry profitability and dictate strategic fit. These dimensions, or forces, are composed of:

1. Degree of industry rivalry (consisting of brand identity, diversity of competition, exit barriers, product differences, among other factors);
2. Threat of entry (focusing on capacity of new firms to enter markets, emerging regulatory influence, exploitable economies of scale, access to distribution, and expected retaliation);
3. Threat of substitutes (composed of buyer propensity to seek out available alternatives and the relative price performance of substitutes);
4. Buyer power (bargaining leverage of consumers, buyer information, price sensitivity, brand identification, and the key factors);

5. Supplier power (involving supplier concentration, forward integration and cooperation among alternative suppliers, and cost ratios exhibited by suppliers).

Porter's model is diagnosed and applied in greater detail in subsequent chapters, but his key insight signifies that the success or failure of a firm's strategy is a function of these intersecting five forces. A strategic plan is thus positioned to be congruent with those clusters of firms that share optimal advantages in regard to each of these five forces. A strategic plan which fails to attain congruence with industry rivalry, entry, substitutes, and buying or supplying power is destined to degenerate, states Porter. His impact on the world of strategic planning since the 1980s cannot be overstated. Porter's critics contend that these forces are less significant than he imagines, and that strategic plans should emphasize primarily the exploitation of the firm's intrinsic strengths. Should a firm's strategic management stress benchmarking to achieve its goals, or simply aim at bringing to fruition its internal expertise, resources, knowledge, and market intuition? This debate has become a cutting-edge issue in strategic planning of the 1990s [21].

1.5.2 "Co-opetitive" Strategic Management

In classic, systems, and five-force modeling, one observes a consistent theme of industrial competition as the guiding format of planning. Indeed, viewed in philosophical terms, strategic planning adheres to a metaphor of confrontation or war in all three instances. A firm's survival is therefore predicated on its capacity to defeat its competitors in the marketplace. Its appraisal of success is based on profitability, productivity, sales growth, cost containment, and prevailing market share. A contrarian model has emerged recently that challenges elements of each of these models [22].

Termed "co-opetition" by its authors, co-opetitive strategic management suggests that newly reconfigured economies are revising traditional industry structures. This phenomenon is especially apparent in those sectors of the economy embracing networked systems of production and distribution. Thus, a new method of strategic planning must be devised in response to seminal economic forces. A networked economy is one characterized by the service sector as the dominant form of commerce and employment, and one which increasingly pivots on rearrangements of producers, consumers, and suppliers. Simply, we are now witnessing a chaotic reordering of industry and competitor rivalries: it is sometimes more advantageous to pursue cooperative ventures, rather than direct conflict, with competitors. In marshaling new business affiliations (mutual promotional agreements), alliances (cooperative nonequity agreements), and partnerships (equity co-ventures), a firm may maximize its sales growth, profitability, productivity, cost containment, and market share. This alternative method of strategic management represents a profound departure from those models which preceded it. A visual depiction of co-opetitive modeling can be summarized in Figure 1.3, illustrating a hypothetical audit of potential strategic business formations available to a communications firm.

The authors of "co-opetition," Adam Brandenburger and Barry Nalebuff, have elaborated their theory with this commentary:

> A player is your complementor if customers value your product more when they have the other player's product than when they have your product alone. A player is your competitor if customers value your product less when they have the other player's product than when they have your product alone. A player is your complementor if it's more attractive for a supplier to provide resources to you when it's also supplying the other player than when it's supplying you alone. A player is your competitor if it's less attractive for a supplier to provide resources to you when it's also supplying the other player than when it's supplying you alone. Friend and foe. . . . There are win-win and win-lose elements in relationships with customers, suppliers, complementors, competitors . . . think co-opetition.

In this alternative arrangement to strategic management, what is most decisive to a firm's success is its capacity to isolate "complementors" and competitors among producers and suppliers. It is the ability to identify these prospective configurations and reconfigurations among complementors and competitors that decide future profitability. Having projected these attractive arrangements, it is then incumbent upon the firm to negotiate favorable working relationships in the form of affiliations, alliances, and partnerships; these working relationships are fragile and subject to rapid change. We can confirm that this is a pattern of strategic behavior of growing significance following the 1990–1991 recession [23]. For reasons discussed in Chapter 6, telecommunications firms are increasingly sensitive to both the opportunities and threats associated with this alternative strategic form.

1.5.3 Strategic Management Through PEST

One method of facilitating strategic management is the use of PEST in policy formulation. PEST (political-economic-sociological-technical variables) imposes on the strategist an evaluation of the key factors embedded in the

Co-Opetitive Business Formation

	Affiliation	Alliance	Partnership	Intrapreneurship
Customer Segment				
Supplier				
Complementor				
Competitor				

Note: A co-opetitor represents a potential ally among customers, suppliers, complementors or competitors. An industry rival becomes a co-opetitor if an agreement is arranged such that both parties realize value-added benefit, reduced pricing, improved service, or enhanced market share; e.g., a firm may choose to negotiate an alliance with a competitor to launch a new product; a partnership with a supplier to reduce distribution costs; an intrapreneurship with a former rival to innovate product; a promotional affiliation with a company not connected to its own sector.

Figure 1.3 The "co-opetitive" model (hypothetical audit of potential strategic business formations).

competitive environment. An effective PEST compels strategic planners to systematically weigh the dynamic interaction of these variables so as to formulate alternative scenarios about the future of the industry. The crystallization of these "alternative visions" advances strategic design by clarifying the direction the firm ought to undertake. Simply, one cannot designate an optimal plan in the absence of deciphering the complexities defining our future. There is little disagreement that political discontinuities, economic change, sociological transition, and technological innovation intersect to continually reconfigure the environment governing strategic design.

One proponent of PEST has observed that the instinct to apply this forecasting method should be extended to industries outside the telecommunications domain. The analysis of the prospective behavior of these peripheral characters is equally susceptible to the application of PEST in terms of the competitive environment:

> . . . a telecommunications-based provider should never just consider its immediate environment but also those of its suppliers, customers, and analogous areas too . . . the main strategic issues identified by telecommunications companies in general are: competition, deregulation, falling costs and prices, downsizing, adding value, globalization, market focus, strategic alliances, and what to do about the Internet [23].

Juxtaposed to the use of PEST in competitive analysis is the measurement and forecasting of tertiary consequences (third-party effects connected with transactions between producers and consumers). The integration of competitive with tertiary analysis is discussed in greater detail in Chapter 5.

1.6 A Note on Menu-Based Strategic Management

In considering the foundations of strategic management, we assess four principal methods or approaches to the craft: Traditional (or Classic); Systems; Five-Force; and Co-opetition. We therefore now operate in a paradigmatic vacuum. At this moment, no consensus exists about that model of strategic management appropriate to all firms. Indeed, what distinguishes the current competitive landscape is the distinct lack of agreement about that model or models most consistent with industry success.

We are left with an expansive but uncertain menu: it is the discretion and intuition of the strategist which will guide and apply each model. A compelling case can be made that no one method truly maximizes strategic

advantage during this chaotic period of telecommunications history. Perhaps a newly-created model, or hybrid thereof, will characterize state-of-the-art strategic design for the industry. The body of knowledge hereafter outlined attempts to provoke innovative applications of each of these models for strategic planners and the challenges they are likely to encounter.

In this volume, the author culls a set of strategic management designs by leading telecommunications firms—among them, AT&T, Nextel, America On-Line, SBC, Bell Atlantic, Microsoft—and examines their planning functions in terms of the four perspectives outlined above. This method of evaluation, it is intended, will assist planners in exploring the range of planning options for their own firms. Also discussed is the feasibility of designing and applying potential planning hybrids drawn from this menu.

References

[1] Wright, P., M. Kroll, and J. Parnell, *Strategic Management: Concepts and Cases,* Upper Saddle River, NJ: Prentice-Hall Co., 1998, pp. 3–7.

[2] Shaw, J., *Telecommunications Deregulation,* Norwood, MA: Artech House, 1998, pp. 55–65.

[3] Joshi, P., "Clash of Phone Titans: No Answer Yet Who Will Gain From Mergers," *Newsday,* March 2, 1998, pp. 7–10.

[4] Jauch, L. and W. Glueck, *Business Policy and Strategic Management,* New York: McGraw-Hill Co., 1988, pp. 375–383.

[5] Taylor, W., "Fast Company, Control in an Age of Chaos," *Harvard Business Review,* Boston, MA, November 1, 1994, p. 64.

[6] The development followed the collapse of planned economies in Eastern Europe, the former Soviet Union, portions of Africa and Asia, as well as the conversion of China to a quasi market-based economy.

[7] Tetenbaum, T., "Shifting Paradigms: From Newton to Chaos," *Organizational Dynamics,* April 1, 1998, Vol. 26.

[8] Mintzberg, H., J. B. Quinn, and S. Ghoshal, *The Strategy Process,* Upper Saddle River, NJ: 1998, pp. 22–48.

[9] Kelly, K., *Out of Control: The Rise of Neo-Biological Civilization,* Reading, MA: Addison-Wesley, 1994.

[10] Trunick, P., "New Demands for Tomorrow's Manager," *Transportation & Distribution,* December 1, 1998, pp. 18–19.

[11] Brown, S. L., and K. Eisenhardt, "Product Development: Past Research, Present Findings, and Future Directions," *Academy of Management Review,* April 1, 1995, pp. 336–343.

[12] Trebling, H. M., "The New Telecommunications: A Political Economy of Network Evolution," *Journal of Economic Issues,* March 1, 1996, pp. 314–316.

[13] Dent, H., *The Great Boom Ahead,* New York: Hyperion Press, 1993, pp. 3–26.

[14] Fiegener, M. K., "The Control of Strategy in Dynamic Versus Stable Environments," *Journal of Managerial Issues,* March 22, 1997, Vol. 9, pp. 72–86.

[15] Sheridan, J. H., "The Best vs. the Rest: A Glimpse at the Strategies, Strengths and Management Styles Behind the Numbers," *Industry Week,* August 18, 1997, Vol. 246, pp. 70–79.

[16] Brown, C., "Human Capital, Social Capital, and Firm Dissolution," *Academy of Management Journal,* February 16, 1999, pp. 125.

[17] Fahey, L. and R. Randall, *The Portable MBA in Strategy,* New York: Wiley Press, 1994, p. 376.

[18] Herman, R. and J. L. Gioia, "Making Work Meaningful: Secrets of the Future-Focused Corporation," *The Futurist,* Vol. 32, December 1, 1998, pp. 16–22.

[19] Robbins, S., *Organizational Behavior,* 8th ed., Upper Saddle River, NJ: Prentice-Hall Inc., 1998.

[20] Darden, J., "Business Planning as Pedagogy: Language and Control in a Changing Institutional Field," *Administrative Science Quarterly,* Vol. 21, February 13, 1999, pp. 23–45.

[21] Porter, M., *Competitive Advantage,* New York: Free Press, 1998.

[22] Brandenburger, A. and B. Nalebuff, *Co-opetition,* New York: Doubleday Press, 1996, pp. 18–35.

[23] Mi, F., "Consolidation: 'Bully Effect' Driving Acceleration of Consolidation in Telecom," *Cambridge Telecom Report,* March 23, 1998, p. 19.

2

Strategic Management and Deregulation: Opportunities Precipitated by the Telecommunications Act of 1996

Two overriding factors will unquestionably drive the immediate future of strategic management in the telecommunications industry: deregulated market structures and technological change. The Telecommunications Act of 1996 has provided statutory impetus to the inevitable restructuring of markets in which communications providers must now confront a plethora of unexpected competitors. This chapter explores the strategic significance of interacting regulatory and technological forces during the period immediately following deregulation, and suggests a framework in which to monitor and measure future strategic alternatives for the industry. The fusion of deregulated markets with technological innovation will define future business opportunities within the telecommunications industry, and all other sectors which support it.

By "opportunities," we refer to emerging markets, consumer bases, and product lines likely to surface in response to the relaxation or elimination of regulatory constraints and sanctions. A discussion of the efficacy of telecommunications regulation, and the debate enjoined by proponents and opponents, is embedded in this chapter. There is little doubt, however, that the industry has embarked on an era of turbulent change, and that business opportunities will advance or recede in all sectors at an increasingly breathtaking pace. How a firm anticipates and responds to these market dynamics will outline its future strategic options.

2.1 Operational Definition of Contemporary Telecommunications

The conventional meaning of telecommunication is the transmission or reception of audio, data, or video signals through the electromagnetic spectrum, wireline, or light. This accepted, standard definition held greatest relevance and efficacy during the period immediately preceding deregulation. Telephony, radio and television broadcasting, and other forms of wireless communication constituted principal forms of telecommunications usage. As long as government regulation differentiated these communications services, and specified their separate domains, this working definition of the term guided and organized the industry.

As outlined below, a new "operational" definition must be elaborated during the era marked by passage of the Telecommunications Act of 1996. This statute carries a number of important directives that will dictate industry prospects. The vital economic distinction between the pre- and post-Act eras compels the adoption of a new definition of telecommunications: today, the artificial regulatory boundaries dividing the telephony, broadcasting, wireless, and other communication sectors have fallen. As a result, telephony firms may now invade the broadcasting industry; cable television firms may enter telephony, radio, or other sectors; long-distance telephone companies may participate in any future communications venture, and so forth. Most significantly, the computer industry, particularly software development, now superimposes its presence over every form of communications services and products. Hence, telecommunications may now be operationally defined as the electronic transmission, reception, storage, or organization of information. This information is found in data, audio, or visual form. Three principal and discrete, but nevertheless interacting, sectors—telephony, broadcasting, and computer—now dominate modern telecommunications [1].

2.2 Telecommunications Deregulation: The Cutting-Edge Debate

The communications industry began its deregulatory awakening in 1982 [2]. Cracks in AT&T's monopoly status had surfaced as early as the 1950s in the form of legal judgments levied against alleged uncompetitive practices. It was, however, the Modified Final Judgment (MFJ), presided over by Justice Harold Greene in 1983, which asserted that AT&T would be required to reconstitute its organizational scheme. The MFJ provided that while AT&T would be permitted to retain its holdings, its regulated monopolistic structure would be reshaped to encourage competition. The agreement consolidated 22 Bell operating companies into seven independent regional units (including US West, Ameritech, Pacific Telesis, Bell Atlantic, Bell South, Southwestern Bell, and NYNEX). This divestiture also allowed the newly reconfigured AT&T to engage in businesses formerly barred by regulatory constraint. In effect, AT&T could now participate in markets formerly closed off from new ventures. The telecommunications giant would be prevented from dominating local markets through its system of deregulated subunits. As we will note later (refer to Chapter 8), this legal judgment had far-reaching effects, particularly with regard to unforeseen strategic externalities and their impact on industries unrelated to telephony.

Competition, innovation, creativity, and expanded choice were the collective objectives of post-1984 deregulation in the communications industry. The combination of enhanced competition, technological advance, and symbiotic business development between the computer, telephony, and broadcasting industries—otherwise known as convergence—were to be the vehicles of spurred price reduction. Contentious and provocative debate would follow the post-1984 period of divestiture, but the principle of deregulation as a tool for enhanced competition had been established. The nation had experimented with deregulation in aviation and trucking in 1978, petroleum and natural gas in 1981, cable television in the early 1990s, and the time was ripe by the mid-1990s, some felt, for a philosophical repudiation of regulatory intervention. A growing number of politicians in both parties, as well as many respected economists, had coalesced to instigate this opposition to regulatory ethos. Some faulted the fundamental logic and hypothesis undergirding this strategy for enhanced competition, but it had become unmistakable by 1994 that Congress was moving in the direction of rewriting the Communications Act of 1934. In February 1996 Congress passed, and President Clinton signed into law, the Telecommunications Act of 1996.

The newly rewritten communications statute, whatever its strengths and limitations, was to provide a new commercial framework for the delivery of telecommunications services and products. The 1934 Act had focused on the development of the radio and telephony industries; its regulatory emphasis was aimed directly at encouraging the growth of these sectors while safeguarding the electromagnetic spectrum for national defense. In contrast, the 1996 Act had explicitly acknowledged the critical importance of divergent and convergent communications sectors. Accordingly, telephone service, for example, had been redefined as to its growing role in delivering Internet and other networked services; local service was to be deregulated so as to speed competitive alternatives for both traditional and nontraditional telephony services.

The 1996 Act was also aimed at revising technical and administrative rules guiding development of new broadcasting services, such as wireless and wireline delivery of interactive television. Cable television service was to enter systematic deregulation, as an alternative competitor to telephone and satellite providers. Provisions in the Act also asserted that Federal Communications Commission (FCC) officials were to hold predominance in regulatory rulemaking over their state and local government counterparts. This guiding principle would later trigger legal disputes among private firms, the federal government, and state and local government administrators. In the years immediately following passage of the Act, continuing disputes in telephony regulation slowed the implementation of deregulation.

Viewed in economic and strategic terms, what emerges in the 1996 Act is a cumbersome deregulatory procedure, applied unevenly in the converging sectors of telephony, broadcasting, and peripheral computer services. The 1996 statute had reaffirmed the principle of universal service, first installed in the Communications Act of 1934, and superimposed over it the power of the federal government to remove or relax barriers to convergent and lateral entry by established firms. The FCC would no longer arbitrarily prevent veteran companies from expanding market share through horizontal and vertical acquisition. Subject to a review of competitive criteria, anti-trust considerations were to be decided on an individual (or case-by-case) basis. In addition, telephony, cable, and broadcasting firms could invade each others' territories without fear of regulatory constraint. This contingency was intended to goad convergent sectors into intensified competitive behavior. As expressed by many firms, particularly large enterprises, this new freedom of market interaction was an unprecedented step in the—albeit disputable—direction of unfolding competition.

A critical conceptual footnote must be acknowledged about the definition of convergence as defined by the author. While the broadcasting and telephony industries—regulated under both the 1934 and 1996 statutes—remain guided by public administrators, the same comment cannot be made about the converging computer sector. Indeed, with the exception of certain controls on the export of sensitive (national security) hardware and software technologies, the computer industry remains unregulated. There is a clear positive correlation between the degree of government regulation and the speed of innovation associated with any industry: the more invasive the regulation, the slower industry innovates, creates, and enhances the quality of deliverable services. This is the inevitable price of government mandates in pricing, security, safety, or other regulatory aims. As a result, we now see a highly dynamic, nimble computer industry intersecting with the comparatively less agile, adaptable telephony and broadcasting sectors. The innovative influence of computers—as communications tools—now supercedes the collective influence of broadcasting and telephony in driving the nation's telecommunications infrastructure. Passage of statutory reform has only accelerated this far-reaching result, a consequence now manifestly obvious with the proliferation of networked information.

Other features of the Act include a key provision which repeals prohibition on communications manufacturing by the Regional Bell Operating Companies (ROBCs). Congress had thus envisioned an expanding manufacturing market bolstered by the elimination of an artificial barrier to entry of these major competitors. The local carriers could also participate in electronic commerce and publishing. Additionally, major broadcasting corporations would be allotted spectrum loan privileges for the delivery of high-definition (HDTV) frequencies. Again, substantial freedom would be granted to the private sector in organizing, dispensing, and supervising the use of the electromagnetic spectrum to stimulate new commercial ventures. The statute is also explicit in forbidding the FCC to regulate the Internet, now vital to triggering development of a wide array of electronic commercial ventures. This feature of the statute holds profound ramifications for aggregation, consolidation, and synergistic business forces—factors discussed in detail in this volume as to their strategic significance.

2.3 The Telecommunications Act of 1996: Debate and Assessment, 1996–2000

The objective of the 1996 statute, as pronounced by the executive and legislative branches, was increased competition in the converging communications sectors. The presumption was that deregulation was a necessity in achieving this objective. Deregulation was intended to increase the number of competitors and expand the array of services while reducing prices, simultaneously spurring new economic opportunities so as to reduce unemployment and enlarge the tax base. Has the newly revised communications statute attained the goals of its authors?

The contentious debate following statutory reform often degenerates into ideological confrontation. Those traditionally defending government regulation as a tool for consumer protection contend that the Act has been a failure if one compares its promise relative to performance. Their argument rests primarily on a critique of: a) rising consumer prices for cable television service across the nation since 1996; b) the absence of genuine competitive alternatives for local telephone service in most regions of the nation; c) the lack of immediate penetration and competition exerted by established players: many telephony firms, for instance, have been content to remain in their own domains before incurring risks associated with providing cable television service [3]. Examined exclusively in these terms, the Act appears to have fallen short of the major goals projected for it. Prices have generally risen beyond the rate of inflation in the four years immediately following passage; in addition, many firms are clearly reluctant to supply local telephony either through the use of established local networks (via lease) or through the construction of wireline alternatives. Viewed in this narrow, and perhaps myopic context, the Act may be labeled a disappointment.

If the method of evaluation is broader, inclusive of a consideration of all forms of telecommunications activity, and adjusted for the long-run, an alternative judgment can be proposed. The 1996 rewriting of the 1934 communications law is a 100-page statute. It is complex and reflects the political realities of an attempt to appease divergent interests among large and small firms, consumer groups, attorneys, politicians, consultants, economists, and expert witnesses. Moreover, the telecommunications industry is a capital-intensive market structure: its evolution and impetus for growth in other sectors of the economy is a function of patient building of customer bases. Applying purely economic criteria to the task, the 1996 Act has effectively propelled change at different rates in various areas of the sector. There has been little competitive development in cable television and local telephony since 1996. However, there is substantial empirical evidence to suggest that vigorous consolidation has reoriented market structures—the size, scope, competitive characteristics—in long-distance delivery, wireless products, Internet development, satellite offerings, and broadcasting services. At least some of these synergies are traceable to the statute. The ability of these providers to enter, depart, reenter, and reconfigure their product lines was effected by key provisions of the law.

One might reasonably characterize the Act as representing a progressive first step on the "macro-level," but failing on the "micro-level." The imperfections of a law designated to satisfy self-interested constituencies ensured inadequacies in regulating any given sector within the industry group. It was thus preordained that some professionals would dismiss the law as ineffectual or counterproductive. On the other hand, the law transformed the strategic mindset of the private sector and stimulated greater expectations on the part of the public: whatever its regulatory deficiencies, prospective improvement in communications products, services, and delivery was made inexorable. The speed of innovation, competitive offerings, or pricing reductions may not have satisfied critics, but the bulk of evidence in the 48 months following deregulation suggests adaptation not otherwise attainable. Additionally, the pace of new communications enterprises—measured through accelerated venture capital formations—was spurred, in part, by adoption of the law [4].

2.4 Economic Backdrop

An understanding of the evolution of the American economy during the years immediately preceding and following passage of the Telecommunications Act of 1996 is essential when assessing future opportunities for the industry. American

macroeconomic developments—employment patterns, price stabi
profitability, business investment, consumer behavior, corporate
government fiscal, and monetary policy—have contributed dec
pace of telecommunications production and delivery (refer to Appendix, Section D).

This section elaborates the juxtaposition of the American economy to the international economic system now poised for transition. What emerges from a detailed analysis of the American economy in its present manifestation is the role of telecommunications in setting its course. By extension, we infer that we will later witness similar patterns of economic development if telecommunications precipitates comparable dynamic change in both the public and private sectors. While this hypothesis is contentious, there exists mounting empirical evidence that telecommunications is now the linchpin to sustained productivity in all sectors of the economy.[1]

2.4.1 Economic Backdrop: Pre-Telecommunications Act of 1996 Era

The American economy in the year 2000 reflects broadly divergent economic forces which had initially surfaced during the prior decade. As the United States shifted from the 1980s to the 1990s, we note in retrospect a major restructuring of American business: high debt accumulations, lagging productivity, organizational restructuring, and labor surpluses had conspired to precipitate a pronounced recession extending from 1990 to 1991. A mathematical gauge of the American economy is given by the macroeconomic expression [3]:

$$C + I + G + Ex - Im = GDP,$$

where C = consumer spending; I = business investment; G = government spending; Ex = exports; Im = imports; GDP = gross domestic product.

This macroeconomic equation holds that the value of all goods and services generated by the economy during a single calendar year (GDP) is directed initially and primarily by consumer spending. (Note: consumer spending is generally more significant in driving the American economy than that of business investment, government expenditures, and exports in stimulating national economies.) Investment in new plant and equipment is essentially caused by consumer demand projected by business and industry. Government spending is enjoined when transactions between consumers and producers are

1. American gross domestic product now approximates $8.5 trillion. The broadcasting, telephone, and computer industries combine to generate in excess of $1 trillion; hence, the telecommunications sector stimulates 15% of national wealth each year. At present rates of growth, the industry will double before 2010 and instigate nearly 20% of GDP at that time (www.bls.gov).

consummated. Exports stimulate capital in-flows for the economy, while imports provoke capital out-flows. The logic of this tool holds that American consumers—in their collective consumption—drive or restrain economic expansion by transmitting market signals through the equation. In effect, business adopts its plans in relation to anticipated spending, while the capacity of government to finance its activities hinges principally on the interaction of business and consumers. Exports equivalent to or in excess of imports stimulate economic growth, while a negative trade imbalance retards prospective expansion.

The American economy had exhibited sustained economic growth following 1982 and extending through 1990, despite experiencing a chronic trade imbalance (that is, imports consistently exceeded exported goods and services throughout the decade). During the second half of 1990 the economy slowed, then contracted into 1991 (because a recession is defined as six or more months of contracting gross domestic product, this development constituted the bottom of the business cycle). The origins of the recession remain a matter of dispute. Some maintain that the economy stagnated in the midst of high debt accumulations which diminished consumer spending. Others saw the recession as the product of higher tax rates installed by the federal government during 1990. Still others concluded that demographic changes dictated restrained consumer and business spending. Some analysts saw the recession as a natural cyclical transition within the business cycle. Whatever the cause or causes of the 1990–1991 recession, high unemployment, slowing corporate profitability, mounting bankruptcies, and expansive consumer, business, and government debt materialized at the beginning of the decade. Further exacerbating these problems were high interest rates. Simply, as unemployment rose, corporate profits shrunk, debt grew, and the aggregate level of entrepreneurial opportunities stalled.

We note several indisputable lessons from the 1990-1991 recession and its implications for the telecommunications industry during the coming decade. First, of the nine recessions punctuating the post-World War II period in the United States (the other eight recessions occurred in 1949, 1954, 1958, 1960, 1970, 1974–1975, 1980, 1981–1982), the 1990–1991 contraction was marked by the absence of major rehiring by corporations which had laid off employees. In other words, corporations deliberately withheld rehiring a major fraction of their workforce—the common post-recession pattern of corporate behavior—through 1995. Second, American labor productivity in these firms, particularly in service industries, remained comparatively unchanged during the post-recession period. Seemingly, corporations choosing not to rehire their former workers did not suffer a decline in relative rates of efficiency; at least in the short-run, these companies apparently persuaded themselves that they could sustain desired output without returning to prior levels of employees. Formerly, firms, upon return to prosperity, rehired workers and leaped to higher rates of productivity.

Lastly, corporations during the post-1991 period felt compelled to "reengineer" their organizational structures in response to major technological change, particularly with respect to the prospective incorporation of information systems [4]. Extraordinary improvements in computer software and hardware engineering had impelled businesses to invest heavily in these technologies; their motive, to be sure, involved gaining higher productivity at lower costs of production. During this systematic reengineering—that is, restructuring organizational schemes to accommodate the use of these emerging technologies—many firms had concluded that they could produce comparable output while simultaneously reducing the number of personnel. Indeed, it was precisely the effective implementation of innovative telecommunications and information technologies which led these firms to conclude that rehiring personnel represented redundancy; in the short-run, many companies found that they could sustain higher profitability with fewer numbers of workers at little strategic risk. As we will later note in culling the empirical evidence, this tactical view of human resource management may now be regarded as myopic relative to the long-run [5].

2.4.2 Economic Backdrop: Post-Telecommunications Act of 1996 Era

The Telecommunications Act of 1996, the first major overhaul of statutory communications law since 1934, was heralded by its legislative sponsors as a critical engine to American economic growth as the new century beckoned [8]. While such a hypothesis was a contentious and provocative one, and still open to alternative interpretations, we can profile domestic economic performance in the years following deregulation. Again applying the macroeconomic expression identified above, we note accelerated economic expansion (measured in terms of gross domestic product) in 1997, 1998, and 1999. Deregulation, the prime impetus to passage of the 1996 statute, may or may not have influenced economic progress in the period following adoption. We can, however, empirically confirm the following with respect to the forty-month period following the adoption of regulatory reform in February of 1996 [6]:

- Consumer spending rose markedly;
- Business investment in plant and equipment, particularly with respect to purchases of telecommunications and information technology equipment, rose significantly;

- Government deficits abated, then reversed course to generate surpluses beginning in 1998 (the first time the federal government had collected more revenue than expended since fiscal year 1969);
- Unemployment and inflation simultaneously declined: by late 1998 the United States attained essentially full employment at a 1% rate of inflation for the first time since the 1950s;
- Corporate profitability rose dramatically during the last half of the 1990s; Stock market valuations had doubled by 1999 (as measured by the Dow Jones Industrial Average);
- Surpluses of jobs relative to available labor began to accrue by 1999 in many professional occupations.

The debate enjoined during the late 1990s regarding the efficacy of telecommunications deregulation remains heated and divisive. There are two noteworthy footnotes, however, to the citations indicated above:

1. Proponents of regulatory relaxation in the communications sector had promised that these contingencies would unfold once the 1996 statute had been signed into law.
2. American economic growth had accelerated in recent years despite the fact that American exports had stagnated during the concurrent global slowdown.

With regard to the latter development, it is significant that the American economy grew at a time when foreigners had cut their spending for American goods and services (implying that a return to global prosperity would further stimulate economic growth as the new century unfolded). With regard to the former factor, advocates of the 1996 telecommunications law naturally took succor in seeing the economy perform precisely the way they had forecasted before 1996. Whether the economy had performed well because—or in spite of—regulatory modifications remained, however, a source of dispute.

An effort is made in this chapter to define the economic consequences of telecommunications deregulation and to set the parameters of the interpretative debate about those effects. Whether Americans have seen true deregulation since 1996 continues to be a matter of dispute, as elaborated below. Deregulation has clearly been more aggressively asserted in certain areas of communications services as opposed to others. Tabular and graphical data about the performance of the business cycle during this period is provided at length in the appendix. It is instructive to note, however, that growth in the computer sector, and its juxtaposition to telephony and broadcasting, is strongly correlated to growth generated elsewhere in the economy [7].

2.5 Telecommunications and the American Economy: The Year 2000 and Beyond

The strategic management process inevitably demands that planners make empirically-based judgments about the future of the business cycle. Succeeding chapters isolate the critical distinction between internal and external "environmental" analysis in strategic planning. The key determinant for external environmental analysis—factors beyond the control of a firm—is the contour of the business cycle (consult Appendix, Section E). The American business cycle, validates symmetry, but is staggered with respect to its peaks and valleys. In other words, we cannot predict with precision or confidence the stages of economic growth that are punctuated by recessions. Simply, the duration of economic growth sandwiched between recessions is not spaced at predictable intervals.

Note, for example, that five-year intervals divide the 1949 and 1954 recessions; yet, a period of less than two years separates the 1958–1960 and 1980–1981 recessions. Nearly a decade spaced the 1960 and 1970 recessions; eight years of growth unfolded between 1982 and 1990. America did not experience a recession between 1991 and 2000 (the longest period of sustained economic growth during peacetime). What should planners infer about the role of the business cycle in driving a firm's prospective sales growth? To what extent do these data provide a portrait of post-2000 U.S. economic growth? What forecasts for domestic and global growth can professional planners reasonably rely upon as we move through the coming decade?

While we cannot predict with certitude the periodic contractions of the economy, we can draw credible inferences regarding those forces likely to drive economic expansion through 2010 and the period immediately beyond. The GDP equation above affirms that consumer spending is decisive in: a) spurring economic growth; b) expanding corporate profitability; c) driving new business (and stimulating the creation of new product lines). In this regard, empirical evidence confirms [8, 9]:

- Demographic patterns drive spending, sales, and innovation;
- Investment capital at low interest rates is critical to product creation (the development of fundamentally new product lines) and innovation (the incremental improvement of veteran product lines);
- Increase in real wages (driving income, savings, and debt accumulation) is necessary to instigate major increases in capital investment by both business and government;
- Productivity is the key to wealth creation and improvement in real wages (real wages are defined as compensation adjusted for inflation).

Hence, sustained economic growth in any historical period is the product of positive demographic patterns (i.e., steadily rising numbers of new consumers) accompanied by expansive capital investment, growth in wages, and subsequent improvement in productivity. Positive patterns in demographics, capital investment and real wages theoretically yield higher productivity over the long-run; accrued efficiencies in output enable firms to consistently increase wages. It is impossible, in short, to envision substantial economic progress in the absence of higher productivity. Any projections about the economy in the year 2000 and beyond pivot on the interaction of these four variables. What do these dynamic and interactive variables suggest about the coming decade?

The United States now finds itself in an extraordinary positive confluence of all four variables. A nexus of positive demographics, investment, wage growth, and productivity is now arising as the new century begins. This intersection of economic forces is not unique in American history, but the sheer size of the "Baby Boom" generation—some eighty million Americans born between 1945 and 1965—promises exceptional consumption for domestic business opportunities [10]. Peak consumption for the typical American occurs at the age of 49: highest spending, savings, and debt consolidation occur during this period of the life cycle. Consider that a person born in 1945 is fifty-five as the new century begins; a person born in 1957, the peak year of births during this era, becomes forty-three in this year; an American born in 1965, the tail-end of the Baby Boom, turns 35 in 2000.

As a result, succeeding enormous waves of consumers will be entering peak consumption through 2014 [10]. The Baby Boomers, the largest generation in American history, are likely to sustain near-term economic expansion well past 2010. Slowing birth rates—and thus consumer growth rates—from the mid-1960s through 1980 suggest that the American economy, barring unforeseen circumstances, is likely to recede dramatically before 2015 [11]. This deduction does not preclude recessions during the coming decade of prospective growth. Indeed, America has never experienced a period longer than a decade without experiencing a recession. However, strategic planners need not be concerned that there are insufficient numbers of customers available to purchase their firm's products. This decisive external environmental factor should spur strategic planners on to creating enticing communications services and products during the coming decade.

2.6 The Telecommunications Act of 1996: Distilling Strategic Significance

The Appendix (see Section A) elaborates those elements of the 1996 law which are of greatest concern to market planners. The law introduces important features, through imposition of key "sections," which instigate strategic considerations:

- Relaxation, phasing out, or elimination of FCC regulatory powers;
- Permission, implicit and explicit, to engage in merger-and-acquisition activity, or formations of alliances, partnerships, or other business affiliations;
- Creation of selected areas of heightened competition;
- Stimulus for technological innovation and new business formations;
- Promotion of economic and technological convergence of cable, telephony, broadcasting, and computer industries;
- Diminution or submission of state regulation to federal encroachment.

Some features of the Act have greater strategic import than others, of course; other components have their greatest impact in terms of time horizon (short versus long-term consequences); still others hold the promise of future regulatory contentiousness that cannot be forecasted with precision [12]. What is most revealing about the construction of the statute, however, is its generalized and sometimes vague phraseology. It is the ambiguities associated with these generalities which hold the promise of continued litigation between local and long-distance providers, between large and newly emerging firms, between the federal and state governments, between all government jurisdictions and large providers in telephony, broadcasting, and cable enterprises. Despite these discontinuities—problems not dissimilar to regulation encountered in other industries—the Act establishes a procedure for mediating disputes.

Critical portions of the Act, in terms of reordering telecommunications strategic management priorities for the future, can be identified as [1, 17]:

- *Section 103:* Public utilities as defined under the Public Utility Holding Act are now permitted to diversify into telecommunications and other services through single-purpose subsidiaries.
- *Section 201:* Advanced television licenses are assigned to incumbents.
- *Section 202:* Relaxes or modifies restrictions on the aggregation of broadcast outlets as owned or controlled by single entities, corporations, or conglomerates.
- *Section 251:* Elaborates a new model for integrated interconnection. The RBOC's are required to provide equal access to interexchange carriers and other information service providers. This element of the law is arguably the most vital short-term issue with which the FCC

will have to contend in orchestrating eventual full deregulation. In the absence of systematic interconnection of networks, full deregulation of the industry is impossible.

- *Section 252:* State agencies retain the right to approve interconnection agreements but acquiesce to the FCC with regard to policymaking authority if their departments fail to act expeditiously in approving these agreements.
- *Section 253:* State and local governments are prohibited from adopting legislation aimed at preventing interstate and intrastate services. Large corporations engaged in product and service delivery, as well as information content, are concerned with this portion of the statute.
- *Section 254:* Universal service is reaffirmed but subject to changing interpretation based on on-going advances in technology and other emerging trends. The creation of a Federal-State Board to determine the efficacy of universal service policy is provided. Additional responsibilities are imposed regarding discounted services extended to educational and health care organizations.
- *Section 256:* Defines infrastructure interoperability on a voluntary basis; that is, the FCC retains responsibility for directing diverse interests to develop interoperability standards and coordination.
- *Section 257:* The FCC isolates and eliminates market entry barriers to the formation or creation of entrepreneurial ventures in telecommunications or derivative services. This is tacit legislative support for the promotion of multiple market entrants, a spur to the stimulation of optimal competition in all communications markets. This feature of the statute is likely to be one of the most significant with respect to the future of long-term strategic planning in the industry.
- *Section 259:* Local exchange carriers are required to share their networks with competitors that have satisfied criteria for universal service subsidies.
- *Section 271:* An RBOC is allowed to provide interLATA service outside its defined region. Interconnection requirements through a "checklist" must be met and approved by federal and state authorities. The RBOCs must offer intraLATA toll dialing at the commencement of interLATA service.
- *Section 273:* Equipment manufacturing prohibitions imposed by the MFJ are eliminated. Therefore, RBOCs may now manufacture and market equipment, provided interLATA service authorization is secured via the checklist identified previously.

- *Section 274:* Electronic publishing by the RBOCs is allowed, provided such commercial ventures are implemented through separate subsidiaries or non-exclusive joint ventures.
- *Section 301:* Rate regulation sunsets (terminates) on March 31, 1999. Full deregulation of industry thus theoretically commenced within three years following passage of the law.
- *Section 302:* Telephone companies may now penetrate the cable subscription market.
- *Section 303:* Local franchising regulators are prevented from prohibiting cable television providers from delivering multiple services; local government power is thus diminished in regulating this industry.
- *Section 401:* The Act mandates that the FCC forbear applying any elements of sections of the law in industries that do not require intervention in order to facilitate competition. The section of the Act discourages arbitrary intervention by regulatory authorities. This portion of the Act also disallows states from applying federal statutes and regulations that the FCC itself has chosen not to implement.
- *Section 402:* Beginning in 1998, and within the parameters of successive two-year cycles thereafter, the FCC is required to review and evaluate all regulations and remove those determined to be irrelevant, immaterial, unnecessary, or redundant. This section of the Act will later be vital for strategists bent on challenging emerging rules and regulations incompatible with their "convergent strategies." This item is discussed at length in Chapter 7.
- *Section 403:* The FCC is given the authority to eliminate any provisions of the 1934 Act that are deemed to be irrelevant. The powers extended through Section 403 are essential in carrying out the intent of Section 402.
- *Section 552:* All forms of telecommunications services—broadcasting, satellite, cable, telephony, and other distributors—are encouraged to designate a fund aimed at developing new technologies to facilitate parental control of video information.
- *Section 702:* The RBOCs assume the responsibility for the confidentiality and privacy of proprietary information they transmit.
- *Section 706:* The FCC and its companion state regulatory commissions are required to expedite deployment of emerging telecommunications services for all Americans. The FCC retains the right to design new incentives for accelerated advancement of critical technologies.

- *Section 707:* Creates a Telecommunications Development Fund designed to generate capital for low-interest rate loans for small businesses in the industry.

Other sections of the Act bear pertinence, directly or indirectly, for the strategic management process. The items indicated above, however, are those which directly affect pricing, promotion, distribution, or emerging product design. In short, firms cannot hope to compete assertively or aggressively in the absence of an assessment of the extent to which these features of the 1996 statute directly impact their response to market forces.

2.7 The Telecommunications Act of 1996: Redefining Strategy in Deregulated Markets

The impact of deregulation has been uneven with respect to the evolution of the communications industry. A contemporary definition of telecommunications had been asserted earlier in this volume: the transmission of video, voice, or data over wireline or wireless networks. These networks, in turn, are governed increasingly by the interaction of the telephony, broadcasting, and computer sectors. Thus, a communications firm can no longer afford to rely on a strategic management process exclusively focused on the parameters of its own sector: a broadcaster cannot define its strategic plan without considering the potential competition from a local exchange carrier; a telephone company cannot permit itself to ignore opportunities or threats emerging from cable television enterprises; a computer firm cannot dismiss an unfolding avenue in internet telephony or content development, and so forth. The shifting business landscape—the product of technological progress, relaxed government regulation, intensified global competition—requires a broadening of strategic planning focus. The modern telecommunications firm is, in short, primarily a provider of communications products or services potentially distributed to market through [12, 18]:

- Multiple technical platforms;
- An expanding array of business formations, including mergers, acquisitions, intrapreneurial and entrepreneurial ventures, alliances, affiliations, and partnerships;
- Co-opetitive competitive structures, in which successful firms learn the process of adapting to transitional agreements with firms whose objectives temporarily conform to their own;

- Global penetration, particularly with respect to markets newly exposed to competitive market structures;
- Dynamic marketing mechanisms, culled from data warehousing and mining, which facilitate customized products for both businesses and consumers.

Subsequent chapters explore cutting-edge methods in designing and refining strategic management processes to accommodate these criteria. It remains salient that those firms who have demonstrated greatest success in the early years of post-Act regulatory reform are those most amenable to these emerging tactics [19]. Deregulation did not originate these competitive changes, but has effectively accelerated them.

2.8 The Telecommunications Act of 1996: Landscape of New Business Opportunities

There are several methods by which one can gauge the impact of the 1996 statute on emerging business developments in the industry. New business opportunities can be broadly categorized as:

- Seminal: newly-created enterprises, triggered by technological breakthroughs, which aim at supplanting veteran service and product lines and often instigate substantial competition within the industry;
- Derivative: enterprises which seek to exploit, but not necessarily originate, improved technologies or production processes so as to improve veteran product lines;
- Supportive: firms which develop only after seminal or derivative enterprises are built, and which seek to serve these interests through production, distribution, or marketing.

Each category is intrinsically entrepreneurial, dynamic, and subject to volatile change; indeed, a change in one category triggers a transformation of each counterpart. Empirical evidence in the post-1996 Act period suggests an expansion of new enterprises within each of these classifications [14]. These entrepreneurial activities represent, in part, a response to deregulation, and, as indicated previously, a reaction to technological and industrial cross-fertilization promulgated by the statute [15]. The economic principles of evolving seminal, derivative, and supportive enterprises are discussed in the following chapter.

There have been serious attempts to measure the current and prospective impact of new enterprises as spurred by telecommunications deregulation and technological innovation. One manifestation of projected new business opportunities is an annual list of ten emerging communications opportunities published by *Telecommunications* magazine [16, 20]. Table 2.1 enumerates the annual lists of new "telecom" business opportunities in the years immediately preceding and following the 1996 statute.

The assimilation of data reveals both convergent and divergent opinion about those communications technologies most likely to impel future change in the industry. Significantly, the year-to-year listings elaborate the growing emphasis on Internet usage, wireless services, and technical apparatuses required to facilitate storage and rapid distribution of information. What is most striking, perhaps, is the extraordinary proliferation of new technologies: the dramatic rapidity of innovation renders discontinuity throughout these lists. No single technology retains its place or prominence on a year-to-year basis. One can discern obvious technological propensities in wireless and wireline data traffic, Internet commercial usage, multimedia development, computer programming, and similar trends, but experts have great difficulty in forming a consensus about dominant state-of-the-art tools. The absence of agreement reflects precisely the uncertainty embedded in current strategic management processes. Clarity of planning begins in successfully forecasting these emerging methods of communication [19].

The data arrayed do not suggest that technological innovation, in the short-run, has been significantly influenced by adoption of the 1996 Act. What has changed in the period immediately following passage of regulatory reform is the nature, number, and design of new business formations intended to exploit those technologies. The Telecommunications Act of 1996 is regularly denounced by some as having fallen short of the goals initially projected for it. This debate will surely endure through the coming decade, but it now appears irrefutable that the statute has intensified the complexity of modern strategic management for the industry.

2.9 Telecommunications Deregulation and the Global Economy

The transformation of American regulatory policy in communications holds great significance for the global economy. Several developments in the industry's restructuring have already punctuated global economic development in the four years following passage of the statute. First, we can detect a dramatic increase in the number of mergers, acquisitions, partnerships, affiliations,

Table 2.1
Year's Ten Hottest Technologies in Telecommunications*

Rank	1994	1995	1996	1997	1998	1999
1	Digital Wireless	Plug/Play Internet Use	Java Prgmmg	Web Broadcasting	Multi-Layer Switching	Service Mediation
2	Remote Networks	DBS	VO Frame Relay	Remote Acc Servers	Erbium Doped Amps	Metro DWDM
3	ATM	PCS	Virtual LANS	Extranets	Packet Voice	OC768SONE
4	Gigabit Networks	Carrier Frame Relay	Cable Modems	Internet Telephony	LMDS	HDSL2
5	LAN Switch Hubbing	ATM/LAN	Gigabit LANS	ENDS	WLL	Unified Messaging
6	Super-Intelligent LAN	Videoservers	Internet Appliances	Website Management	Virtual Pvt Networks	Policy-Based Management
7	Voice/Data Integration	Wireless Data	Personal Sat Phones	IP Switches	Web Call Centers	MPLS
8	Reengineered Security Systs	Data Crypts	Intranets	WDM	XML	SmartAntenna
9	Neural Netks Management	Digital Cash	ANMs	Digital Sub Lines	TMN Technology	Optical Switching
10	Multimedia Info Delivery	Real-Time App Sharing	Data Mining	POTS Modems	Java Net Management	Wireless Location

*Note: Projections drawn from a variety of experts throughout the telecommunications industry, including consultants working in the field of specialization indicated. Listing employs qualitative methodology and culls expert opinion each May during the respective years Identified above. Data was culled from [21].

alliances, and intrapreneurial ventures within the industry since early 1996 (documentation certifying this trend is supplied in Chapter 8). This pattern of industry restructuring is now being replicated in other regions of the world, particularly in Europe. Second, the convergence of telephony, broadcasting, and computer technologies (in products and services) has spurred productivity and enhanced competition in industries ancillary to the communications sector [21]. Third, that fraction of American trade comprised of telecommunications is rising at a significantly greater percentage than other sectors of the domestic economy. Together, these developments suggest that many economists and strategists may have underestimated the impact of deregulation on communications and, in turn, the growing tertiary importance of this sector for higher value-added services throughout the economy.

Simultaneous to the American experience with deregulation is the experiment now underway in various regions throughout the globe. The pattern of industrial reconfiguration now unfolding in the United States is running concurrent to a number of experiments now engaged elsewhere. Deregulation, reinforced by privatization of public communications infrastructure, is asserting an impact in many nations (refer to Chapter 9 and the Appendix, Section E for technical analysis). One can embrace or deplore the inevitable restructuring prompted by deregulation, but the pace of global change now unfolding is unlikely to be restrained or redirected in the near future.

References

[1] Shaw, J., *Telecommunications Deregulation*, Norwood, MA: Artech House, 1998, pp. 4–9.

[2] The Bureau of Labor Statistics at http://www.bls.gov.

[3] Schiller, B., *The Economy Today*, New York: McGraw-Hill Co., 1997, pp. 34–36.

[4] Case, K., and R. Fair, *Principles of Economics*, Englewood Cliffs, NJ: Prentice-Hall Co., 1992, pp. 851–857.

[5] Gates, W., *Business at the Speed of Thought*, New York: Warner Books, 1999.

[6] Whiting, R., and J. Sweat, "Profitable Customers," *Information Week*, March 29, 1999, pp. 44–52.

[7] Griffin, C. E., "Electric Avenues: Telecommunications Reform Breaks Down Old Barriers and Creates New Opportunities," *Entrepreneur*, June 1, 1996, pp. 134–137.

[8] U.S. Federal Reserve at frb.org; Also, Bloomberg for analysis of post-1996 telecommunications sector at http://www.bloomberg.com.

[9] Dent, H. S., Jr., *The Great Boom Ahead*, New York: Hyperion Press, 1993, pp. 3–46.

[10] Dent, H. S., Jr., *The Great Jobs Ahead*, New York: St. Martin's Press, 1995.

[11] Dent, H. S., Jr., *The Great Boom Ahead*, New York: St. Martin's Press, 1995.

[12] Heldman, P., *Competitive Telecommunications*, New York: McGraw-Hill Co., 1997, pp. 3--72.

[13] Pressler, L., "Telecommunications Act Oversight," *Congressional Testimony*, February 25, 1999.

[14] Frost and Sullivan Report, "Telecommunications Deregulation to Act as a Spur for Growth in Intelligent Networks Market," *M2 Presswire*, October 1, 1998.

[15] Noll, M. A., "The Telecom Act of 1996: Two Years Later," *Telecommunications*, July 1998, pp. 12–14.

[16] Perrin, S., "Ten Telecom Start-Ups to Watch," *Telecommunications*, July 1998, pp. 36–42.

[17] Brands, H., and E. T. Leo, *The Law and Regulation of Telecommunications Carriers*, Norwood, MA: Artech House, 1999.

[18] Flanagan, P., "This Year's Ten Hottest Technologies in Telecom," *Telecommunications*, May 1998.

[19] Aaker, D., *Strategic Market Management*, New York: John Wiley Sons, Inc., 1998, pp. 58–77.

[20] *Telecommunications* annual review of emerging technologies, May 1994–1999, at http://www.telecommagazine.com.

[21] O'Keefe, Susan (ed.), Masud, Sam, and Doug Allen, "The Ten Hottest Technologies," *Telecommunications*, pp. 26–33, May, 1999.

[22] "Innovation in Industry," *The Economist* (Special Reprint), February 20, 1999, pp. 5–28.

3

The Economics of Telecommunications Convergence: Telephony, Broadcasting, and Computer Integration

Strategic forecasting, planning, and management cannot be successfully formulated in the absence of economic analysis. Fundamental tools of microeconomics and macroeconomics must be effectively applied to the marketplace in order to ascertain whether a firm's strategy is congruent with the competitive environment. These questions are among the factors which the telecommunications industry must confront in the new millennium:

- To what extent does the business cycle influence the demand for telecommunications products and services?
- Does the prospective proliferation of new competitors in the global marketplace stimulate or suppress the supply and demand for American communications services?
- To what degree will competition intensify during the next decade, and can these forces be accurately projected?
- What dynamic forces will drive the competitive and oligopolical market structures for each sector of the industry?
- What variables will guide the economics of competitive interaction between the telephony, broadcasting, and computer industries?
- How will changes in telecommunications influence other sectors of the economy, and how will this resultant adaptation guide future demand for communications products and services?

- How should communications firms evaluate their own competitive positioning during the next decade?
- What communications products will likely dominate the future competitive marketplace, and what strategies will guide their distribution to consumers?
- Which synergies will define future telecommunications competition?
- Will traditional strategic management techniques be useful in the newly emerging marketplace and, if not, how should they be redesigned to respond to the unique needs of this industry?

It is the dynamic complexity of these questions which increasingly disorients communications firms as they devise strategy [1]. This phenomenon is no less apparent in other sectors of the domestic and global economies, but the rapidity of technological change which typifies the telecommunications enterprise amplifies the priority for strategic clarity. A first step toward strategic illumination lies in understanding the determinants of global economic transformation, accompanied by the tools necessary to gauge that change.

3.1 Telecommunications and the U.S. Economy, 2000–2010

The American economy has grown substantially during the past decade. Following the recession of 1990–1991 (the most recent of nine recessions punctuating the post-World War II period), the U.S. economy has experienced declining rates in unemployment, inflation, and interest rates while concurrently enjoying higher corporate profitability, rising stock valuations, and spectacular increases in both consumer and business investment. This era has represented a remarkable period in American economic development. Americans have become wealthier while the world concurrently receded in recession and depression during the late 1990s [2]. Government deficits have evaporated, businesses have become more productive, and manufacturing substantially more competitive. American firms have increased output while restraining price inflation and reducing the unit costs of production [3].

If these patterns of economic development have characterized the 1990s American economy, what reasonable macroeconomic projections might be made with respect to the coming decade? As is often the case, economists have developed no absolute consensus about our probable long-term economic future. We may therefore speak only in terms of probability as we assess emerging business opportunities during the next ten years. We do have at our disposal a set of macroeconomic determinants that assist in providing an educated forecast (see Appendix, Sections D and E).

Data asserted in Chapter 2 generally suggest that American economic expansion, though always subject to temporary downturns, may well extend through 2010. The resurgence of productivity, positive demographic factors, reduction of business and government debt, and the restoration of price stability imply favorable business conditions for all industries. The issue confronting strategic planners in telecommunications is a deceptively straightforward one: to what extent will the communications industry share in this prosperity, and how do these forces define the future of the sector's competitive landscape? With which firms should veteran telecommunications players align themselves in this economic environment? Does a period of sustained national economic growth necessarily expand the demand curve for communications services? If it does, is there any assurance that the established telephony, broadcasting, and computer sectors will all benefit from rising consumer consumption of these services? Are newly emerging firms poised to seize traditional market share from veteran companies?

A positive economic forecast based on the variables indicated above obviously does not guarantee a favorable path for either established or entrepreneurial firms. However, the collection of data—reflecting enhanced productivity, heightened entrepreneurialism, expansive consumption, strong demographics—does suggest the following:

- Inflation is likely to remain very low, reducing the risk of capital acquisition and investment. Thus, business demand for communications products and services is likely to expand significantly during the coming decade, particularly as the world economy emerges from recession after the year 2000.
- Low inflation dictates low interest rates over the long-run. Again, this factor increases the probability that both business and consumer demand for communications services will rise significantly: the purchase of these products defines positive future demand for telecommunications services, particularly for the telephony industry in the short-run. The dramatic reduction in and elimination of government deficits during the late 1990s strengthens this forecast, as government at all levels—federal, state, and local—no longer absorbs capital from the financial markets.
- Unemployment is likely to remain low during the next ten years. This development is a function principally of the decline in new workers entering the workforce after 2000. With a decline in the aggregate

numbers of Americans seeking work, wages are likely to rise in real terms (wage increases minus inflation equal increases in real income). The capacity of Americans to consume will rise, and it is not unreasonable to infer that demand for advanced communications will accelerate, particularly in regard to those services to which consumers become "dependent": these might include Internet usage, wireless data and voice service, and other products hardly yet conceived. It is instructive to recall that basic telephone and cable television service were both regarded as luxuries at the time of their inception. The strategic conversion of "luxury" items to "staples" has now become a principal priority of the modern communications firm.

- The telecommunications industry increasingly displays a propensity toward inelastic demand: that is, individuals, households, and businesses must use communications services irrespective of downturns in the business cycle. This development suggests that consumers require such services independent of cost. Such a predilection does not mean the elimination of price discrimination among consumers—that users will not seek out competitive pricing—but implies that, once adopted, consumers regard such services as necessities (see above). If this tendency asserts and sustains itself during the coming decade, then communications firms may become impervious to fluctuations associated with the business cycle; demand for veteran products and services may not be directly tied to changes in income, productivity, wealth, or national economic policy. The advance of electronic commerce (Internet-based transactions) represents a new conduit for business, and is fueling recent global economic growth. This activity started in the United States in 1994, and has accelerated at a spectacular rate. By some estimates global "e-commerce" will generate $1 trillion in gross revenues by 2002, and perhaps double again before 2010 [4]. One cannot overstate the significance of this transitional era in American economic development. The federal government acknowledged the criticality of this phenomenon in a report issued in 1999 [5]:

> The digital revolution is also changing the respective roles of government and the private sector. In the 19th and for much of the 20th centuries, governments played a key role in helping build or actively regulate much of the country's infrastructure. The federal government made extensive land grants to encourage private capital to expand the nation's rail network. Government subsidies were used to stimulate the development of an airline industry. Federal and state dollars combined to build and maintain the interstate highway system. In communications, the government granted a virtual monopoly to a single company and

> regulated the industry after its breakup. Most power companies have been regulated monopolies at the state or federal level.
>
> The federal government funded and developed early versions of the Internet for national security and research purposes. It will continue to provide funding for research and development on future Internet and high-performance computing technologies. However, most of the capital to build the computing and telecommunications infrastructure is being provided by the private sector.
>
> The pace of technological development and the borderless environment created by the Internet drives a new paradigm for government and private sector responsibilities. Creating the optimal conditions for the new digital economy to flourish requires a new, much less restrictive approach to the setting of rules.
>
> Governments must allow electronic commerce to grow up in an environment driven by markets, not burdened with extensive regulation, taxation or censorship. While government actions will not stop the growth of electronic commerce, if they are too intrusive, progress can be substantially impeded.
>
> Where possible, rules for the Internet and electronic commerce should result from private . . . action, not government regulation. Governments do have a role to play in supporting the creation of a predictable legal environment globally for doing business on the Internet, but must exercise this role in a nonbureaucratic fashion. . . . Competition in telecommunications and broadcast industries should be encouraged so that high-bandwidth services are brought to homes and offices around the world and so that the new converged market place of broadcast, telephony, and the Internet operate based on laws of competition and consumer choice rather than those of government regulation. . . . There should be no discriminatory taxation against Internet commerce. The Internet should function as a seamless global marketplace with no artificial barriers erected by governments. As with any major societal transformation, the digital economy will foster change and some upheaval. The Industrial Revolution brought great economic and social benefit, but it also brought about massive dislocations of people, increased industrial pollution, unhealthy child labor and unsafe work environments. Societies were often slow in responding to these negative side effects. Similarly, the digital economy may bring potential invasions of privacy, easier access by children to pornographic and violent materials and hate speech, more sophisticated and far-reaching criminal activity and a host of other as-yet unknown problems.
>
> The private sector and government, working together, must address these problems in ways that make the Internet a safe environment while not impeding its commercial development.

These conclusions by the federal government are highly revealing. Many public officials now acknowledge not merely their inability to "manage" electronic commerce, but implicitly concede that national economic growth and competitiveness are vitally tied to the course of a digital economy. The migration from a conventional to digital economy creates winners and losers, as well as new efficiencies, but also discontinuities and dysfunctions. These effects are inevitable when consumer choice fuses with deregulation [6]. The communications industry thus instigates precipitated threats and opportunities for all sectors of the macro-economy.

3.2 Telecommunications and the World Economy, 2000–2010

There is no concrete method to chart the course of the entire world economy, and the role of telecommunications in influencing it, for the coming decade. We can make reasonable inferences about the prospective supply and demand of communications services, however. Assuming that the pattern of post-World War II growth rates is sustained through the coming decade, we can presume global economic growth of 3–8% on an average annualized basis [7]. If the world economy grows only moderately during this period, we might see growth at 4%; a high growth rate (beyond 6%) is conceivable if global trade and productivity accelerate.

Two hundred nations comprise the world economy. Each finds itself in a slightly different stage of economic development. For telecommunications firms seeking optimal strategy, these varying stages of development connote a pecking order with regard to investment and marketing. It is probable, and indeed empirical evidence presently confirms, that multinational communications corporations will pursue investment initially in the most advanced industrialized societies [8]. It is in these regions that profitability will be maximized in the short-run. We may also infer that these firms will pursue business-to-business marketing strategies, while corporate customers are more likely to purchase high-end services, thus promoting economies of scale for the most aggressive carriers. Thereafter, households with high disposable incomes and established patterns of communications consumption, will represent the most attractive customer base. Such consumers, given the uneven distribution of global wealth, are scattered in all nations and signify viable markets only when they can be successfully aggregated. Concentrated groups of business and household users, sharing these patterns of behavior, will lure communications investment at the front-end of the next decade [9].

Nominally, it may be argued that the pattern of U.S. communications investments during the 1990s—in mergers, acquisitions, affiliations, partnerships, alliances, and start-ups—will be replicated elsewhere in the decade to come. There are several reasons for this phenomenon. Experiments in deregulated monopolies, coupled with privatization, are seen with increasing frequency throughout the world as the new decade begins. The collapse of command-control economies—principally those in the former Soviet Union and Eastern Europe—have greatly aided this trend, of course. By some estimates, free-market economies have consequently absorbed in this decade an increment of more than one billion consumers for prospective communications products and services. Elementally, it is not feasible to participate in the world economic system without conforming to the rules adopted by the largest players.

Second, the encroachment of information technology in all areas of human endeavor necessitate speedier, more adaptable organizational structures which might consolidate market clusters [10]. The most expedient way of efficiently distributing services to global markets lies, in the short-run, in exploiting the advantages of large, multinational corporate firms: large players can leverage vast amounts of capital so as to quickly build required infrastructure [8]. This step must naturally precede the development of those service enterprises, large and small, which must rely on such infrastructure to attract their customers.

Finally, the surge in networked economies—linking common interests across the globe without regard to geographic boundaries—has impelled governments to accede to intrusive corporate interests. Any government of national, regional, or local origin effectively weighs a trade-off in responding to this development: an attempt to forestall the vanguard of multinational corporate consolidation runs the risk of inciting consumer and business opposition. This opposition is a function of local interests which sustain competitive disadvantage if not permitted access to state-of-the-art communication services. Global governments, for better or worse, now find themselves drawn into the broader web of multinational corporate investment typifying consolidation in the late 1990s. The pattern of corporate investment and consolidation is explored further in Chapter 8.

3.3 The Emerging Economics of Telecommunications

Economists and market researchers have discerned the emergence of new industry structures governing the creation and distribution of telecommunications products and services. In effect, innovation in communications products have precipitated economic changes in other industrial and consumer sectors of the economy; this resultant economic transformation has influenced subsequent innovations in market planning. The telecommunications industry is fostering an explosion of product proliferation which has introduced a metamorphosis in economic and strategic analysis. The planning of strategy must now consider these emergent economic developments:

- *Aggregation:* the method by which firms seek to expand and retain a customer base;

- *Complementation:* the synergy of two or more firms cooperating to develop new products or cultivate new markets;
- *Consolidation:* the procedure in which competitors merge, acquire, or otherwise transform their company structures to achieve aggregative or complementary strategies;
- *Disintermediation:* the unfolding loss of traditional distribution markets, in which firms and their employees lose access to their customer base via the encroachment of new firms which directly link producers with consumers;
- *Tertiary analysis:* an economic tool, based on the study of externalities, aimed at forecasting the development of new markets which are directly precipitated by the advance of technology; and
- *Convergence:* the circumstance in which a traditional sector of communications is either invaded by, or chooses to invade, an unrelated sector as a result of technological innovation.

The following sections isolate the economic determinants of aggregation, complementation, consolidation, disintermediation, tertiary analysis, and convergence. One must bear in mind that the economic and strategic significance of these tools remain in flux, and are influenced by variables as yet not fully understood. We can empirically verify the unfolding presence of these economic phenomena, but not arrive at a certitude about their long-term meaning for strategic planners.

3.4 The Economics of Aggregation

To aggregate—to attract and retain—customers has become an overriding concern for communications firms during the past decade. The strategy of market aggregation is, in principle, no more significant for telecommunications than for other sectors of the economy. However, new methods of estimating and projecting the economic value of communications customers have attained heightened status in recent years. There are two primary reasons for this development: (a) in the face of deregulation, customers—households and businesses—are becoming resourceful at pursuing alternative providers of both products and services; (b) the continued expansion of Internet services has promulgated many new firms whose own valuations pivot on the dynamics of aggregation [11]. Simply, communications firms are groping for a method of estimating their own true prospective value in a changing marketplace. Traditional methods (among them price-earnings ratios, subscriber revenues, and cash-flow management) of company valuations are often inadequate in and of themselves to provide such a determination.

How does a telecommunications firm assess its own value amid mounting competition stimulated by deregulation and the convergence of the telephony, broadcasting and computer industries? A common basis of valuation is required, and such a quantitative method must conform to the new contours of technological innovation. In this environment, we are slowly witnessing a consensus among market forecasters: the economics of *reach* and *loyalty* define present company value [12]. "Reach" is a quantifiable variable indicative of the degree to which a firm's message penetrates markets. In other words, that fraction of the market to which a firm's message and image are exposed assert an essential element of underlying value. "Loyalty" refers to that percentage of a customer base who may be regarded as continuing customers [13]. The interaction of these two variables increasingly correlate with the perception of a firm's value; the stock and venture capital markets in recent years reaffirm the growing significance of economic aggregation. A firm which can demonstrate (see Figure 3.1) growing valuations via these two variables is rewarded for such success, whether publicly or private-held. Further, the capacity to float bonds or borrow from the capital markets—often necessary for expansion—is considerably eased

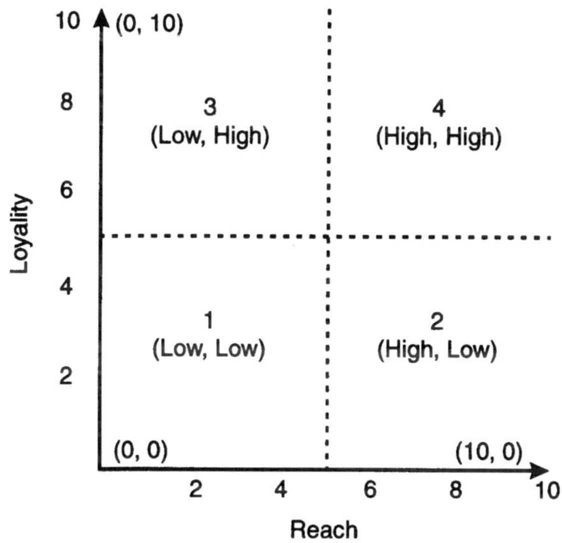

Figure 3.1 The economics of "reach" and "loyalty" hold that a firm's valuation (or market capitalization) is optimized when high rates of customer reach and loyalty intersect. Optimal valuation is attained in Quadrant 4 (High, High) REACH/LOYALTY.

through valuations that rise in response to strategies which exploit loyalty and reach. Costs of running a business are reduced through lower interest charges, and profitability is increased through the attainment of this strategic objective.

Economic aggregation, the measurement of loyalty and reach, is likely to also be a critical factor in instigating mergers and acquisitions, as well as business alliances and partnerships. A firm adept at effectively designing strategy applied to both loyalty and reach is often rewarded for its long-term prospects, even if short-term profitability is nonexistent. The capacity to perpetually engage the customer's attention represents extraordinary long-term value to communications carriers, content providers, and other distributors of information [14]. Investors presume that perpetual customer exposure to a firm's message signifies primary, secondary, and tertiary value for the long-term. These issues are discussed further in Chapter 4.

3.5 The Economics of Complementation

The process by which two or more firms cooperate by exploiting their counterpart's competitive advantage may be referred to as *complementation.* In dynamic industries, where change is impelled by dramatic transformation of social, cultural, political or economic systems, or through technological innovation, we can confirm the rapid reconfiguration of business enterprises [14]. Such reconfigurations—new organizational schemes—are intended to penetrate new markets, expand market share, and overcome prior obstacles to market entry. In the case of telecommunications, political, economic, and technological change have coalesced to fuel such synergies.

It is no longer feasible for a single telecommunications firm to dominate all avenues of infrastructure, just as no content provider can direct or influence the course of broadcasting. The disintegration of AT&T's monopoly, the elimination of arbitrary dichotomies between the principal communications sectors, and the competition—domestic and global—promulgated by deregulation, have combined to necessitate cooperation of competitors. This is not to say that these firms do not regard themselves as industry rivals or long-term opponents. To the contrary, the shifting competitive landscape can alter the very reconfigurations of those business firms which surfaced in response to such change.

It remains salient, however, that a broadcaster may itself be in need of software support only supplied by the computer sector; the telephony provider may be compelled to absorb a cable firm with access to markets it cannot presently secure; the computer firm may wish to facilitate the objectives of leading firms in other sectors to assure a ready demand for its product. The multiplicity of scenarios is expansive, ever-changeable, and demands far-sighted strategic management. These matters of "co-opetitive" dynamics are explored in greater detail in Chapters 7 and 8 but, as elaborated in Chapter 1, there exists a menu of criteria isolating the economic dynamics at work.

The economics of complementation hinge on these simple criteria:

- A competitor is a complementor if consumers value a company's product more when value absorbed from that competition makes that product more desirable than all others available at that time. If your cooperation with another firm leads to an expanded consumer base—whether or not that competitor resides in your industry sector—it is foolish to ignore a prospective alliance, partnership, or affiliation.
- A competitor represents a prospective complementor, after an alliance is forged, if suppliers are more likely to provide a continuity of supply at favorable prices than if the same firm operated in the absence of such an alliance.
- A competitor is a complementor if this is the only mechanism available to facilitate immediate improvement in price, quality, service, or post-purchase satisfaction.
- Complementation is a viable strategic priority if these criteria are met and acceptable standards of benchmarking—methods of accurately evaluating competitive performance—can be established. These benchmarking methods can include productivity performance, consumer satisfaction, market share percentages, profitability, cost reduction, and similar measures.

The principal authors of "complementation," Adam Brandenburger and Barry Nalebuff, have simplified their elaborate theory in this way: ". . . there are win-win and win-lose relationships with *customers, suppliers, complementors, competitors* . . ." and these relationships continually shift [15]. Identification of the time to forge or disband complementary relationships is partly intuitive, but also partly empirical science. Employing these criteria to define optimal moments for engineering such relationships refines strategic alternatives.

3.6 The Economics of Consolidation

The extraordinary proliferation of telecommunications mergers and acquisitions following the passage of regulatory reform in 1996 is noteworthy in terms of size, scope, and complexity. Consolidation, the product of merger and acquisition, represents a strategic attempt to combine the resources of symbiotic partners while reducing unit costs of production.

Consolidation is inevitable in the early stage of deregulation in virtually every industry. In trucking, aviation, petroleum, natural gas during the late 1970s and 1980s, in banking, financial services, and electric utilities during the 1990s, we can verify this repetitive experience. Economists debate the value and purpose of consolidation from a social and political perspective. The motive for such a strategy, viewed exclusively in business terms, remains unmistakable: lower production costs coupled with greater market share yields higher profitability and competitive advantage.

Integration, *vertical* and *horizontal*, is the tool of such consolidation, and both procedures have been seen with increasing frequency in the communications industry since 1996. Indeed, this trend has been manifested in both domestic and global mergers in the past several years. Mergers are defined as a combination of two or more firms; an acquisition represents the purchase of one firm by another. Vertical integration is defined as the ownership of networked production or distribution of a product of service. A firm engaged in vertical integration chooses to purchase or merge with those enterprises that support its delivery of products or services. Vertical integration is expressed as "forward" or "backward": if a cable television firm purchases a manufacturer of set-top boxes for consumer selection of programming, its strategy is supportive and therefore backward. If a motion picture studio purchases a television network so as to enter a new market for distributing its product, its strategy may be termed "forward." It is forward rather than backward integration which incites anti-trust investigation and enforcement.

Horizontal integration occurs when one firm purchases another so as to expand market share. Viewed benignly, such integration can enhance economies of scale; if reduced production costs result in lower prices, both producer and consumer may benefit. If reduction in competition yields higher prices and less innovation, sometimes the result of such strategy, prospective improvements in production, and distribution are stifled.

These standard definitions of horizontal and vertical integration do not fully reveal the economic complexities of consolidation. The telecommunications industry, given the converging telephony, broadcasting, and computer sectors, is stimulated today by shifting domains of content and infrastructure. In this respect, increasingly sophisticated attempts to deliver multi-dimensional content to consumers dictate advanced infrastructure; innovative infrastructure, in turn, enables content providers to lure more consumers to high value-added services. The interests of content providers and producers synergistically blend in each of these three sectors. Thus, the wave of mergers and acquisitions *within* each sector will evolve over the longer-run to a comparable trend *between* these sectors. Consolidation may paradoxically provoke intense competition between and among these converging sectors, although this scenario remains pure conjecture at present. Table 3.1 offers a hypothetical projection of horizontal and vertical integration, and the objectives and motives of such strategy, through 2010 for the communications industry.

3.7 The Economics of Disintermediation

American business has now entered the first stage of what might be termed "economic disintermediation." Disintermediation is the process by which mid-level management and distribution processes are effectively contracted or eliminated. The advent of an Internet-based economy has facilitated, for the first time, direct transactions between producers and end-users. Consumers, through electronic commerce, are now able to circumvent formal distribution channels to command lower prices, simultaneously accelerating rates of transaction. The result is that mid-tier managers and other workers—traditional "middlemen"—are vulnerable to job loss. Those firms who choose not to participate in the digitized economy lose competitive advantage and market share, in addition to prospective clientele. Economic disintermediation is becoming a fact of contemporary business life, and cannot be avoided or evaded in light of advances in communications technology [16]. While advances in telecommunications technologies have imposed this restructuring of modern commerce, so too have they introduced similar opportunities and restrictions in this industry.

Table 3.1
Projected Horizontal and Vertical Integration, 1997–2007

	Objective/Motive			
	Front-end (1997–2002)		Back-end (2002–2007)	
M/A Partners	Horizontal	Vertical	Horizontal	Vertical
Service providers	Immediate expansion of market share	Procurement of marketing expertise to support sales	Diminution of competitor market share	Diversification (global and domestic)
Content developers	Access to networks and technical support	Procurement of complementary expertise in emerging markets	Long-term competitive positioning	Access to global markets

In the transition from an intermediated to networked economy, a firm now faces the challenges of *managed disintermediation:*

- Applying the economics of intermediation, disintermediation, and reintermediation;
- Identifying shifting customer tastes amid proliferating choices in the networked economy;
- Training sales and marketing staff to specify emerging market opportunities;
- Specifying prospective opportunities for disintermediation in the distribution chain;
- Planning sector strategy during the networked transition;
- Elaborating new methods of reintermediating value in the distribution chain;
- Isolating appropriate forecasting methods for networked market objectives;
- Leveraging technology to maximize market exposure at minimal cost;
- Differentiating product value in networked distribution chains;
- Pricing products and services during transitional reintermediation.

In this reconfigured networked economy, both individuals and organizations must learn the techniques of reintermediation [17]. Consider the following strategic scenarios: (1) How does a computer reseller insert itself back in a distribution chain when other firms have seized its market through direct producer-to-buyer sales? (2) How does a television broadcaster respond to the proliferation of niche cable channels, resulting in displacement of traditional mass audiences? (3) How do telephony carriers retain traditional revenues in the face of Internet telephony? (4) How do veteran traditional telecommunications providers in telephony and broadcasting respond to dramatic innovations in software development which replicate or simulate standard services formerly the domain of large infrastructure competitors?

These contingencies are among recent developments now seen in the communications industry. Scarcely a month has passed since the passage of regulatory reform that dramatic evidence of disintermediation has not been seen in the American economy. The motive for disintermediation lies in cost reduction; decline in production costs leads to more competitive pricing; competitive pricing, for those firms able to exploit such advantages, leads, of course, to rising market share. As noted in subsequent chapters, however, those firms which successfully practice the art and science of disintermediation may later be subjected to the same tactic. Those who disintermediate can themselves be disintermediated. Strategic design aimed at exploiting and responding to disintermediation is discussed in Chapters 7 and 8, along with the issues described above.

The century-long trend in adding intermediaries—intermediation—is now giving way to the technological turmoil and opportunity of disintermediation. Conceiving tools of reintermediation now beckon all firms, and the professionals who manage them. In this regard, both the domestic and global economies shift to a new paradigm for those businesses seeking the benefits of strategic disintermediation [18]:

> The key managerial challenge of the emerging global infostructure will be to reconfigure organizations to create and maintain high profits in a low transaction cost environment . . . These costs include those of searching for and identifying products, drafting, negotiating and safeguarding the terms of a sale or purchase, payment and settlement, and the costs incurred to enforce contracts or to correct and resolve contract disagreements. As communication becomes inexpensive, and software agents, technology and transaction templates are better defined, the transaction costs incurred by firms and individuals will fall. This in turn will make markets more efficient.
>
> Firms will have to reconfigure their strategies and organization to respond to inexpensive communications, lower transaction costs and reductions in profit opportunities from market inefficiencies in the emerging global infostructure. As management and economics researchers have noted, successful firms focus their strategies along a few salient dimensions for competitive advantage.

For communications firms, the heightened priority of cost reduction, and the organizational development requirement to instigate and sustain it, is likely to remain predominant during the coming decade. Savings produced via cost reduction can be reinvested back into value-added services necessary to strengthen customer retention. The proliferation of new business formations seen after passage of the Telecommunications Act of 1996 certify the inevitability of this transition. Chapter 8 outlines the multiplicity of motives surrounding these creative business restructurings, and the strategic path major competitors have now carved for the immediate future.

3.8 The Economics of Tertiary Analysis

That body of economics which examines the unintended consequences of business transactions is known as *externalities*. Externalities may be categorized as positive or negative, depending on the perspective of the observer [19]. If, for example, the adoption of new communications technology leads to increased use of telecommuting—a tactic whose primary purpose it is to improve productivity—secondary firms will necessarily be created to support it. These newly emerging firms, in turn, will generate other businesses as a consequence. If telecommuting suddenly permeates a community, rents may fall, restaurants nearby might fail, and government tax revenues could recede; these tertiary consequences might be grouped as "negative." At the same time, demand for home office furnishings and equipment might rise exponentially, automobile insurance outlays might fall (as a result of fewer accidents on the roadway), and environmental quality might improve; these resultant tertiary effects would be construed as "positive."

Viewed in practical business terms, every transformation of the economy associated with such technologies constitutes threat and opportunity. Economist Joseph Schumpeter has referred to this phenomenon as "creative destruction." New technologies not only disrupt veteran industries, their unintended effects can often re-shape virtually every sector of the economy. If a firm can successfully forecast with some precision these unintended consequences, new markets can be cultivated to its competitive advantage. Every metamorphosis precipitated by technological innovation inevitably leads to these subordinate tertiary consequences. Applying a systems perspective to the analysis of tertiary consequences, we can arrive at some broad principles underlying these changes. A paradigm for such analysis is posited in the Figure 3.2.

The model identifies the unfolding market effects of telecommunications innovation. When a new product is introduced, serving its constituency, we note a first-ordered tertiary effect. The development of network television, as a case in point, introduced sales and maintenance professionals necessary to sustain the industry. A second-ordered effect was the transformation of the motion picture industry, as this sector coped with a declining market for movies. A third-ordered effect was the reconfiguration of the advertising industry, which suddenly responded to a mass audience now bound to daily television viewing. As each ordered-consequence is assessed, we note subsequent tertiary consequences (fourth-, fifth-, and sixth-ordered effects) for competitors, complementors, suppliers, and distributors. The economics of tertiary effects, and consequent markets for products and services, should become a concern of strategic planners in communications firms in the years to come. The very

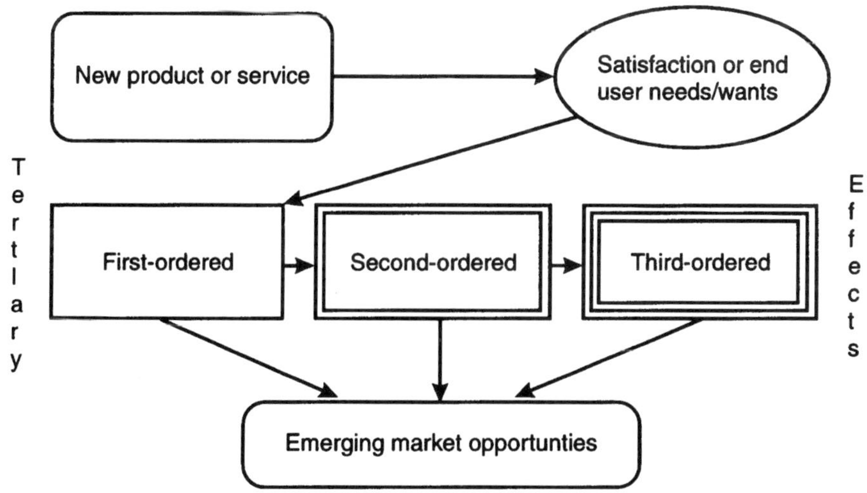

Figure 3.2 The telecommunications tertiary effect.

innovations introduced by communications firms may result, in some instances, in future competitive threats for the industry.

3.9 The Economics of Convergence

The decision of a communications firm to penetrate an unrelated sector of the industry—that is, launch new product lines or cultivate emerging markets outside its own domain—can be construed as sector convergence. Ordinarily, the expression "convergence" is used as a euphemism for the dynamic interaction among the broadcasting, telephony, and computer industries [20]. This recent phenomenon is directly attributable to deregulation, which effectively diminished or eliminated the artificial boundaries dividing these sectors.

Technological innovation demands that new economic tools be applied to estimate the business effects of convergence. The strategic significance of these overlapping sectors, and the economic changes they promulgate, are as yet only vaguely understood by economists and business strategists [21]. While little methodological progress has been made in forecasting the dynamics of economic convergence, the parameters of the problem are becoming apparent [22]:

- Technological innovation has accelerated the process of product substitution in recent years, thereby dramatically contracting development schedules and expediting product release dates. Firms are thus compelled to "obsolete" their own product lines at ever-increasing rates of acceleration.

- There is increasing evidence that it may be more important to be "first to market acceptance" rather than "first to market" relative to new product lines: a firm must anticipate, more than ever, the risks associated in investing in *product development versus market research.* Amid dynamic convergence between these three sectors, does a firm prosper more if it is the originator of new technology . . . or should the same firm permit (or even encourage) its competitors to make such an investment while its strategizes a procedure to capture market share after replicating that product?
- To what degree does synergistic interaction between these sectors affect prospective demand elasticity for communications services? In other words, as these sectors both compete and cooperate, which products and services represent future platforms for the industry? Will the computer industry dominate telecommunications services in the future, or will consumer demand be linked primarily to veteran firms whose focus has been an improvement in traditional infrastructure? Will content development drive advances in infrastructure, or does infrastructure define eventual communications usage and distribution?

The following chapter explores economic tools that can be applied by strategic planners to estimate future consumer demand, the elasticities of that demand, and its probable determinants. One is cautioned to recall that the principles and models of macroeconomics and microeconomics are being reshaped in the post-Act era. Information usage—and the products and services which support it—follows a logic of its own. Unlike all other economic resources, information—and the infrastructure which undergirds it—proportionately increases relative to its consumption. The more we exploit knowledge—the key economic resource of the modern economic period—the more expansive our need for data storage, distribution, and interpretation [23–25]. Knowledge and information cannot be depleted, an inescapable and overriding tenet of the information era.

References

[1] Hope, J., and T. Hope, *Competing in the Third Wave,* Boston: Harvard Business School Press, 1997, pp. 109–128.

[2] Dent, H. S., Jr., *The Roaring 2000's,* New York: Simon and Shuster, 1998, pp. 14–27.

[3] "Report on the U.S. Economy," Business Intelligent Unit, New York: DeLoitte, Touche, in cooperation with DeLoitte Consulting, January 15, 1999.

[4] Kambil, A., "Electronic Commerce: Implications of the Internet for Business Practice and Strategy," *Journal of Business Economics*, October 1, 1995, pp. 27 (7).

[5] "The Emerging Digital Economy," U. S. Dept. of Commerce Report, Washington, D.C.: GPO, 1999.

[6] Shaw, J., "Future Scenarios for the Telecom Industry: A Ten-Year Forecast," *New Telecom Quarterly*, Austin, TX: Technology Futures Inc., October, 1996, pp. 25–33.

[7] Schiller, B., *The Economy Today*, New York: McGraw-Hill Co., 1997, pp. 370–377.

[8] Buckley, P., and M. Casson, "Models of the Multinational Enterprise," *Journal of International Business Studies*, March 22, 1998, pp. 21 (24).

[9] Baron, D. P., "The Nonmarket Strategy System," *Sloan Management Review*, Vol. 37, September 22, 1995, pp. 73 (13).

[10] "Telecomms Deregulation to Act as a Spur for Growth in Intelligent Networks Market," Frost and Sullivan Report, M2 PressWIRE, October 1, 1998 (press announcement).

[11] Martin, C., "The Strategist," *New Media*, March 1999, p. 20.

[12] Wagner, C. G., "Digital Engine Powers New Economy," *The Futurist*, Bethesda, Maryland: World Future Society Press, April 1999, p. 20.

[13] Golum, R., "E-Commerce Crazy Economics," *Investor's Business Daily*, March 26, 1999, p. 12; For information on new methodologies to estimate the value of Internet-based marketing penetration visit industrystandard.com.

[14] Schwartz, E. I., "Brands Aren't Everything," *The Standard*, May 10, 1999, p. 41.

[15] Brandenburger, A., and B. Nalebuff, *Co opetition*, New York: Doubleday Press, 1996, pp. 36–39.

[16] Scott, D. L., "Disintermediation," *Wall Street Words*, January 1, 1998, p. 22.

[17] Wilder, C., "Special Report: Emerging E-Commerce," *Information Week*, October 20, 1997, pp. 94.

[18] Weiner, E., and A. Brown, "The New Marketplace," *The Futurist*, Vol. 29, May 1, 1995, pp. 23–29.

[19] Case, K. E., and R. Fair, *Principles of Economics*, Englewood Cliffs, NJ: Prentice-Hall Co., 4th ed., 1998, pp. 444–458.

[20] Shaw, J. K., "The Emerging Business Economics of Telecommunications," *New Telecom Quarterly*, April 1997, pp. 9–19.

[21] Rosenbush, S., *Telecom Business Opportunities*, Newport, RI: Aegis Press, 1998, pp. 35–68.

[22] Asthana, P., "Jumping the Technology S-Curve," *IEEE Spectrum*, June 1995, pp. 59–54.

[23] Davis, S., and C. Meyer, *Blur: The Speed of Change in the Connected Economy*, New York: Addison-Wesley Press, 1998.

[24] *Designing Tomorrow's Organization*, Economic Intelligence Unit, (Special Report) DeLoitte, Touche Consulting, 1998.

[25] Tapscott, *Blueprint to the Digital Economy*, New York: McGraw-Hill Co., 1998, pp. 96–110.

4

Strategic Management and the Consumer: Analysis of Emerging Demand

Effective strategic management is predicated principally on the analysis of dynamic supply and demand. The interaction of these forces defines price, outlines production schedules, spurs future innovation, and specifies the determinants of consumer preference. Armed with this information, practitioners gain a keener insight into managing strategy and enhancing competitive positioning. While the capriciousness of the American consumer remains a preoccupation of strategic planners in all sectors of the economy, empirical evidence mounts which enables us to profile the nature of telecommunications consumption. There are significant variations among providers in the telephony, broadcasting, and computer industries, but recent research reveals a continuity in demand for particular services [1]. Before strategists weigh the organizational appropriateness of the proposed strategic models described in this volume, a nominal understanding of emerging consumer demand must be attained.

Economists have confirmed the variables responsible for consumer demand, and describe these factors as determinants. The determinants of demand are assigned by the computation of [2]:

$$D = T + I + E + \text{(other factors)}$$

where D = demand; T = taste; I = income; E = expectations; and other factors include the availability and price of alternative goods or services. We refer to this expression as the "demand equation."

Thus, the three main determinants of demand include consumer taste (or preferences), prevailing income, and the expectations (or prospective satisfaction) consumers hold for that product. These causative elements are supplemented by competitive options consumers have before them at the time of transaction. The credibility of these variables as decisive in setting the pattern of consumer demand have been documented over and over again in recent decades, and are regarded as beyond methodological dispute. Superimposed over these individual patterns of consumption are environmental factors which inspire the interaction of group and individual; these contributing influences include social, political, and cultural dynamics.

4.1 Strategic Management and the Emerging American Consumer

Data supplied in Chapters 1 and 3 confirm the transitional character of the American consumer following the 1990–1991 recession. It is during this era, and particularly after 1996, that we can empirically confirm a striking improvement in per capita American income, consumption, productivity, and wealth creation [review Appendix, Section D for a detailed analysis of these data) [3]. While the confluence of these improvements clearly suggests the capacity to purchase an expanding array of goods and services of various kinds, they in no way imply a wider distribution of that consumption. In other words, apart from the social implications of concentrated wealth, marketers should be aware that the immediate future holds that present consumer segments will prevail. Aggregate wealth is likely to mount significantly in the decade ahead, cementing current consumer niches (whether denoted by income, demographics, educational status, regional origin, and so forth).

An understanding of the broad dynamics of consumer demand as they inspire change during the coming decade is a necessity for strategic planners. We can discern through the empirical evidence the contours of these demographic, economic, and social forces. These "environmental dynamics of consumer demand" are united with the maturation of the Baby Boom (that generation ordinarily defined as having been born during the twenty years following World War II). It is certifiable fact that peak consumption, investment, and savings are variously attained by Americans moving through the ages of 35 to 55 [4]. Eighty million Americans reside within this demographic, all within working age, all now approaching peak productivity. For the coming decade, the American and global economies will be governed principally by their interaction. In addition, a transformation of consumer demand will be ignited vis-a-vis the environmental dynamics described in Section 4.2.

4.2 Environmental Dynamics of Consumer Demand

In recent studies cutting across all sectors of the economy, and including the communications sector, we can specify the growing significance of these environmental dynamics and the nature of consumer demand:

- The shift from producer to consumer control: The vast proliferation and ubiquity of information available to consumers has effectively demystified consumer decision-making at the point of transaction. Consumer access to, and storage and distribution of, information have compelled producers to respond expeditiously to consumer needs. Customization of options has led some firms to contemplate particle marketing: treating each household as if it were a market unto itself [5]. Such one-to-one market research has been made a reality through advances in inventory management, technology, and data warehousing. This environmental dynamic is directly connected to advances in reduction of production costs.
- Production cost reductions: new management techniques in inventory control, market research, and networked supply distribution have combined in dramatically reducing production costs. Just-In-Time Inventory Management has activated Just-In-Time Labor Management. These twin J.I.T. phenomena portend an immediate future of minimized inflation because production cost reductions have accrued to those firms adapting to this production shift. Firms most agile in adjusting to this change have secured greater market share—exploiting competitive advantage—at the expense of those unable to respond [6].
- Mass customization: As noted in Chapter 1, we have witnessed during the past two decades a rapid increase in the number of consumer market niches. In many industries, this expansion of consumer choice, evolved from mass segmentation, has led to the theoretical possibility of customization: that circumstance in which consumers dictate production control by specifying all attributes of a product or service. In effect, such an environment is one in which the consumer tells the producer what to make, when to make it, and when to distribute it. Consumers exercise increasing influence relative to product pricing. One might reasonably infer, as many observers have, that smaller firms are thus positioned to strip away succeeding layers of a corporation's consumer base—that a trend toward customization represents a market opportunity for smaller firms able to customize a product line. However, the evolution of this strategy naturally leads to the concept of mass customization: that environment in which a major firm is able

to customize its market for every consumer of its product or service [7, 8]. Product customization can now be aligned such that a major producer can permit customers ample latitude in outlining the characteristics of same. Market segmentation has thus transitioned from niche marketing to customization (particle or "one-to-one" marketing) to mass customization (see Figure 4.1 below).

- *Product life-cycle contraction:* advances in technology have now intersected with sophisticated market research. Firms today have a keener sense of the desires, interests, and needs of customers. These two factors have enabled companies to introduce new products at dizzying rates; indeed, many marketing and sales departments now report that they have little time to study a new product line, such that they can adopt appropriate strategy. Products now come on line so rapidly that marketers must innovate new techniques to educate consumers as to their value [9]. Predictably, obsolescence is now instigated by the very firm which originated the product; this phenomenon intensifies obsolescence promulgated by the competition, of course.
- *Value of time and convenience:* There exists mounting evidence that consumers, more than ever, are sensitive to the interacting value of time and convenience. Products and services aimed at maximizing the time available for customers to pursue other tasks—the implicit value

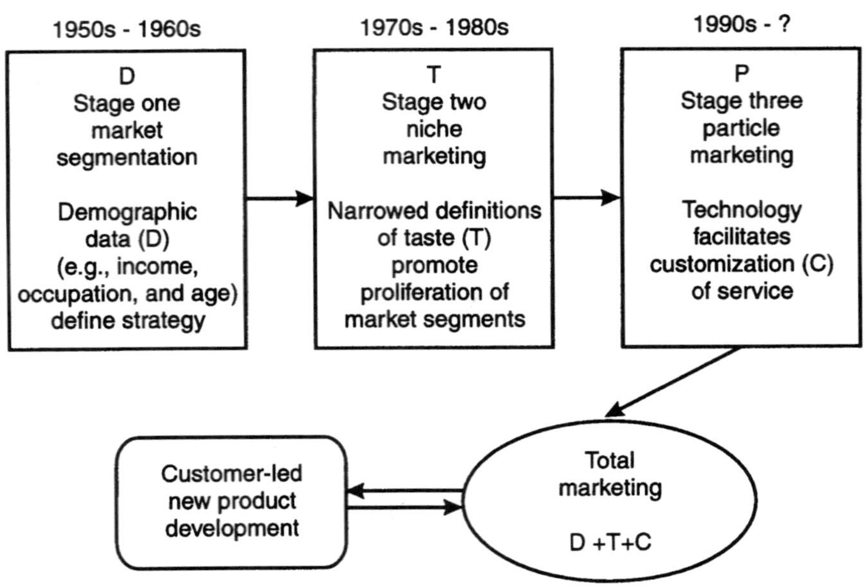

Figure 4.1 Customer-led customization model.

of opportunity—substantially improve both the quality-of-life and material well-being of customers. New methods are emerging which estimate the value that consumers place on opportunity costs, whether or not they are conscious of these implications.

- *Sensitivity to discounting:* A heightened sensitivity to price-cutting for both goods and services is now manifestly evident in the economy [10]. There is probably greater reluctance for consumers today, both domestic and global, to accept standardized pricing than at any time since the end of World War II. To engineer lower prices, firms have had to necessarily pursue cost reduction relentlessly, as described above.
- *Customer service and post-purchase satisfaction:* Reinforcing all environmental factors is the requirement that customers now demand greater attention to their needs not only at the time of transaction, but during the post-purchase phase of customer service [11]. With an expanding array of customer choice pivoting on expansive global production, consumers can now be fastidious about spending and the producers with whom they align. It is empirically verifiable that it is cheaper to retain a profitable customer than recruit a new one. As a result, progressive organizations are beginning to invest heavily on customer retention, shifting resources in many cases from simple "customer prospecting" to "customer loyalty."
- *Quality:* durability, reliability, and predictability become the watchwords of customer satisfaction during the current era. Benchmarking quality performance—against global standards—becomes an indispensable corporate tool during this competitive era. Identifying, monitoring, and measuring the highest standards of the most competitive player now ascends to a firm's foremost priority. It is this standard that the informed consumer will expect of his provider.
- *Value-added design:* In an age when firms lose their capacity to manipulate or influence price, the point of differentiation in product or service design becomes the layers of value that producers continually provide. In essence, we refer here to "functionality." What does the customer need and want? What will the customer need in the future? The ability of a firm to successfully evaluate and forecast these answers determines its prospective competitive edge in the market.

Together, these environmental factors set the current and future parameters of consumer behavior. A firm must treat these variables as dynamic and ever-changing, and accept the growing reality that a permanent shift has unfolded in the modern American economy: producer power has given way to consumer choice amid mounting supply. This theme is about to be converted into a mantra for the communications industry.

4.3 Components of the Determinants of Demand

The determinants of demand described above can, in turn, be divided into a set of active components. Dissecting the determinants of demand provides a keener understanding of what drives consumers to make their decisions and, in particular, what may eventually influence their choice of communications menus.

4.3.1 Problem Recognition

The ability of a consumer to discern choices in the marketplace may be termed "problem recognition." The evidence, in industry after industry, suggests that problem recognition constitutes a critical factor in consumer choice after initial exposure to a new communications product or service [12]. Consumer decision-making becomes more refined as purchase involvement increases. As consumers gain experience with a menu of services, they become more sophisticated in the use of a product, and most fastidious with respect to future expectations. More significantly, consumer ability to differentiate products heightens and sensitizes. In short, the capacity to distinguish the value of products in relation to their optimal use denotes problem recognition. It is not unreasonable to infer, owing to the evidence elaborated above, that consumers are increasingly discriminating in this regard.

Market researchers have labeled consumer decision-making, at the time of product adoption, via three discrete categories: habitual, limited, and extended. Habitual decision-making is defined as that category in which consumers consistently purchase the same item. Such purchases eschew extensive information-collection in favor of the security and continuity of trusted brand names. Memory and experience guide such behavior. Resistance to price and dissatisfaction with service or quality can undermine consumption dictated by habit. For firms generating products or services falling within this category, every effort must be made to ensure quality control. Habitual behavior will shift to limited decision-making proportionate to the level of dissatisfaction: consumers will seek out alternatives to the degree their level of dissatisfaction rises. Extended decision-making—that category in which consumers spend substantial time pursuing and evaluating product alternatives—is energized when price sensitivity magnifies differences among competitors, and post-purchase satisfaction with veteran product lines deteriorates.

4.3.2 Information Search

One of the essential elements of consumer decision-making concerns the method by which a consumer obtains information in order to formulate selections. Such information can be dichotomized between internal versus external sources. Internal information, or stored information, is culled from personal experience, memory, and passive recall. External information is knowledge drawn from individual interaction with consumers, promotion, government databases, or direct product inspection. Customer pursuit of external sources of information, in light of widespread, dispersed information via the Internet now becomes a major concern of strategic planners in all industries. One manifestation of the increased importance of external information in recent years is the heightened significance marketers assign to brands [13]. Branded images resonate in memory and offer reassurance in a world of mass production and resultant chaotic choice.

A serious analysis of a consumer's information search yields important inferences about the prospective development of telecommunications markets, domestic and global. The proliferation of products and services now underway—from telephony to Internet usage, from wireless services to interactive television, from carrier operations to content producers—may overwhelm and bewilder the consumer in the short-run. The pattern of product acceptance in other industries suggests that, as time passes and product acceptance is manifested, communications firms can expect consumers to invest additional in external information searches. Fastidiousness in product selection accompanies this evolution in consumer preference. As a result, it becomes incumbent upon the firm to establish its market position and image early, to reinforce this linkage with product standards among the highest formed in its sector, to pursue customer satisfaction vigorously so as to build loyal clientele. The telecommunications sector has generally endorsed the concept of "bundled packaging" with accompanied consumer customization to realize these objectives [14]. This strategy is examined and evaluated in detail in the following chapter. It is essential to bear in mind that it is consumer preference, derived from newly developing external sources of information, that are dictating a number of these strategic priorities.

4.3.3 Lifestyle and Attitudes

Firms must naturally be sensitive to shifting developments in consumer lifestyles and attitudes as cultural, social, and political contexts evolve. There is considerable evidence that the time required for market segments to experience and respond to market developments is contracting. Indeed, the trend toward shorter product life-cycles, confirmed repeatedly in empirical studies, is one manifestation of this development in the communications industry [9]. To remain "ahead of the curve" a company must anticipate these developments so as to exploit market opportunities when they arise.

Among the dominant factors influencing consumer behavior are lifestyle and the attitudes associated with the human life-cycle. Identifiable independent variables influence consumption with the passage of age, and the lifestyle one adopts. Empirical evidence asserts that as one ages lifestyle generally transitions to a more conventional, conservative pattern; conformity to social norms and dictates of family life outline social, political, and cultural attitudes [5]. Such transformations are accompanied by rising consumption, as noted in Chapters 1 and 3. Consumption peaks at about age 49 and generally falls incrementally thereafter [15]. Lifestyle and attitudes define the capacity and nature of consumption. One cannot separate age, lifestyle, and attitudes from the pattern of telecommunications consumption likely to evolve during the next decade.

To gauge the impact of lifestyle and attitude on future configurations of telecommunications demand, market researchers apply psychographic methodology. Psychographics is a method which estimates consumption according to categories of shared attitudinal characteristics. By grouping consumers into such classifications—labels based on psychological need, age, and peer groups—firms can designate the core of demand. In an era of intensifying competition, those firms able to accurately apply these methodologies will retain comparative market advantage. A visual description of psychographic methodology is provided in Figure 4.2.

4.3.4 Motivation, Emotion, and Personality

Much of consumer behavior, and thus purchase and post-purchase decision-making, is a function of economic intangibles. Principally, these intangible, often indefinite, but always influential variables include individual motivation, emotion, and personality make-up [16]. Voluminous literature has been compiled to document empirical tests relating the variance with which motivation determines consumption. A. H. Maslow's seminal research suggests that all human beings shift through five stages of motivation before achieving full satisfaction in life: these include, in order of transition, physiological, safety, belongingness, esteem, and self-actualization. These transitions determine varying tastes and rates of consumption. Complementary research by others has identified the importance of curiosity, self-expression, affiliation, and modeling. Examined in totality, human beings—despite their general desire for independence and self-assertion—are highly influenced by the actions, motives, and behavior of others. This phenomenon is pertinent to an analysis of the adoption rates of new services now projected for the communications sector.

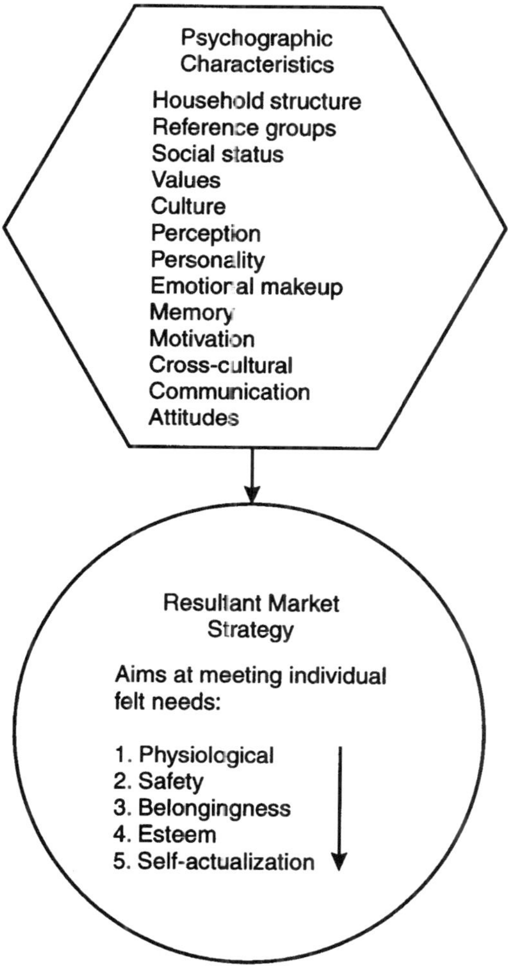

Figure 4.2 Psychographic determinants of consumer behavior and decision making.

The use of surveys, focus groups, and individual and group testing allow a firm to ascertain the extent to which motivation precipitates consumer choice. Emotion, a subset of motivation, can drive the purchase of goods and services but is less likely to influence purchase and post-purchase decision-making processes. The ambiance likely to be characteristic of telecommunications consumption in the post-deregulatory era is not a function of discretion drawn from emotion; it is a function of felt need and reliance on continued innovation. Thus, personality will be given a higher priority in future telecommunications market research, particularly with respect to the construction of strategic plans [17].

Personality is the collection of traits defining the individual. The economy and industry are moving to an era of customized menus, as explained in Chapter 3. Simply, advanced technologies now afford a firm the luxury of designing a communications palate specialized and attuned to the taste of individuals. Passage of the Act effectively accelerates the competitive thrust to enlarge market share. Those firms unwilling or unable to meet this challenge are likely to suffer declining market share and possible extinction; the link between personality and consumer demand is magnified as a result of deregulation. Furthermore, new market research efforts in "data warehousing" and "data mining" facilitate competitive opportunities scarcely known previously in the industry.

4.3.5 Evaluating Consumer Preferences

The process of judging a product's traits—price, quality, and other objective characteristics—constitutes objective criteria with respect to gauging and measuring consumer preferences. Other characteristics, such as reliability, dependability, and prestige, are more subjective in nature but nevertheless influence consumer behavior. Under such circumstances, it is common for consumers to rely on branding—or corporate image and reputation—to differentiate alternatives. How do consumers decide which product or service to adopt when inexperienced with available competitive alternatives? While this question elicits diverse reactions within the industry, strategists must contemplate its implications. The effectiveness with which a company addresses this issue will be an important factor in attaining long-term market penetration. In a world of proliferating choice in telecommunications, particularly as firms align in forming partnerships to deliver services, the identification of evaluative criteria become an immediate economic and methodological question. Benchmarking the industry's strategic response to this question becomes a prime methodological priority, as discussed in the following two chapters.

We can assert a general methodology to the process by which consumers evaluate alternatives: the subjective measurement of qualitative and quantitative information, internal and external, which consumers use to differentiate alternatives. How consumers perceive available selections on the basis of branding, image, reputation, and prestige constitutes the first and foremost area of inquiry for market research. Second, the comparative importance of each of these four variables—as well as others—defines the use of evaluative criteria necessary to construct a marketing plan.

Among information compiled by telecommunications firms are databases detailing the patterns of aggregate household consumption. These data are formulated in cyclical, seasonal, temporal, and other categorical descriptions. These are reliable data about which credible forecasts can be made as to regular and periodic demand for communication services. Before passage of the Act, such forecasts were often generated with a high degree of confidence, even certainty about prospective aggregate demand. Regulation cemented a "gradualist" environment: innovation, in relative terms, was less dynamic, less radical, and less disruptive in nature. Deregulation is in the process of encouraging counterintuitive developments in consumer tastes, as new technologies bring unanticipated change to telecommunications as consumer expectations in unrelated industries compel telecommunications providers to continuously upgrade their menus. This phenomenon is even more pronounced by content providers in broadcasting, electronic commerce, and multimedia [18].

Consumption of telecommunications services generally falls within four definable areas of adoption: consumer applications, household life-cycle, and organizational demand. With respect to household life-cycles, there exist certain influential variables which impinge on demand. These variables include cultural and social class factors, patterns of socialization, family structure, role specialization, segmented decision-making, and the changing stages of the household life-cycle [19]. In prior periods, the decision to adopt communication products would have been relegated to a single member of the household. Deregulation promises to complicate the process of forecasting demand amid proliferating choice. Many emerging services retain unique value to each family member.

It is reasonable to anticipate that emerging wireline and wireless services, particularly in terms of interactive applications, will be customized to the need of each member of the household. Proliferating competition assures that every individual will approach the key economic decision-maker in the household for those services of greatest personal value. The household decision maker, therefore, will have to contend with multiple choices on the one hand, and the heightened demands of others to formulate a menu of services for the entire unit. For strategists, it will be vital to focus on the habits of each individual to estimate household consumption patterns [20].

Decisions regarding communication product adoption were formerly the domain of a single individual, group or committee in many firms. The convergence of the telephony, computer, and broadcasting industries, and the industries that support them have had the effect of encouraging individuals throughout an organization to investigate product alternatives. We thus note a parallel situation in contemporary organizational consumption: expanding alternatives have overwhelmed even telecommunications experts in many companies. Individuals without such expertise are nonetheless cultivating knowledge in a single technology which they find essential to their work.

In designating marketing strategy to meet the consternation of organizational buying during this era, communications firms might consider the following tactics:

1. Isolate purchase decisions in this fashion: (a) new task adoption, (b) straight re-buying, and (c) modified re-buying. The most complex of these decisions involves the deployment of new communication services—adopted to solve emerging problems—while replacement decisions ordinarily are less provocative in their organizational implications.
2. Identify the organizational style of a firm: styles can generally be classified as command, cooperative, or competitive [21]. In firms where command decisions are made, product adoption is simplified from the viewpoint of the vendor; on the other hand, post-purchase servicing is typically more facile in organizations that adhere to cooperative styles of decision-making. Telecommunications firms clearly must consider the strategic and tactical implications of the organizational style governing product and service agreements. This issue becomes a difficult one with which to contend in organizations noted for shifting from one style to another over time.

4.4 Monitoring Consumer Behavior

In monitoring, measuring, evaluating, and forecasting consumer behavior in the telecommunications sector, firms may integrate a set of traditional and state-of-the-art tools. Traditional methods of monitoring changes in consumer preferences, tastes, and patterns of consumption include qualitative and quantitative applications. The use of such techniques remains a fixture in basic market research and in the extrapolation of data used to predict prospective demand. These tools have recently been augmented by sophisticated new methodologies based on data warehousing, data mining, and interactive applications [22].

Data warehousing is the process of accumulating information about all aspects of consumer behavior, including consumer preferences in style, taste, pricing, and post-purchase service expectations. Database management and the use of object-oriented technology to integrate diverse aspects of multiple market databases now permit organizations to identify nearly every detail impinging on consumer demand. Data mining concerns the procedure of filtering voluminous data in a warehouse; by isolating and applying the patterns of behavior unique to each consumer, unprecedented inferences can now be drawn about current and prospective demand. In short, the stage has been set for the marketing of customized telecommunications services to individuals, households, and organizations. Monitoring consumer behavior is thus now more efficient, despite the complexities of expansive competition and technological discontinuities.

4.5 Diffusion Rates for Products and Services

The cadence by which veteran and new product lines are introduced to and adopted by a market is defined as a diffusion rate. Diffusion rates describe the rapidity with which consumers purchase communication products. Estimation of diffusion rates is a critical link in predicting aggregate future demand for a firm's products and is performed in assessing individual, household, and organizational life-cycles.

Diffusion rates are categorized in three ways, as described in Figure 4.2. We note rapid diffusion, standard diffusion, and slow diffusion as typical descriptions of unfolding rates of adoption. There are important strategic reasons associated with a firm's identification of its prospective product market opportunities. When viewed in the context of unfolding consumer behavior, diffusion illuminates critical strategic deliberations:

1. The entry to market by competitors is accelerated after product introduction in cases of rapid diffusion. Simply, rapid consumer demand quickly inspires competitors to emulate, simulate, or substitute relative to successful product lines.
2. The appropriate strategy to exploit consumer demand is to inaugurate a new product and rapidly improve its quality as it "rides" the diffusion wave. In this respect, the firm certifies a reputation for creativity and innovation in emerging product lines.
3. Each of the three illustrations describes an S-curve phenomenon. That is, consumer adoption, and therefore diffusion, is slow at the outset as a new product gains acceptance; thereafter, growth becomes exponential as consumers influence the uninitiated to adopt the same product or service lines.
4. Generally, rapid diffusion is contingent on new product introduction. Standard and slow diffusion rates are characteristic of innovative product lines.
5. In rapid and standard diffusion rates, the time required to generate 10% market penetration is comparable to the period necessary to shift from a 10% to 90% rate. In other words, market acceptance is slow at the outset and climbs rapidly when nonusers witness the benefits gained by those who experiment.

S-curves, as tools used to assess prospective consumer behavior, have various limitations and contingencies. We may infer that their most effective application occurs in those instances in which broad practical and social acceptance seems likely, in which economies of scale can reduce manufacturing costs, and when such services perform a manifest need or otherwise substitute for existing product lines in efficient or cost-effective ways.

Much research has been performed in recent years to classify, differentiate, and segment consumers [16]. The effort here is aimed at determining what marketing strategy is necessary to appeal to every "type" of consumer. The presumption is that certain groups of consumers share characteristics that then enable a firm to standardize its approach to the market. Marketing is defined as the pricing, promotion, and distribution of goods and services.

A starting point for the description and analysis of consumer behavior involves the placement of consumers along the S-curve. Figure 4.3 positions consumers according to four primary categories of behavior, empirically based, along the S-curve. Early adopters are those interested in experimenting with new products; these individuals have the discretionary income to engage in such experimentation and enjoy establishing precedent or creating new images. Early majority consumers are those who perceive benefits tangibly demonstrated by those who initiate such purchases but who are unwilling to be the first to adopt such innovations. Late majorities, riding the S-curve to 90% market penetration rates, join the early majority as prices drop and applications become obvious. Laggards move into the market after social acceptance of new products has been established, prices approach nominal levels, and the stigma of being "left out" becomes palpable [23].

It should be noted that many firms further segment these categories—early adopters, early majorities, late majorities, and laggards—in many additional segments. However, all such breakdowns belong to one of these four typologies. Significantly, exponential growth in sales can only be achieved when a firm meets the criteria outlined above and when pricing diminishes to accommodate each successive category shifting into the market. We may reasonably presume that quality and service improve as consumers drive the S-curve, although there exists considerable variance from firm to firm in this regard.

4.6 Emerging Developments in Consumer Behavior

In the era of rapid, unsystematic change unfolding since regulatory reform, firms will inevitably reorient their strategy to serve and retain customers. Amid the discontinuities and competitive thrusts promised by converging telephony, broadcasting, and computer sectors, what trends are likely to transform strategic response to consumer behavior? The following conjecture is the product of both empirical verification and speculation [24].

4.6.1 Shift to Bundled Services

The initial years following a deregulated communications market are likely to be characterized by a consumer desire for "bundled services." In moving from monopolistic or oligopolistic markets—in which clear lines were drawn between the telephony, broadcasting, and computer industries—to a market-based system, consumers may be so inundated by choice that consternation results. The lack of experience in deciding upon a local telephony provider, for example, could conceivably instill in consumers a desire for simplicity and predictability. By extension, it is logical that a firm that can provide all fundamental services—telephony, television, Internet, and wireless access—can readily secure market share. It is precisely in this direction that companies such as AT&T have moved [25]. Such a propensity now accounts, in part, for the wave of mergers and acquisitions that followed enactment of the Act.

4.6.2 Heightened Consumer Awareness

As underscored in this chapter, it is a reasonable inference that, as time passes, greater consumer involvement in accessing sources of alternative information is probable. As a result, a bundled strategy holds effective tactical implications for the short-run but cannot substitute for competitive pricing, quality, and servicing over the long-term. For segments of the early adopters, early majorities, late majorities, and laggards, "bundled" market tactics represent an expedient measure, but the competition that unfolds thereafter could instigate new innovations in product design and delivery. Consumers, within the next few years, may well become savvy purveyors of telecommunications services as they link market knowledge with personal need.

4.6.3 Particle Marketing

Particle marketing represents the pricing, promotion, and distribution of services to customers on a one-to-one basis. In other words, to fully engage particle marketing a firm must customize services for every customer in its defined market. This is an unprecedented development and has particular significance for the telecommunications industry. Because the telecommunications industry depends on the provision of intangible goods, instantaneous transmission of information is a realistic objective. When consumers exploit the dynamics of particle marketing, they are empowered to provide feedback to providers in a way unimaginable in most industries.

A contentious, but reasoned conclusion would hold that an informed consumer contingent, once sensitized to proliferating options, would be willing to abandon the stability of bundled packages for personalized services adapted for changing tastes. One can only speculate about the efficacy of sophisticated market networking to build and sustain market share; it seems likely, however, that more affluent consumers will explore options that permit personalized menus. Thereafter, a "learning curve" for other customer segments could well dictate the future outline of communication services. Particle markets, using methodological options now in their infancy, are predicated on a pattern of deregulation reducing market concentration. The development of an industry oligopoly, should deregulation not achieve the aim of its advocates, would minimize the probability of this scenario. In the meantime, a competitive environment encourages a firm to extend its organizational resources to advantage; strategic positioning is thus aimed at establishing an image of accommodation to residential and business need.

It should be emphasized that a firm's ability to monitor changes in consumer behavior is crucial to exploiting the emergence of particle markets. Such factors as strategic marketing, organizational restructuring, process reengineering, and the realignment of customer support services will be vital in superseding the challenges of competitors who perceive the same opportunities.

4.6.4 Relational Marketing

It is manifestly obvious that some customers are more profitable than others. High-end users requiring a multiplicity of applications pay more for services, and are thus more valuable to a firm, than those whose communication requirements are minimal or incremental. As a result, some market segments generate greater sustained revenue streams than others. A common precept of contemporary strategic planners holds that all market segments should be pursued vigorously. Paradoxically, a firm may be pursuing such an objective in a way that damages its long-term profitability. The pursuit of all market segments, expensive and time-consuming, may deprive a firm of its best customers by diverting resources necessary to sustain the loyalty of its own customers.

We can assert in this context a tension between the need to capture new customers versus the maintenance of existing consumer bases. The process by which a firm devotes a majority of its resources to nurturing and cultivating its best customers may be described as "relational marketing." The relentless pursuit to satisfy a firm's best customers yields benefits, albeit at a price. Effective relational marketing, and its ramifications for strategy, can maximize revenues but alienate potential customers whose need to purchase products may be as yet deferred. On the other hand, a firm which deliberately eschews unprofitable customers in favor of the quest to gain profitable ones is sending its costly clientele to unwitting providers. There are thus twin benefits to relational marketing when successfully applied. Clearly, much research needs to be undertaken in this emerging field.

The application of relational marketing to strategic planning will vary widely in the communications sector. Content providers will typically see less immediate application than carriers; cable operators will sense greater benefit than broadcasters; software developers may be more responsive to its potential relevance for their sector, than hardware manufacturers. In most cases, relational marketing, and its potential to minimize cost while optimizing revenue, will have greater initial relevance for those areas of communications more regulated than others. Carriers assume greater costs in conducting their business than do their content counterparts, owing to regulatory constraints. Over the longer term, relational strategies may be eased in favor of broader efforts to saturate every prospective market segment. In effect, customer "prospecting" may reassert itself a decade from now in those firms which currently aim to exploit the economics of relational strategy.

4.7 Shifting Patterns in Consumer Preference: Domestic and Global

Social and cultural values decide, to a significant extent, the varying patterns of economic consumption. Therefore, consumer demand projected for telecommunications services must be examined in light of changing tastes. Changes in tastes—or consumer preferences—are influenced by shared social and cultural norms among a variety of segmented groups. To fully anticipate changes in demand, and to forecast with precision their impact on sales and post-purchase satisfaction, one must: (a) define social and cultural values that distinguish each segmented group, and (b) link these values to patterns of demand for each service [23].

Mechanically, one approaches the problem of forecasting changes in consumer preferences by stratifying (organizing) groups according to age, sex, gender, ethnic background, socioeconomic status, and other characteristics [26]. By applying correlation analysis to the identification of "strength of relationship" between each of these independent variables and resultant demand, we gain a keener sense of prospective consumption. A crucial element affecting the communications industry in the post-World War II period has been the transformation of American behavior precipitated by dynamic changes in technology, politics, economics, and social interaction. As a result of dramatic change, consumer behavior has been altered fundamentally because conflicts emerge between and among stratified groups. Differences in social and cultural values magnify differences in consumer attitudes. It is therefore vital to determine the marketing strategy congruent with changing, segmented groups. A paradox, and business opportunity, implicit in this dynamism concerns the potential for greater understanding and acceptance between and among the same groups. Such understanding may hold an economic ramification as private alliances might emerge between stratified groups who learn to employ improvements in communications and information infrastructure to generate new business.

An important footnote to this discussion concerns the economic influence of women in American society during the past two decades. An increasing fraction of the workforce consists of "working women," a contingent whose need for advanced communications services parallels that displayed by men. With outsourcing, telecommuting, and home-based businesses now transforming segments of the private sector, one cannot overemphasize the significance of such restructuring. It will therefore be vital for a firm to forecast demand for each stratified group according to preferences exhibited by men and women. In other words, there are likely to be differences between men and women as to their consumption of services within a wide range of bundled alternatives.

Just as consumer tastes shift with the passage of time and the dynamic interactions described above, so too do consumer preferences vary throughout the world. We can identify two overriding considerations in behavioral change affecting the demand for global communications services. The first concerns the cultural influence of American entertainment, education, technology, and public policy. The United States is the largest exporter of intellectual property in its broadest form: the contribution of ideas and knowledge to the design of new enterprises [27]. As a result, when intellectual property is fused with advanced communications, American influence on global consumer preferences sustains a profound impact. Many people throughout the world simply emulate, in varying degree, American tastes and attitudes. The exercise of such influence is connected directly to the adoption of emerging technologies aimed at linking diverse peoples to American dissemination of information. The demise of Cold War political structures has intensified this development.

Concurrent to this phenomenon is the growing influence of "globalization"—one market, not many, in which arbitrary geographic distinctions disappear in the face of producer and consumer interaction. Despite governmental attempts to control, dictate, or regulate production and employment within their own boundaries, consumers—now equipped with sophisticated forms of transmitting and retrieving information—assert their independence in purchase decisions.

References

[1] Russell, C., "The New Consumer Paradigm," *American Demographics*, April 1999, pp. 21–34.

[2] Tucker, I. B., *Microeconomics for Today*, St. Paul, Minnesota: West Publishing Co., 1997.

[3] U.S. Dept. of Commerce, Bureau of Labor Standards and the Conference Board at www.bls.gov; Also consult www.yardeni.com for hyperlinked monthly data interpretation.

[4] Dent, H., *The Great Boom Ahead*, New York: Hyperion Press, 1993, pp. 21–46.

[5] Shaw, J., "Future Scenarios for the Telecom Industry: A Ten-Year Forecast," *New Telecom Quarterly*, December 1996, pp. 25–33.

[6] Senge, P., "The Art and Practice of the Learning Organization," *The New Paradigm in Business*, Los Angeles: Perigree Books, 1993, pp. 126–137.

[7] Gates, B., *Business at the Speed of Thought*, New York: Warner Books, 1999.

[8] Gates, B., *The Road Ahead*, New York: Viking Press, 1996.

[9] Valery, N., "Innovation in Industry," *The Economist*, February 20, 1999, pp. 5–28.

[10] "Remarks by Chairman Alan Greenspan," *Federal Reserve Board Speeches*, Washington D.C.: Federal Reserve Documents, October 16, 1996.

[11] Kambil, A., "Electronic Commerce: Implications of the Internet for Business Practice and Strategy," *Business Economics*, October 1, 1995, pp. 27–34.

[12] Mowen, J., and M. Minor, *Consumer Behavior*, Upper Saddle River, New Jersey: Prentice-Hall Co., 1998, pp. 347–368.

[13] Pettis, C., *Technobrands*, New York: American Management Association, 1995.

[14] Sirois, C., "Telecom in the New Millenium: A Shift in Power," *Telecommunications*, May 1999, p. 18.

[15] Dent, H., *The Roaring 2000s*, New York: Hyperion Press, 1998.

[16] Kotler, P., *Marketing Management*, Upper Saddle River: Prentice-Hall Inc., 1998.

[17] Harari, O., *Leapfrogging the Competition*, Washington, D.C.: American Century Press, pp. 95–116.

[18] "Old Medias Internet Boom," *The Economist*, May 22, 1999, pp. 53–54.

[19] Bartholomew, M., *Successful Business Strategies Using Telecommunications Services*, Norwood, MA: Artech House, 1997, pp. 85–120.

[20] Imparato, N., and O. Harrari, *Jumping the Curve: Innovation and Strategic Choice in an Age of Transition*, San Francisco: Jossey-Bass Press, 1994, pp. 53–78.

[21] Arlen, G., "Choosing the Right Path," *Convergence*, May 1996, pp. 39–48.

[22] Bartholomew, op. cit. Also see Edelstein, H., "Mining for Gold," *Information Week*, April 21, 1997, pp. 53–70 for an incisive account of recent developments in data mining strategy.

[23] Bernhardt, K., and T. Kinnear, *Cases in Marketing Management*, New York: McGraw-Hill Co., 1997, pp. 349–416.

[24] Wilde, C., "Telecom Analysts Take Stock of the Market," *Tele.com*, May 15, 1997.

[25] Data on new business formations can be secured through www.datasecurities.com and www.ventureeconomics.com.

[26] Salak, J., "Delivering the Goods," *Tele.com*, May 1, 1997, pp. 56–65.

[27] Edvinsson, L., and M. Malone, *Intellectual Capital*, New York: Harper Business Press, 1997, pp. 65–74.

5

Strategic Management and the Producer: Analysis of Emerging Supply

One manifestation of the Telecommunications Act of 1996 is the transformation of the industry's market structure. Before 1996 the telecommunications industry consisted of a set of subdivided markets in which consumers purchased products and services under conditions ranging from substantial competition to regulated monopoly. Passage of the Act reconfigures industry market structure by relaxing or eliminating constraints that formerly existed in telephony and broadcasting. The third tier of modern communications, the computer sector, remains unregulated and is now asserting its influence.

Industrial economics—the study of competitive characteristics of industries—holds that any sector can be categorized as perfectly competitive, competitive, monopoly, oligopoly, or monopolistic competitive. These five market structures continue to characterize elements of the economy, although regulated monopolies are rapidly disappearing in the United States. By classifying telecommunications goods and services according to the market structures in which they are distributed, we may deduce the short-term, mid-term, and long-run competitive influence anticipated by authors of the Act. Market structures set the environment for competition and, thus, ultimately provide a context for projecting capital investment, business opportunities, and profitability. The identification of transitional market structures, adjusted for time horizons, clarifies emerging competitive threats while refining organizational objectives.

5.1 Analysis of Market Structures

Market structure is an expression referring to attributes of interaction between competitors in the marketplace. Characterizing the physical interaction of firms necessarily involves a consideration of three principal functions: (a) the transmission of information between companies, the key element of which is pricing; (b) physical barriers to entry; and (c) the number of firms and the market share they control or influence [1]. These dynamics, and the speed with which they introduce new communications services to market, will be a function of the restructured marketplace redefined by the Act. Deregulation has altered the rules of telecommunications competition primarily by eliminating cross-ownership restrictions between firms operating within the converging framework of the telephony, broadcasting, and computer industries. Deregulation has thus transformed structural characteristics of the market and reshaped strategies by which products are distributed to consumers [2]. The primary difficulty in gauging deregulation, and its resultant influence on market structure, lies in the uneven influence it exercises within the three communications sectors. As a practical matter, deregulation will have its blast effect in those areas heretofore governed by regulatory authorities. In theory, if not always in practice, deregulation shifts the management of transactions from public oversight to private interests.

Rules guiding the sale of communications services are contingent upon the market structure in which products are distributed. A spectrum of rules running the gamut between perfectly competitive and monopoly structures define the market. Within these parameters, variations in market structures—competitive, oligopoly, monopolistic competition—play a vital role in the immediate future of the telecommunications industry. The spectrum of market structures and adjoining rules are noted in Table 5.1.

Market structures can be defined in the following way as applied to the communications industry:

- Perfect competition: optimal number of firms jostling in the market, characterized by dynamic changes in price, quality, and service options, with extremely limited market share held by any single firm.
- Competition: large number of firms moving in and out of markets, with no firm able to set market price.
- Monopolistic competition: a market structure in which many firms sell segmented, differentiated product lines, often characterized by some producers offering a single product but providing substantial variations of choice within that line.

5.3 Analysis of Market Structures and Strategic Decision Making

Strategic decision-making, in economic terms, is the process by which a firm plans for the anticipated competitive response to its pricing, promotion, and distribution of goods or services [1]. Strategic planners measure competitors' behavior prior to launching their own plans. This procedure is a complex and delicate one and is based substantially on the market structure governing sector behavior. In competitive markets, strategic planners operate under the presumption that rivals can respond adroitly to tactics launched by one or more companies. In oligopolical structures, the speed with which change occurs is muted because market share (and thus pricing influence) is controlled by a limited number of competitors.

Strategic decision-making will become a watchword for the early years of telecommunications deregulation. With little experience in marketing, competitive research, and strategic decision-making in competitive markets, large players, in the early years of deregulation, have operated at something of a disadvantage relative to those with such background [7]. In competitive market systems, one may generally assume that market players respond rationally; therefore, their behavior is theoretically predictable and thus can be modeled.

Among the critical strategic issues confronting telecommunications strategists in the early years of deregulation are:

1. Developing market plans that enhance customer relations (hereafter identified as relational strategies);
2. Recruiting and educating new customers for value-added product lines;
3. Retaining customers amid proliferating competitors and choices;
4. Identifying lifetime projected values for customer bases;
5. Computing break-even and roll-out response rates for each product line;
6. Enhancing customer loyalty for those product lines less vulnerable to radical variations in or departures from S-curves;
7. Extending customer-base reach through innovative promotional strategies; and
8. Pursuing the creation of strategic alliances, affiliations, mergers, acquisitions, partnerships, and intrapreneurships aimed at expanding customer "reach" and "loyalty."

Predictably, these patterns of strategic behavior have already been witnessed during the first four years immediately following passage of regulatory reform. What is less clear, and indeed quite nebulous, is the degree to which these patterns of strategic behavior are affecting the supply of products and services. We can measure the physical production of telecommunications goods, domestic and global; we can gauge the affect of such production in terms of both imports and exports (these items are statistically elaborated in Sections D and E of the Appendix). What is quite intangible is the extent to which the supply of telecommunications services has been influenced by the interaction of the telephony, broadcasting, and computer sectors. In particular, to what extent has the emergence of Internet-based commerce outlined the future of such services? Is the supply of services limited or constrained in a physical sense? Are we on the cutting-edge of an explosion of such services, or are we creeping toward a finite array of service production? These intriguing questions are examined in greater detail later in this chapter.

The initial years of deregulation are likely to bring limited competition to some areas of telecommunications, while others may experience intense competitive interaction. There is empirical evidence which supports both contingencies during this initial phase of deregulation. The laws of industrial economics state that strategic decision-making will follow the contour of market structure in question; intense competition will expedite strategic decision-making while oligopolical competition will reduce the value of rapid response. Strategic planners in every sector of the industry will inevitably confront these dilemmas. It is reasonable to infer, given industry convergence, that to the extent the computer industry invades its companion sectors, it will stimulate imaginative strategic decision-making that might not otherwise materialize [5]. The need to embark on innovative strategic modeling is now a requirement for industry planners; incentives have been reconfigured (note Chapter 6) to assure this denouement.

One approach to evaluating the impact of deregulation on various market structures is to review questions that strategists are themselves addressing relative to the Act's implementation [6]. If we isolate five sectors in the communications industry—local telephony, long-distance telephony, wireless voice and data, Internet applications, and video services—these observations arise:

- Local telephony: When will local carriers open their networks? How competitive will local resale and loop unbundling be in the near future? How facile will local providers be in launching bundled packages? Will local exchange carriers merge, acquire, align, partner with prospective rivals to develop their product lines, or instigate an alternative strategy to achieve their aims?

- Long-distance telephony: How will long-distance and international exchange carriers position themselves vis-a-vis RBOCs? How expeditiously will IXCs penetrate markets other than standard telephony? What role will Internet telephony play in reconfiguring profit motives, production, and future strategic planning for long-distance providers? Will domestic competition be enhanced after industry consolidation of international communications players is formalized?
- Wireless voice and data: How will incumbent cellular providers respond to its prospective competitors (paging, SMR, PCS, wireless local area networks)? Should the wireless industry immediately diversify product lines or refrain from such strategy until customer needs are effectively determined?
- Internet applications: Will Internet usage define future business opportunities, including production, consumption, distribution? Will multimedia applications supplant other technologies in voice and video transmission? Will interactive applications introduce new competitive threats to other areas of the communications industry? Does prospective Internet ubiquity form a new market structure likely to influence other segments of the industry? What long-term impact does flat-rate pricing have on local and long-distance telephony networks?
- Video services: Does the interaction of DBS, CATV, MMDS, and other video services reformulate pricing? What will be the impact of ILEC's entry into this field? Is there a market for interactive television? Does HDTV have any short-term effect on the video market structure?

Superimposed over these provocative questions is the profound and baffling impact this collection of technologies will have on the economy generally, and the extent to which a restructuring of the economy will, in turn, demand new products from the communications sector. The synergies borne from technological innovation will assuredly foment new products and services that can be scarcely conceived a decade out. Anticipating business opportunities inspired by the economy's reaction to these innovations may well become a preoccupation of strategists in the near future.

One can reasonably conclude that each of these five sectors can be characterized by a defined market structure, as illustrated above [5]. Local telephony represents the least competitive environment in telecommunications. Federal and state regulations have been only gradually phased out; this factor, combined with RBOCs' resistance to a restructuring of their markets, has forestalled greater competition. Long-distance telephony, the first market to undergo the rigors of competition extending back to 1984, remains a competitive environment. Wireless voice and data communications are in the infancy of growing competition; some projections suggest that as many as a dozen wireless providers may enter the market in metropolitan regions by the late 1990s. Internet usage is growing exponentially worldwide and is presently the most dynamic and competitive communications market, partly because regulatory constraint is essentially nonexistent. One must dichotomize networking—with its reliance on traditional telephony structure—from emerging service provision. The potential power of the Internet to supplant traditional wireline services makes it the greater imponderable in contemporary telecommunications.

The telephony, broadcasting, and computer industries price their products in varied market structures, as defined above. We must differentiate three factors as being decisive in price determination. These include: (1) current market structure, based on deregulatory transformation; (2) newly emerging variables; and (3) the response of competitors in forming respondent strategy. Strategists are impelled to describe, prescribe, and project the collective pricing influence of all competitors in the market. Description, prescription, and projection—defining current reality, prescribing strategic response, and forecasting the effect of that response—are embedded in a firm's strategic scenario. Short-term instabilities associated with deregulation make these procedures exceptionally difficult; the economic stability many anticipate for a later "industry shake-out" will ease the burden and forecasting of strategic pricing. For reasons described above, "industry shake-out" is not likely to be effected until after 2002.

The incorporation of description, prescription, and projection modeling as applied to pricing allows a firm to simultaneously take into account both its own actions as well as the competitive response of others. In competitive market structures, one inevitably concludes that all three variables operate in a context of greater fluidity than in oligopolical environments. Defining price in a market occupied by a handful of competitors is comparatively stable and, thus, intrinsically more predictable. The dynamics of technological change must be accounted for in the process of projection because a sudden innovation could render existing price strategy obsolete at any moment. Indeed, this phenomenon represents the rule more than the exception.

Market structures define whether a firm acts as a "price-setter" or "price-taker." In competitive markets, a firm is not able to set its own price with indifference to customers. In such an eventuality, the company behaves as a price-taker, adopts prices determined by the intersection of supply and demand among producers and consumers. In markets characterized by oligopoly or monopoly forces, the firm tends to behave as a price-setter, thus exercising marginal or significant influence over prices consumers pay. It may be inferred that, historically, firms in the telephony and broadcasting industries have typically acted as price-setters rather than price-takers, albeit under regulated circumstances. The computer industry, in both hardware and software development, has evolved, for example, in competitive markets, and simultaneously encouraged competition and innovation in telephony and broadcasting not otherwise attainable.

5.4 Economic Tools for Telecommunications Strategists

Because a firm's plans are impinged upon by a variety of market forces, a set of microeconomic tools should necessarily be applied to strengthen strategic forecasting and policy formulation. There exists a body of evidence in strategic planning which confirms that an understanding of supply and demand dynamics, elasticities, profit maximization, substitution analysis, and indifference curves are essential devices to calibrate strategy and probable competitive response [9]. While planning must address variables outside the domain of economic analysis—politics, social trends, and technological innovation, among them—there is little doubt that the tools described below can be of great assistance in clarifying the competitive landscape and informing the strategic management process.

5.4.1 Dynamics and Determinants of Communications Supply and Demand

The marketplace brings consumers and producers together. Transactions are measured in price. Prices are defined by the interaction of supply and demand. Some markets are local in nature, while others are regional, national, or international in scope. The geography of markets is an important characteristic in the evolution of telecommunications networks and the services they support.

We have established that in the telecommunications triad of the telephony, broadcasting, and computer sectors there exists significant variation in competitive structures. While the computer industry remains fully unregulated, its counterparts are significantly regulated even after enactment of deregulation. Within telephony and broadcasting, some elements of the Act have modified or relaxed regulations, but government influence over local telephony has remained entrenched. Whatever the underlying public interest motive, the fact remains that market supply and demand are influenced by government edict. We must therefore diagram intersecting supply and demand according to the market activity in which that enterprise is engaged. In areas characterized by competitive market transactions, as illustrated by the principles above, we diagram prevailing supply and demand as illustrated in Figure 5.1. We refer

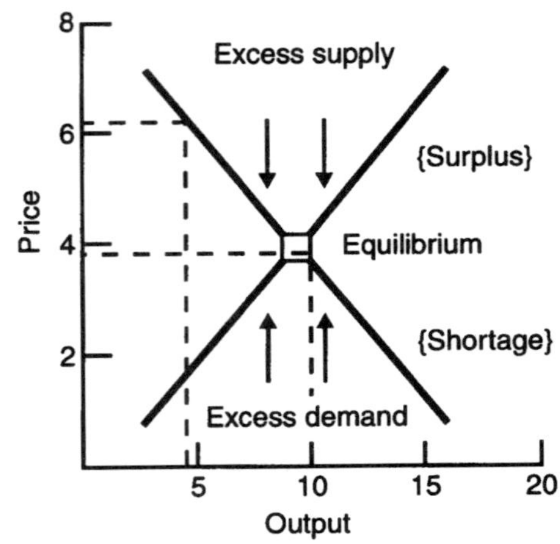

Figure 5.1 Dynamics of market supply and demand. Prices set above the equilibrium point generate surpluses, while prices set below precipitate shortages. Price reductions in the form of "sales" lower excess supply to the equilibrium point over time; price increases mitigate excessive demand until equilibrium is attained. Shifts in lines of supply and demand tend to be more dynamic in instances of excessive supply or demand, with competitors entering and leaving the marketplace at accelerated speed.

simply to the setting of prices in such environments as "competitive market price determination" [10].

The interaction of supply and demand defines price, with supply and demand curves changing constantly. Price will remain consistent if government asserts its regulatory powers with that aim in mind. Price stability will be sacrificed in the short-run if deregulation is manifested in the way its sponsors intended, but symmetry between market supply and demand will be achieved in the longer run. The effect of this congruence would, in theory, lead to equilibrium price, that situation in which buyers want to buy the same quantity that sellers are prepared to sell. Therefore, if a seller sets a price higher than equilibrium, surpluses occur; if a seller sets a price too low, excess demand ensues and shortages inevitably follow. Surpluses, in short, beget subsequent declines in price; shortages induce higher prices to the point of equilibrium.

It can be inferred that the use of supply and demand analysis is equally appropriate for content providers and carriers, though there is variance relative to product line. Those communications firms that supply content for infrastructure—telephony, television, and computers—thrive or die by the prevailing dynamics of supply and demand. Those who manage, control, design, or create networking infrastructure operate, however, in a fundamentally different environment. Costs of generating content for infrastructure are labor-intensive; costs of developing that infrastructure are capital-intensive and typically require a longer time horizon to gain profitability. The cost of labor is decisive in content development; the cost of capital is pivotal in infrastructure development. Thus, economies of scope (cost savings through multifaceted service provision) coupled with economies of scale (the decline in per-unit costs as production rises) are essential for network managers. Reducing network costs per subscriber will likely be a preoccupation of firms in the immediate future. It is precisely this objective which is fueling the wave of mergers and acquisitions in the industry following 1996.

Large network providers hold a principal short-term advantage during the early stages of deregulation: firms best able to exploit the fundamental principles of economies of scope and scale are strategically positioned to engage in vertical integration by absorbing competitors [10]. There exists an intrinsic advantage for large network providers—the addition of each subscriber increases the value of the network for veteran users [11]. These considerations encourage vigorous marketing, thus leading potentially to expansive market share. Carriers in the short-term are concerned with market share and competitive threat; the law of supply and demand takes over in the long-run to define market winners and losers. Despite the comparative advantage of size, and thus financial leverage which some firms enjoy, the risks of committing long-term capital while network innovations advance signify major disincentives.

Aligned with capital risk is the complexity of user-churn, the frequency with which consumers move from provider to provider for their communication needs. With the prospect of proliferating competition, with new technologies supplanting veteran product lines, user-churn denotes threat and opportunity. One must simultaneously evaluate the risks of capital and user-churn while developing long-term strategy. In the absence of predictable cash flow, it is not practical to raise or borrow the capital required to build and maintain state-of-the-art infrastructure. It is for this reason that customer-led customization is now regarded as an emerging strategic priority by some firms.

5.4.2 Telecommunications Price Elasticity

The measurement and evaluation of changes in quantity demanded in relation to changes in price is called price elasticity. Companies must estimate how responsive consumers will be with respect to variations in pricing within the same product line. Elasticity is calculated by dividing the percentage change in quantity taken by the percentage change in price. In defining elasticities at various pricing points, a firm is able to compute gross revenues in relation to the number of units sold. Thus, analysis of elasticity serves two overriding objectives: one, to gain a keener sense of the sensitivity of consumers to current and prospective changes in price, and two, the forecasting of revenues in the face of competitor pricing.

The calculation of elasticity will yield one of three results in each application: a value greater than 1, equal to 1, or less than 1. A value less than 1 defines inelasticity, meaning demand is not responsive to changes in price. A value greater than 1 specifies elasticity, suggesting that demand fluctuates with changes in price. Unitary elasticity occurs when the computation equals 1. If price elasticity of demand yields a value greater than 1, then a rise in price will generate less revenue. If the demand for a product is said to be inelastic, then an increase in price will produce more revenue. In situations where demand is unit elastic, the percentage change in quantity equals the percentage change in price.

There is a spectrum of demand for communications products that is variously elastic or inelastic. Telephone service for most people is generally inelastic, since it is regarded as a staple of modern living. Cable television service has apparently assumed the same role for many Americans in recent years. For workers involved in occupations where mobility is crucial, cellular telephones are regarded as a necessity. On the other hand, interactive television service remains comparatively exotic, and thus pricing is highly elastic. The inelasticities associated with many consumer staples reflect potential opportunities for firms to increase prices. Yet, under conditions of emerging competition, the capacity of firms to raise prices while controlling market share is limited. Inelasticities of demand for certain services inspired proponents of regulatory reform, who sought to prevent higher prices associated with oligopolistic market structures.

5.4.3 Defining Profit Maximization in Telecommunications

A telecommunications firm optimizes profit by increasing sales and minimizing cost. The difference between revenue and costs yields profit. In competitive market structures, a company would compute its marginal revenue and marginal costs and identify that level of output at which the two are equal. Marginal revenue is defined as the change in total revenue precipitated by a one-unit change in output level. Marginal cost reflects the increase or decrease in total costs a firm bears as a result of producing an additional increment of output. Prices are defined through the intersection of supply and demand, and the calculation of marginal costs and revenues specifies output. The extent to which a firm sells its outputs maximizes its revenue; the difference between revenue and costs thus determines profit.

The measurement of profit is vital to outlining a firm's strategic vision. The computation of profit, the demonstration of profitability, are essential in conveying a firm's credibility to stockholders. This issue is explored in detail in Chapter 6, as the linkage between profit, stock performance, and the related quality of strategic management are connected.

The computation of profit margin provides communications firms with a valuable tool in justifying their strategies to potential investors. Profit margin is the ratio of income to sales and is measured in two formats. Gross profit margin is computed by the percentage return that a company earns over the cost of goods or services sold. It is measured by dividing gross profit (sales minus costs of goods sold) by total sales. The gross profit a firm earns essentially covers operating expenses, including administrative expenses, taxes, and interest. Net profit margin, or return on sales, reflects the percentage of net income generated by each sales dollar. The standard computation of net profit margin results from dividing the income statement figure for net income after tax by total sales [10].

These calculations provide investors and competitors with important clues about the value-added desirability of communications products. In an environment of rising competition, the firm which produces a product that provides value-added attributes is likely to generate higher profit margins. In this sense, the determination of profit margins is significantly more important than gross profits.

5.4.4 The Telecommunications Substitution Effect

The extent to which consumers choose from among alternatives so as to satisfy their needs is defined as the "substitution effect" [12]. The forecasting of dynamic substitution will grow in importance in the telecommunications industry in the years ahead. Multiple factors explain the dynamics of substitution, but most economists and marketers agree that the differentiation of elastic from inelastic preferences for segmented markets reveal much about the future of communications needs [13].

Most observers will not disagree that food and milk are, in principle, inelastic staples. It is reasonable to conclude that video cameras and vacations represent preferences best characterized as elastic, with consequently wider price fluctuations. Consumers do not alter their patterns of consumption of staples, regardless of changes in price. Price elasticity is related to variation in income for luxury items. In the communications palate, the issue of substitution becomes a complicated one amid rapid technological change. The need of individuals and organizations to communicate rapidly has grown exponentially in recent years, but consumers will always search out lower cost alternatives—

in effect, lower-cost substitutes. How does one measure and project the course of substitution over time? Ephemeral factors make such forecasting highly complex, but in differentiating each communication product relative to another, we may deduce the determinants of market demand.

In contrasting "luxurious" versus "staple" communication products, there will exist substantial variation among income, occupational, and other stratified groups. Where market demand is elastic, consumers will have multiple alternatives available; this might eventually mean employing Internet telephony as a substitute for local or long-distance telephone service. In markets characterized by a wide range of substitutes, the price consumers pay tends to represent a higher percentage of income. Longer time horizons precipitate a propensity toward elasticity; new alternatives are introduced to market, and consumers are able to select the most desirable alternative. Consumers alter their tastes and preferences, as well, in response to emerging substitutes.

The economics of "substitution" can be expressed through cross-price elasticity, or the computation of responsiveness consumers exhibit toward changes in price for a particular product relative to changes in price for available substitutes. For purposes of this measurement, available substitutes may be either identical or complementary in nature. Technological innovation threatens veteran product lines. Forecasting cross-price elasticities will assume a heightened priority for telecommunications firms because it elicits important information about the strategic efficacy of existing products in terms of their pricing. There is expansive empirical evidence which suggests that consumers explore the desirability of substitutes in a way that their predecessors did not. The modern communications provider must maintain close scrutiny—that is, it must engage in competitive (or surveillance) analysis—with respect to those firms instigating such changes [15].

5.4.5 Applying Indifference Curves

Increasingly, consumers have before them an expanding array of communications products and services for a variety of needs. Indifference curves very likely will become an important tool for communications firms seeking an understanding of the extent to which one product may be used for a variety of tasks. Both content and infrastructure providers will seek out the benefits of measuring the degree to which consumers satisfy their communications needs by choosing from among proliferating choices. The indifference curve measures bundled choices from among various alternatives. For each increment of expenditure a consumer is prepared to provide, how will that aggregate sum be allocated for telephony, television, Internet, wireless, and other services? Will a greater sum be extended for one service as opposed to others over time?

Can one service be leveraged such that a smaller sum is required to finance product alternatives? These are among the strategic questions contemplated by planners as they gauge product innovation over time.

Consumers may opt for a simple bundle of telephony and television service, or shift to the Internet for delivery of all services, or select a single provider that can supply a menu of options unique to that palate. The interaction of the determinants of supply, as outlined, guarantee no outcome, and provide no assurance as to the extent to which consumers will augment or simply substitute those preferences.

As cellular phones, Internet service, interactive television, and other services transform the demand for products, how will the consumer reallocate his or her communications budget? Is it likely that the future need for information storage, interpretation, and transmission is so ravenous that today's emerging services will become tomorrow's staples? A general response to these questions cannot be formulated, but the measurement of reallocated communications budgets can be projected through the application of indifference curves. Under these conditions, a firm would design indifference curves for each of its consumer or business-to-business markets. The strategic imperative of indifference curves is also discussed in the following chapter.

5.5 Notes on the Economics of Telecommunications Supply

Space constraints herein prevent a more extensive treatment of the economic tools that are appropriate to evaluating, measuring, and forecasting prospective telecommunications supply, particularly with respect to the long-run. You will note that Sections D and E of the Appendix cumulate yearly data on telecommunications output in the United States. Included in the data are specifications of equipment, infrastructure, and services; also embedded in this material are credible data on the flow of exports and imports to the American market and European markets. Also refer to Section B, which isolates current and future infrastructure capital input in the United States, and its relationship to global supply of communications and information technology.

It should be emphasized that an appropriate complement of economic tools for gauging competitive supply would include:

- Business Cycle Analysis, focusing on the relationship between the business cycle and telecommunications output and distribution;
- Supply and Demand Dynamics, emphasizing the design of demand functions appropriate to market segments, and their connection to incentives for communications producers;

- Computation of market price equilibrium points, identifying the points in both wireline and wireless industries at which market efficiencies are attained (that point at which all available supply is utilized by households and businesses;
- Substitution Analysis, discussed above;
- Elasticity Forecasts, also elaborated above, applied to the prediction of those products or services destined to shift from luxury to staple status. Inserted in the Appendix is a set of primary data websites which provide extensive access to the methodology of telecommunications economics (refer to subsets under economics, finance, and marketing). Note, in addition, the websites listed under the federal government, particularly with respect to the Bureau of Labor Statistics, Department of Commerce, International Trade Administration, Federal Communications Commission, and the National Telecommunications Information Administration.

References

[1] Lidz, T., *The Person*, New York: Basic Books, 1968.

[2] Hawkins, D., R. Best, and K. Coney, *Consumer Behavior: Implications for Marketing Strategy*, 7th ed., Homewood, IL: Irwin Press, 1999.

[3] O'Keefe, S., and S. Masud, "Who has the Winning Strategy," *Telecommunications*, April 1999, pp. 13–16.

[4] "Inquisitive Middle Class—Lost Cause?" Survey Data, PR Newswire, April 14, 1997.

[5] Russell, C., "The New Consumer Paradigm," *American Demographics*, April 1999.

[6] Dent, H., Jr., *The Great Boom Ahead*, New York: Hyperion Press.

[7] Drucker, P., *Management: Tasks, Responsibilities, Practices*, New York: Harper and Row, Inc., pp. 202–214.

[8] Kotler, P., *Marketing Management*, 9th ed., Englewood Cliffs, NJ: Prentice-Hall Co., 1999, pp. 422–428.

[9] Davis, D., *Business Research for Decision-Making*, Belmont, CA: Duxbury Press, 1999.

[10] Schiller, B. R., *The Economy Today*, Homewood, IL: Irwin Press, 1997.

[11] Roddick, A., "A Third Way for Business," *Business Management*, Vol. 127, April 3, 1998, pp. 24–26.

[12] Colander, D., *Economics*, Homewood, IL: Irwin Press, 1996, pp. 531–540.

[13] Frank, R., *Microeconomics and Behavior*, New York: McGraw-Hill Co., 1991, pp. 134–158.

[14] Rosenbush, S., *Telecom Business Opportunities*, Newport, RI: Aegis Press, 1998, pp. 39–82.

[15] Jauch, L., and W. Glueck, *Business Policy and Strategic Management*, New York: McGraw-Hill Co., 1988, pp. 121–126.

6

Applying Strategic Management Models in Telecommunications Firms

Selection and adaptation of strategic management methodology has become a preoccupation of telecommunications market planners in recent years. The proliferation of new and converging technologies, expansion of global competition, deregulated markets, and compression of product life-cycles have conspired to instigate baffling uncertainties as to the identification of that strategic model most congruent with the competitive environment at any given instance. As discussed in the initial chapter to this volume, these challenges to effective planning are nevertheless overridden by the necessity to bring clarity to organizational direction and purpose.

In the current economic environment, strategic management—the design, implementation, and monitoring of strategy—serves two objectives. The first pertains to the need to respond to changes in the competitive landscape, the second a necessity to bring confidence and esprit de corps to a firm's personnel. In the former circumstance, a firm must learn how to position, price, differentiate, promote, and distribute its product to market. In the latter case, staff naturally seek reaffirmation and confidence that the company's goals, resources, and incentives are aligned [1]. Empirical studies confirm that those firms most effective in designing, launching, and adjusting their strategies—as measured by benchmarks in sales, productivity, and profitability—are those which achieve excellence relative to both criteria [2]. The following chapter explores examples of those communications firms most successful in designing and implementing strategy, as adjusted by these two interacting criteria.

Theory and practice elaborated in chapter one clearly delineated four primary models of strategic management: the classical model; the systems model; the competitive analysis model; and the "co-opetition" model. Each paradigm sustains strengths and limitations in its application. The pivotal concern to telecommunications strategists, however, lies in the fact that any model of strategy be susceptible to rapid, flexible, pliable implementation. If a model cannot be adapted quickly to the discordant currents of communications markets—where new products and services are literally launched monthly—it has little value for the contemporary strategist. Accordingly, we might review the suitability of each of these models for the present market environment.

6.1 Strategic Management Models: Strengths and Limitations

Chapter 1 isolates the procedures involved in implementing each of the four primary strategic management models available to strategists. Variations in these models' emphasis and objectives assure that no single strategic paradigm is uniformly applicable in every competitive environment. Therefore, we have a menu of four options available to design and administer strategy. Indeed, communications planners might reasonably customize a fifth model, representing their own hybrid of these four options, and reflecting the unique present conditions of their own firm. We have asserted, and demonstrated, that no single model is appropriate for every firm or, for that matter, every product over the space of time. In a world of expansive change, one requiring rapid strategic response, we might isolate the strengths and limitations of each model, as noted below [3].

It can reasonably be inferred that the classic model is practical, cohesive, and easily understandable to all employees. Clarity of purpose and application is the strength of classic modeling. The model, however, lacks, in the minds of some, a comprehensiveness of analysis, particularly with regard to pricing sophistication and analysis of the competitive environment. Is the classic model appropriate in a world where product differentiation and customization require not merely response to, but an expeditious forecasting of, changes in consumer taste? Classic strategic management is not the optimal method of anticipating those changes in consumer preference likely to emerge in the coming decade. It is precisely for this reason that many strategists have abandoned the traditional method of strategic analysis.

In the systems model of strategic management, adapted by many theorists and practitioners, we note an emphasis on thoroughness and detailed analysis. The systems model, incorporating every conceivable internal and external vari-

Table 6.1
Strategic Management Models: Comparative Strength and Limitations

Model	**Strengths**	**Limitations**	**Speed**	**Practicality**
Classical	Easily understood and applied by all staff	Lacks comprehensive treatment of variables	Can be applied readily to incremental change within an industry	Ease of use makes application most useful in smaller firms
Systems	Most sophisticated approach to global economic and technological complexity	Requires substantial time and expenditure of resources to successfully apply	Cannot be applied quickly to profound changes within an industry	Most suitable in large, complex bureaucratic structures, particularly multinational corporations
Five forces	Most recognized tool to dissect industry and product profitability	Least effective in changing market structures or emerging or seminal product lines	Provided historical data is available, can be applied rapidly to dynamic competitive change	Most useful in oligopolical or monopolistic competitive market structures; less useful if industry is characterized by technological dynamism
Co-opetition	The model most effective in identifying prospective opportunities for alliances/ partnerships	Requires managerial insight, intuition, and industry networking	Highly responsive to changes in industry structure, and relationships among firms	Most effectively applied in smaller firms with aggressive marketing and sales staffs

able known to influence organizational performance, contains the very defect of its strength: its comprehensiveness requires an extraordinary commitment of time and other resources. To be thorough in strategic design requires consensus from all units within the firm. During an era of mounting competition, however, is the time truly available to adhere strictly to the tenets of this model? If consensus is pivotal, does the time required to implement this model provide for rapid product innovation—a presumed necessity during the post-deregulatory era of mounting competition?

Michael Porter's competitive analysis model, noted for its sophistication and precision, represents an evolutionary leap in strategic management. Where the classical and systems models stressed internal processes and procedures, Porter reverted to an evaluation of those strategic priorities associated with the competitive landscape. In other words, Porter's research focused on the elements of a firm's environment: these include, as outlined in Chapter 1, the

- Threat of new entrants, or prospective competition;
- Supplier power, or the capacity of a supplier to influence or control the behavior of a firm's competitors through its distribution and pricing policies;
- Rivalry among existing companies, or the degree to which a firm's competitors are prepared to vigorously compete on the basis of pricing, quality, service, and the intent to expand market share;
- Buyer power, or the ability of customers to drive the definition, form, or reshaping of a company's products or services; and
- Threat of substitutes, simply the emergence of credible alternatives to existing product lines.

To fully employ Porter's model, a company must measure, monitor, and benchmark its performance relative to its competitors according to the five criteria described above. A "strategic audit" is thus recommended for a company committed to this form of strategic management. We note in Figure 6.1 a method of implementing Porter's competitive analysis model.

Porter's arduous analysis, generally accepted by modern strategists, is supported by credible evidence affirming its efficacy [4]. It meets the tests of forward-thinking strategic managers in terms of rigor, comprehensiveness, relevance, and specificity. It delineates when a product-line should be abandoned, when it should be innovated, when competitors are likely to replicate or substitute for it. Competitive analysis isolates the determinants of market change, and identifies methods of product differentiation. Its application is appropriate for both domestic and global strategic management, and it can be adapted to changing competitive environments once fully installed by a firm.

There is, however, one essential limitation in this method of market planning: the model lacks a forum by which a strategist can deduce when and how to negotiate cooperative agreements (in the form of alliances, affiliations, or partnerships) with other firms. Increasingly, deregulated market environments—competitive landscapes in which no communications company can dominate all production and distribution—require the ability to design nimble strategy which can exploit transient economic and synergistic opportunities. Competitive analysis thus represents state-of-the-art strategic management in design, implementation, and scope; it is not, however, an adequate device for

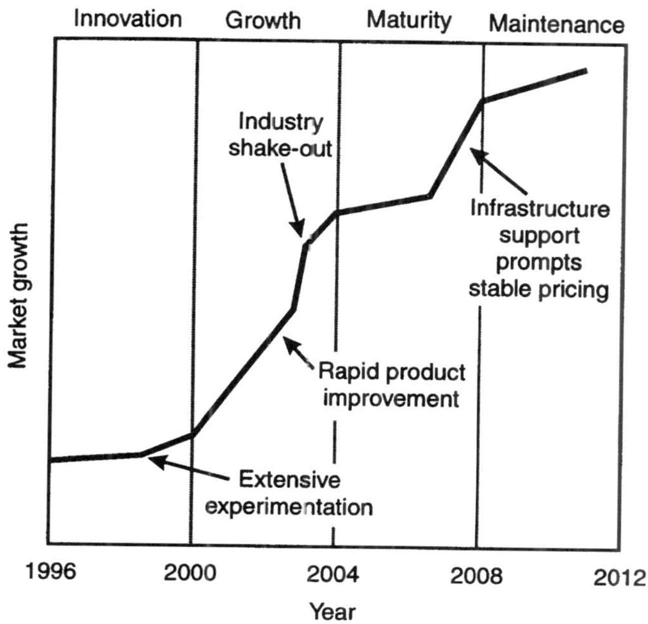

Figure 6.1 SRC/LRS model.

constructing cooperative agreements with other firms whose resources magnify a firm's own competitive strengths [4]. For this contingency, "co-opetitive modeling" prevails as an important tool in the arsenal of strategists.

Co-opetition, elaborated in Chapter 1, provides a mechanism to identify those instances in which a firm can approach both rivals and neutral entities regarding cooperative ventures. The synergies of cooperation outweigh whatever prospective threats might be engaged as a result of such ventures. A firm knows when to adopt co-opetitive measures when: a) an agreement results in customers valuing your product to a greater degree after cooperative ventures are encumbered; b) your own market share increases because customers have greater access to your product; c) when suppliers prefer working with your firm rather than your competitors; and d) when strategic disadvantage accrues to your competitors as a result of such action. In each instance, cooperation results in the expansion of market opportunity through the enhancement of internal resources or the diminution of competitive threat (refer to Figure 1.3 for a review of the mechanics of co-opetition).

In sum, given the array of strategic modeling herein described, contemporary telecommunications strategists in an optimal environment might elect to apply all four methods. For obvious reasons, however—among them the availability of time and financial resources—it would seem prudent to rely on the competitive analysis paradigm, augmented by the judicious use of co-opetition. The precision of Porter's model, supported by a nimble, pliable use of co-opetition, affords the greatest probability of strategic success in today's competitive landscape. It should be noted, however, that nothing prevents the creative strategist from incorporating elements of all four models to designate a practical alternative. Imaginative strategic management is continuously evolving.

6.2 Suggested Guidelines for Application

We can assert some straightforward guidelines for identifying the suitability of these models for each of the three elements of modern telecommunications. In telephony, an industry marked by established structures of competition, and thus far resistant to the dictates of deregulation, one is more likely to succeed at employing traditional or systems modeling. The resistance to competitive change has enabled this industry to adhere, in many cases, to classic strategic design. If strategic response is not dictated by the constraint of time and limited product life-cycles, systems modeling might be most appropriate, particularly with respect to the largest corporations.

In broadcasting—radio, television, and their cooperating sectors—competitive change is more immediate, and therefore a compelling feature of strategic management. Broadcasters increasingly must anticipate not merely the fastidiousness of the consumer, but the dramatic competition promulgated by technological advance. Broadcasters must be active in innovating their product and customizing its features for consumers. In this instance, Porter's model, with its reliance on the analysis of supplier power (a critical feature for content providers), seems most advantageous.

For the computer industry, and its rapidly advancing encroachment on its companion sectors in telecommunications, an imaginative blend of competitive and co-opetitive analysis appears imperative. Indeed, in the following chapter are discussed those firms with a proven track record of success in this sector; these firms are characterized by an extraordinary ability to aggregate a customer base so as to minimize their rivals' strategic progress. Strategic acuity, coupled with nimble and proactive negotiating ability, marks the performance of the successful strategist in this sector.

6.3 Laying the Foundation for Strategic Management

One must distinguish between two critical research mechanisms that necessarily precede a serious application of strategic management. Market research is the collection of data relating consumer behavior to prospective demand for goods and services. Market research, at its core, juxtaposes elements of behavior—attitudes, socioeconomic background, personality, and so on—with changes in consumer preference. Competitive research, also referred to as "surveillance" by some practitioners, represents the review and evaluation of strategic management adopted by one's competitors. Competitive research, in its totality, includes quantitative and qualitative treatment of competitor performances, monitoring of emerging alternative technologies, and assessment of consumer interaction with other firms relative to price, quality, service, and post-purchase satisfaction. Those who employ competitive research techniques seek to determine the extent to which competitors' strategies alter prospective market share and the effect this will hold for the future of the firm [5].

6.3.1 The Tools of Market Research

The tools of market research include segmenting, targeting, and positioning. Segmenting the market refers to the dichotomization of customer bases for promoting a product. Market segments could expand or contract, depending on changes in consumer preference as evidenced by customer surveys, focus groups, and other tests. Targeting involves pinpointing current and future market opportunities, that is, identifying prospective congruence between a firm's resources and its ability to satisfy a market segment. Positioning is the procedure by which a firm sets its product line apart from those of competitors, which is especially vital in industries where impending competition is about to proliferate [6].

Traditional market segmentation calls for the application of the following sequential procedures:

1. Identification and measurement of the market: this procedure demands that researchers group consumers within market segments and apply appropriate methodologies to confirm the extent to which a firm's product line conforms to consumer needs. Can the firm measure a segment's characteristics, desires, and needs?
2. Determination of segment profitability: is the market segment of sufficient size to justify a firm's commitment of financial and other resources?
3. Market communication and distribution: can the firm introduce its product to market and effect channels of communication to monitor prevailing consumer preferences?

4. Elasticity of response: does the market respond to changes or differences in market strategies? Can the firm ensure that the effectiveness of its market strategy can be scientifically evaluated?
5. Market segment stability: do forecasts reveal stability—therefore predictability—of market segments, or is dynamic change expected? This factor is critical in determining a projected modification of market and strategic plans.

The basis upon which market segments are defined includes the manipulation of both descriptive and behavioral platforms. Descriptive platforms consist of variables, the integration of which provides a profile of groupings within defined market segments. Such variables include age, sex, gender, nationality, income, occupation, religion, family size, household life-cycle, and education. Behavioral platforms are comprised of variables including brand loyalty, readiness to buy, social class, lifestyle, usage rates, user status, and other identifying characteristics focusing patterns of behavior impinging on demand. The creative assimilation of descriptive and behavioral platforms is what distinguishes firms in terms of their relative success in market research [6].

Telephony and broadcasting firms will survey, in light of the Act, the effectiveness of marketing plans devised by firms with comparable competitive experiences in other industries. In this way, errors in market segmentation will be avoided and the probability of marketing success will be enhanced. Historical analogy is discussed in greater detail in the next chapter.

6.3.2 Market Targeting in Telecommunications

The selection of target markets is characterized by the identification of groups whose consumption would result in greatest profitability. Ordinarily a firm chooses to introduce its product to those groups within market segments who not only can afford to purchase that item but where profit margins can best be extracted in a competitive environment. Thus, there exists a ranked-order methodology to targeting; the largest and most profitable groups are introduced to new product lines before extended market absorption is attempted [7]. This procedure is both traditional and sensible in the eyes of many, but new marketing techniques may obviate its effectiveness in light of unpredictable competitive threats emerging from technological substitution and new product lines.

The identification and evaluation of target markets lead to strategies appropriate to exploit such groupings. Typically, marketers will employ either concentrated or differentiated strategies to exploit emerging opportunities. A concentrated strategy focuses on a single, most profitable group of consumers within a market segment. A differentiated strategy, sometimes described as "multisegment strategy," fully develops all niches within a segment and promotes its product to each differentiation [8]. Particle marketing constitutes a third approach to the identification of targets: each consumer becomes a market unto him or herself. Some observers refer to particle marketing as "one-to-one" promotion or, alternatively, "atomization" or "one-to-one marketing." The trend toward narrowly defined targets is unique to this decade, but some speculate that the next progression in targeting may be "subatomization": the identification of the predominant independent variables responsible for a customer's behavior. These subatomic units can vary over time and thus require constant monitoring and measurement to be useful tools in the targeting process [10].

We can infer an important underlying impetus to target research for communications firms during the next decade. For established companies previously constrained by the rigors of regulation, it is probable that they will ease into traditional concentrated vs. differentiated methodology before embarking on alternative strategies. Organizational inertia and operational size prevent regulated firms from engaging customers on a one-to-one basis. Those communications firms previously untouched by regulation, and who remain more nimble by virtue of smaller, decentralized marketing units, are better suited to apply particle marketing.

6.3.3 Market Positioning in Telecommunications

Market positioning incorporates a number of organizational and product characteristics to gain a foothold in either established or emerging markets. The emphasis is not on image but competitive advantage: the issue at hand concerns gaining comparative leverage over all other competitors. In the communications market, the procedures by which positioning is accomplished are important because they must be applied in a context of great volatility. Those engaged in positioning telecommunications firms pinpoint:

- Product attributes: The association of advantageous characteristics vis-à-vis competition (cellular phones, for instance, are superior to paging services because of value-added characteristics).
- Solution specification: This service solves or satisfies a basic consumer or business need (wireless data transmission enables mobile workers to integrate e-mail and Internet services into their daily routine, for instance).
- Product user: Firms have positioned themselves by linking personalities (famous spokespersons) to their firm or product line.

- Quality: Studies reveal that an increasing fraction of the population is sensitive to product quality and reliability; communications firms producing hardware, for instance, are attempting to exploit this sensitivity.
- Multifunctionalism: communications firms are now defining their market positions by designing products with versatile functionality. A single tool that can solve multiple problems retains great customer appeal during the early era of deregulation when uncertainty prevails in the minds of many consumers.

Segmenting, targeting, and positioning are the operative words of contemporary telecommunications market research. Deregulation cajoles the incorporation of all three techniques to gain comparative and competitive advantage. Rapid adaptation to these strategies leads to the development of new markets, facilitates expansion of market share, and mitigates the costs associated with retention of customers. Firms operating without a history of regulatory constraint now pursue their strategies in the face of prospective capital infusion by their larger counterparts; funds designated for market and competitive research can be expected to grow exponentially in the years ahead, particularly in the area of content providers who must meet the demands of discriminating consumers.

6.3.4 Recent Developments in Competitive Research

Competitive research focuses on the evaluation of competitors' strategic plans. The emphasis is placed on the segmenting, targeting, and positioning of other firms seeking a larger market share. Those who employ competitive research methodology on an on-going basis seek definition of "market path." The future contours of the market are seen as a function of strategic plans designed at present.

The object of competitive research is to "benchmark" a firm's own record relative to others operating within comparable markets. Such comparisons of performance provide a practical method of measuring progress while increasing marketing accountability. Such data, if carefully and properly evaluated, are likely to be a significant force in creating strategic plans during the early stage of deregulation. The use of benchmarking methodology is contentious because debate exists regarding appropriate methodologies for its application and interpretation. The primary methodology, at present, involves the use of accounting and statistical methods applied across a spectrum of firms operating under comparable conditions.

For telecommunications firms, market and competitive research is now applied in an "age of adversity" [11]. Two dynamics make competitive research in communications challenging in today's environment: the proliferation of competition and technological innovation. In relaxing regulatory constraints, government has simultaneously encouraged not only new competition but technological innovation and substitution precipitated by converging industries.

6.3.5 The Changing Environment of Telecommunications Competitive Research

During the 1950s, mass marketing techniques were aimed at generating demand for single-brand markets; one kind of automobile, one kind of cosmetic, and one type of appliance was promoted for large markets, geared to a period of rising consumption. This pattern continued into the 1960s, with some differentiation in market segments (or market segmentation). A single product was subdivided into a set of multiple niches, distinguished by variance in consumer income, attitude, or taste. The advent of more sophisticated methodologies brought further subdivisions, sometimes referred to as "niche" marketing during the late 1970s and 1980s.

The 1990s have signaled a period ripe for the development of "market customization," that is, the creation of product lines tailored to the specific needs of individuals, households, and businesses. The array of database procedures, combined with technological innovations, now makes this process feasible. Advances in data warehousing, mining, inventory control, and other database management techniques now make feasible what was formerly impractical.

A number of concerns quickly surfaced for marketers within months following adoption of deregulation. Among the most vital concerns is managing customer life-cycle marketing. Firms must now address:

- How to expeditiously seize market share;
- How to maintain that market share in the face of rising competition.
- How to implement a life-cycle strategy (satisfy customers over the long run);
- How to build effective customer retention programs;
- How to exploit greater profitability for each market segment;
- How to forecast emerging customer needs;
- How to identify the most profitable customer life-cycle segments;
- Whether to diversify early or later across geographic boundaries; and
- How to efficiently employ state-of-the-art database strategies to track customer recruitments, retention, and satisfaction.

Life-cycle marketing had become a critical issue for communications firms because limitations in capital restrained the design of customer bases that predictably build regular cash flows. Costs of capital construction can only be justified by cash flows engendered on a month-to-month basis. In regulated industries, such predictability was a foregone conclusion; in the face of competition, capital construction necessarily became riskier. Gaining a foothold on the forces driving consumer behavior became a key to industry success. While life-cycle marketing became pivotal, new techniques of leveraging data to extract further profitability were held as an equal priority [12].

To contend with these developing issues, communications firms in some instances sought unconventional strategies to differentiate themselves from competitors. Strategically, redefinition of marketing channels, cooperative alliances, mass-customization techniques, new affiliations with related industries, and other unconventional methods were explored with increasing seriousness [12]. Customer loyalty became an overriding objective for firms in both wireline and wireless services [14].

6.4 Strategic Management and Deregulation

Recent studies have confirmed the nature of two distinct behavioral trends in telecommunications consumption for both households and business consumers:

1. Customers, desiring simplicity in communication menus, prefer "bundled" packages.
2. A "new media household" engenders interest primarily from high-income consumers in the early phase of deregulation [15].

A bundled package is defined as a set or group of communication services for which consumers select one company offering such provision at a discounted price. Industry consternation and the proliferation of choices are expected to lure many consumers to the simplicity of single providers. This development, if realized, may signify a long-term industry trend. These empirical results suggest that those firms successful in targeting markets blend effective pricing with service. This analysis also indicates that it is incumbent on single providers to expeditiously evaluate the degree to which their targeted strategies succeed in satisfying customers. After price-sensitivity, these studies verify that intangible factors (such as customer-loyalty) define consumer choice.

Significant numbers of households demonstrate a knowledge of emerging communication technologies which enhance the quality and productivity of their lives [16]. These innovators are curious about such technologies and foresee obvious application in education, health, entertainment, and government services. Relatively small in number, this contingent's experimentation is necessary to launch the wave of early-majority, late-majority, and laggard segments characteristic of the consumer S-curves.

Tracking behavioral change in service adoption can now be accommodated by manipulating billing accounts [17]. Immediate answers to these questions can now be accomplished, which was unimaginable until recently. The monitoring of consumer behavior—the application of state-of-the-art techniques to formulate strategy—may well differentiate the success of strategic plans in the short-run. For smaller firms unable to manage the tracking of such data, new outsourcing strategies are available; firms specialized in the evaluation of such data can provide interpretation in an expeditious manner [18].

Effective market and competitive research is tied today to uncovering behavioral patterns embedded in the data. A major development in monitoring consumer demand is the cooperation of firms whose resources are synergistic. Such complementation represents multiorganizational tracking, the interpretation of customer data by firms able to "team" to implement billing and customer care strategies. A firm interested in the value of customer data for its objectives might be willing to perform customer analysis for another company willing to part with its database.

There is a delicate balance between proprietary hold on data versus the need to gain expertise at minimal cost, but this trend is in evidence in the industry [19]. We note the shift toward complementary and supportive alliances characterizing strategic planning during the era of deregulation. No single firm in the industry can independently supply all of its needs while pursuing every available avenue of opportunity.

6.5 Strategic Management and Forecasting

A necessary precondition for the creation of strategic plans is the application of forecasting methodology. Serious attempts to predict market developments, however complex, provide a context in which firms can forge competitive response. In the absence of forecasting, it simply is not possible to define organizational mission, goals, objectives, and future contingencies. In applying forecasting methods a firm does not obsess with absolute predictive accuracy.

For the telecommunications industry, effective forecasting has never been more vital. Deregulation has sped new competition to market, encouraged a rapid proliferation of mergers and partnerships, engendered new incentives for product substitutes, and heightened the anxiety of consumers inexperienced with the proliferation of alternatives from multiple suppliers. It is precisely dynamic market environments which inspire firms to engage in the most sophisticated forms of forecasting methods.

Practitioners of telecommunications forecasting are experiencing extraordinary change in their craft, particularly with respect to establishing systems aimed at anticipating changes in consumer preference and changing attitudes of consumer loyalty. Data contained in Sections D and E of the Appendix have been carefully selected to provide a foundation for successful forecasting of consumer behavior. The following discussion, and the treatment provided in the following chapter, refer to the importance of benchmarked data in sales volume, employee productivity, market share, customer-base growth, and the retention of customers. The collection of telecommunications forecasting methods discussed below is used to anticipate market developments in two essential endeavors:

1. Generating general market data on prospective consumer preferences, sales, emerging competition, and other priorities as defined by the firm; and
2. Specifying the patterns of technology adoption precipitated by variations in product diffusion, mortality, and substitution.

With respect to the assimilation and evaluation of data on general market developments, two dozen qualitative and quantitative techniques are employed in the business environment. The application of these tools holds universal value in all industries [20]. The array of qualitative tools includes historical analogy analysis, visionary forecasting, Delphi methodology, panel consensus, and pre- and post-testing of focus groups [21–23]. The virtue of qualitative methodology lies in its capacity to treat information not susceptible to traditional statistical treatment. Useful information generated through environmental scanning, discussed later in this chapter, is often anecdotal or intuitive in nature, and such methods provide a mechanism to transmit the intangibles of market research. The instinct of qualitative methodologists is to identify patterns of consumer behavior that are inherently repetitive—and therefore cyclical—in nature. Forecasts are thus contingent on patterns of behavior presumed to be repetitive in terms of lifestyle, attitudes, household age, or other subjective standards [24].

Quantitative methods include the use of: moving average, regression analysis, exponential smoothing, Box-Jenkins tools, X-11 time analysis, trend projections, traditional input-output analysis, systems theory, anticipation surveys, econometrics, life-cycle analysis, leading indicator composites, and diffusion indexes [24]. Other quantitative methods, traditional and emerging, are sometimes applied by firms in the forecasting process, but these can generally be regarded as falling into one of these categories. Quantitative methods, with their often arcane and intimidating scientific foundation, are often preferred by strategic planners and marketers. As is the case with qualitative methodology, however, these techniques have their limitations; always, their value lies in the credibility of data inserted into the appropriate method [25].

For communications firms, all of these traditional techniques hold potential value in gaining competitive advantage. As a result, two principles guide the application of forecasting methodology:

1. Practitioners should identify every tool appropriate to the data or information at hand.
2. The intersection of results, when the appropriate methods have been applied, should be treated as the key to effective forecasts.

6.6 Analysis of Diffusion, Substitution, and Mortality

The communications industry is driven by technological change, innovation dictating prospective competitive shifts. Forecasters have experienced their greatest success in defining the future of communications markets through the application of S-curve methodology [25]. New product introduction, and estimation of adoption rates associated with that marketing, have become the overriding priority of communications prognosticators seeking an understanding of the contours of market demand [25].

In predicting the destiny of emerging products and services, forecasters differentiate three economic forces:

1. Product adoption: the rate at which consumers—residential and business—adopt new products. Adoption is visually captured through the construction of an S-curve (see projected telecommunications product development during the years following deregulation as outlined below), tracing the phenomenon of consumers who gradually enter the market as a product gains market acceptance. This phase is captured within the "innovation" stage of product adoption. Often, new product adoption represents an effort by consumers to "substitute" improved products for older, less efficient ones.
2. Product diffusion: a technique employed to specify consumer prototypes likely to purchase a new product. "Innovators," "early adopters,"

"early majority," "late majority," and "laggards" constitute the primary components of the diffusion process. The importance of this method lies in the fact that early adopters do not enter the market until innovators have established precedent. Acceptance by early adopters is necessary to trigger purchase by early majorities, and so on. Many companies dichotomize these diffusion typologies in relation to the peculiar characteristics of their products and the consumers they attract [21–23]. The growth and maturity phases of product adoption, then, are pivotal in defining product success, with resultant subsequent implications for product innovation and eventual "industry shakeout."

3. Product mortality: product breakdown, not obsolescence, drives "product mortality" with regard to telecommunications infrastructure, network, and other equipment. If the "maintenance" stage of product adoption fails to yield substantial product improvement relative to market need, one may expect radical substitutes to be launched, thereby assuring certain mortality for veteran product lines. In the area of content providers, obsolescence represents the key determinant of product mortality.

6.7 Linking S-Curve Methodology with Product Substitution

As we evaluate the path of a hypothetical S-Curve, it is inevitable that we address a critical issue: will consumers be moved to shift from one technology to another? The projection is an awkward one to make, and involves the "jumping of S-curves" as discussed below. A straightforward focus on pricing is inadequate in responding to this issue, and increasingly the most successful firms are able to factor this variable into their forecasts. Strategies sometimes fail precisely because they are unable to contend with the dynamics of this question.

The empirical evidence suggests that market acceptance is often frequently an enigma, and requires an understanding of mass psychology as well as the economics of pending substitution.

The hypothetical point on an S-curve measures the key transition from concave to convex growth, signifying diminished growth rate. These inflection points signal vulnerability to a market leader in a rising consumer market in which competition mounts. In this illustration, we can identify an inflection point for cellular consumption in the year 2000. The cellular industry continues to grow during this period, but PCS and SMR technologies (wireless rivals) accelerate their growth and capture market share. As is typically the case when industry rivals penetrate the market share of their principal rival, we can anticipate intense competition, characterized by falling prices. The S-Curve, in this regard, therefore serves strategists because it indicates when pricing pressures are likely to develop. It also acts as a predictive tool in anticipating that point at which industry rivalry is likely to intensify. Continued technical innovation, accompanied by improvement in quality and service, is likely to be pronounced during this period. While competitors will fight increasingly for greater market share, we can also expect that, as a practical matter, customer retention will become a pivotal concern during this phase of the S-Curve.

The period from 2000 to 2006 will be an era of severe competition during which the market leader will have to innovate in response to aggressive emerging technologies. It is improbable that cellular can capture more than one-third of total market share during this period, and it is a distinct possibility that PCS could significantly erode cellular market share.

Note that as each of the four primary wireless technologies evolves into the next century, the graph designates cellular technology as the first to reach an inflection point: the shift from concave to convex market penetration specifies a market opportunity for the remaining competitors.

6.8 Applying "Jumping the S-Curve" Methodology

When products or technologies in growing consumer markets undergo a change of arc in their growth rates, potential market vulnerability signals product innovation. A mature technology can "jump" an existing S-curve and once again experience an ascending concave arc. The logic implicit in this strategy is obvious, and, for the cellular industry, the key turning point is likely to occur sometime around the year 2000. Yet, a number of technologies have not been improved to accommodate such a strategy. We note these illustrations constitute examples of lost opportunity and eventual product decline.

Companies have frequently hesitated in "jumping" to a new S-curve for a number of reasons. Product innovation that might have resulted in a transition from established to new S-curves has been effectively prevented by:

- Misuse, dismissal, or misappropriation of market data;
- Reticence in assuming the risk of implementing new product lines;
- Inaccurate projections of competition emerging from alternative product lines; and
- Inability to satisfy customers during periods of rising competitor alternatives.

Cellular providers could be undermined by overconfidence during the next decade if they presume that personal communications services, paging, and specialized mobile radio providers can achieve no more than marginal improvements in market share or customer base.

It has been assumed that the most effective way to capture a rising market share in a rising consumer market is to innovate a technology, grow with the market, and dominate pricing in the early years of minimal competition. It is this tactic that has characterized the growth of the cellular industry since the 1980s. The primary disadvantage associated with this tactic are the costs associated with product improvement and market research. Those arguing that such costs are justifiable during an era of an expansive S-curve with its attendant unbridled opportunities for growth in primary, secondary, and tertiary markets.

It has become axiomatic in the computer industry that "first to market acceptance" is substantially more important than being "first to market" [25]. This lesson is not lost on strategists in other sectors of the telecommunications market, where experiments in testing new consumer markets are growing increasingly expensive. A firm prepared to take the risk that it will not be first to market with new technologies, is also a company poised to exploit the knowledge gained from its rivals' market tests—without incurring the costs associated with such experiments. Whether content, service, or infrastructure provider, a firm must balance the costs and benefits measured in terms of new product development versus product innovation. There is no absolute method of defining this balance and, hence, there is substantial latitude in applying managerial judgment and intuition. The human factor in deciding when to launch new product lines thus asserts its influence—scientific forecasting is not sufficient in this regard.

In the case of the wireless industry, as noted in the aforementioned illustration, the ascending concave arc characteristic of the industry S-curve affords competitors an exceptional and unprecedented market opportunity. It is unlikely that such an opportunity will return after the year 2006. This logical methodological conclusion thus clarifies, if not commends, a judgment as to the assumption of risk. Methodology and intuition should be seen as mutually supportive, rather than conflicting, mechanisms in the strategic management process.

6.9 Limitations of Telecommunications Forecasting

The wireless case study demonstrates both the strengths and limitations of telecommunications forecasting as it relates to customer demand. The enthusiasm now in evidence for wireless communications eluded some of the most sophisticated industry forecasters a decade before. We may therefore infer that any communications technology for which there is prospective practical application, and in which ease-of-use is a definable attribute, holds the potential for market penetration. At the moment these two characteristics coincide, pricing becomes the critical determinant to ubiquity of acceptance and use.

It should be added that forecasters reviewing prospects for wireless technology relied almost exclusively on quantitative forecasting methodology to form their conclusions. As a result, they lost the benefit of exploiting the value of qualitative methods that might have unlocked the unfulfilled needs and desires of consumers.

6.10 Connecting Environmental Scanning, Forecasting, and Strategic Management

The contribution of forecasting in clarifying strategic planning options can only be made on the basis of reliable data submitted to the process of environmental scanning. Environmental scanning synthesizes both quantitative and qualitative information according to a set of variables that includes technological developments, social and cultural trends, economic change, political factors, and other elements.

Data warehousing, data mining, and advances in inventory control now impel the need for advanced telecommunications strategic management. If the proliferation of data on customer behavior are employed to greatest advantage, "mass customization" will certainly arise in the industry.[1] It may be possible for large corporations to provide highly personalized service of a kind only known in niche firms. Personalized service occurs when the customer dictates adjustments in customized menus at his or her own discretion and the organization accedes to that design. It has earlier been noted that customized menus of telecommunications services are probable in the future. Corporate strategists should now take this possibility seriously as they design their plans.

6.11 Applying Strategic Management Models: A Critical Review

How does one determine the appropriateness of these models for a given strategic challenge? The procedures, steps, and techniques of analysis outlined

1. For additional background see the first three websites under the market-function portion of the Appendix, Section F.

for each model in chapters one and six of this volume represent a constructive first step. In addition, it is useful to perform this critical audit of the competitive environment in which the planner must design strategy. A consideration of these criteria assist in clarifying that model most congruent with the current competitive environment:

- Is the industry in which the firm competes a network or infrastructure provider, or is its principal function the generation, storage, or transmission of content? If the former characteristics designate the landscape, change is less vexing and the classical or systems model may afford greatest flexibility and efficacy. If content generation or transmission is the mission of the firm (Internet or electronic commerce providers, for example), Porter's model, complemented by co-opetitive strategy might strengthen the firm's competitive leverage and advantage.
- Is the firm supported by personnel dedicated to its mission and loyal to its goals? If this circumstance holds, then the firm can exploit the intrinsic strengths of the systems model. Continuity, stability, and fidelity to the firm's long-term objectives may justify some sacrifice in competitive nimbleness and dexterity. Predictability in organizational behavior can override the loss of transient advantages that some firms might gain which rely on the co-opetitive model. Indeed, recent experience of network providers—telephony, broadcasters, computer firms—confirms that the loss of "intellectual capital" can devastate one firm while serving its rivals (the loss of key personnel to industry competitors compounds strategic discontinuities).
- To what degree is technological innovation dictating industry performance? If product improvement is the overwhelming criterion of strategic performance, then the classical, systems, and even, perhaps, competitive models may not be sufficient mechanisms to design, launch, and adjust strategy. The capacity to negotiate, create, and sustain strategic partnerships becomes the overwhelming concern, and speed to market with new product supercedes the value of strategic continuity otherwise available in the alternative models.

The following chapter examines common attributes of successful telecommunications firms, and the strategic plans they construct. It is feasible to evaluate strategic effectiveness and congruence by subjecting these models to common benchmarking criteria in sales, productivity, profitability, market capitalization, and a host of other variables.

References

[1] Wright, P., M. Kroll, and J. Parnell, *Strategic Management: Concepts and Cases,* Upper Saddle River, NJ: Prentice-Hall Co., 1998, pp. 55–70.

[2] Porter, M., *Competitive Strategy,* New York: Free Press, 1998.

[3] Fahey, L., and R. Randall, *The Portable MBA in Strategy,* New York: John Wiley & Sons, Inc., 1994.

[4] For a thorough review of Porter's thesis and competitive analysis, refer to www.analy-sys.com.

[5] Pettis, C., *Technobrands,* New York: American Management Association, 1995, pp. 190–212.

[6] Hiam, A., and C. Schewe, *The Portable MBA in Marketing,* New York: John Wiley & Sons, Inc., pp. 209–225.

[7] Whiteley, R., and D. Hessan, *Customer-Centered Growth,* Reading, Massachusetts: Addison-Wesley Press, pp. 20–44.

[8] Naisbitt, J., *Global Paradox,* New York: William Morrow Press, 1994.

[9] Russell, C., *The Master Trend: How the Baby Boom Generation is Remaking America,* New York: Plenum Press, 1993.

[10] Daniel, L., "Combining Technologies to Support 1:1 Agenda," *Business Geographics,* November–December, 1996, p. 18.

[11] Edelstein, H., "Mining for Gold," *Information Week,* April 21, 1997, pp. 53–68; Extensive related data can be accessed at: www.egri.com.

[12] Imparato, N., and O. Harrari, *Jumping the Curve,* San Francisco: Jossey Press, 1994, pp. 229–238.

[13] "Communication Preferences Survey Highlights Opportunity to Deliver Packaged Services Deregulated Marketplace," *BusinessWire,* March 31, 1997. Also refer to subsequent reports produced by *BusinessWire* at www.businesswire.com and www.reuters.com.

[14] Flanagan, P., "The Ten Hottest Technologies in Telecom," *Telecommunications,* May 1996, pp. 29–38.

[15] Tice, W., and W. Shire, "One-Stop Telecom: The New Business Model," *Telecommunications,* April 1997, pp. 63–64.

[16] Dholakia, R., N. Mundorf, and N. Dholakia (eds.), *New Entertainment Technologies in the Home,* Mahwah, NJ: Lawrence Erlbaum Associates, 1996.

[17] Weiner, E., and A. Brown, "Emerging E-Commerce Supply Chain: The New Marketplace," *The Futurist,* Vol. 29, May 1, 1995, pp. 15–28.

[18] Guy, S., "Putting Words Into Action: Bundled Services," *Telephony,* January 20, 1997, pp. 36–38.

[19] Brandenburger, A., and B. Nalebuff, *Co-Opetition,* New York: Doubleday Press, 1996, pp. 110–158.

[20] Valentine, L., and C. Dauten, *Business Cycles and Forecasting,* Southwestern Press, 1996.

[21] Shaw, J., "Future Scenarios for the Telecom Industry: A Ten-Year Forecast," *New Telecom Quarterly,* Winter Quarter, 1996, pp. 25–33.

[22] Schneider, K., and C. R. Byers, *Quantitative Management*, New York: John Wiley & Sons, Inc., 1979, pp. 3–52.

[23] Byers, C. R., *Business Forecasting Methods*, Boston: Harvard Business Review, Spring, 1975.

[24] Mowen, J., *Consumer Behavior*, Englewood Cliffs, NJ: Prentice-Hall Inc., 1998, pp. 244–258.

[25] Asthana, P., "Jumping the S-Curve," *IEEE Spectrum*, June 1995, pp. 50–53.

7

Determinants of Effective Strategic Management: Common Attributes of Successful Telecommunications Firms

We have affirmed through empirical evidence the theory, design, and conceptual tools of contemporary strategic management modeling. With regard to the theory and design of strategy, we have isolated four principal models and their adaptation to the telecommunications industry. These include classical modeling, a traditional method of simplifying organizational goals and their implementation; systems analysis, a comprehensive method of linking inputs, process, outputs, and feedback to ensure continuity of planning; competitive analysis, stressing the interacting influence of the five forces directing the interaction of firms in the marketplace; and co-opetition, a procedure aimed at extracting industrial information so as to create synergistic alliances, affiliations, and partnerships.

In sum, we have four available models to design and manage strategy in the current business environment. As suggested in previous chapters, a strategic planner might reasonably add a fifth model, representing a customized hybrid of these models adjusted to the unique environment of his or her firm. A critical, inevitable question surfaces perpetually among practitioners at this juncture—which firms "are doing it right" [1]? In essence, what factors—or determinants—are responsible for competitive success in the current volatile marketplace? What are the characteristics or attributes of successful strategy in the turbulent telecommunications environment? What criteria should be applied in measuring a firm's strategic effectiveness? How does a firm know when it has maximized its full market potential through the cultivation of strategy? What are the turning points in strategic design—how and when does a firm transform its strategy so as to meet competitive developments in the marketplace?

These are among the vital questions now tugging at the shoulder of every strategic planner [2]. These challenges are not unique, of course, to the telecommunications strategist, as outlined in the initial chapter; they indeed cut across every industry sector. There are, perhaps, two distinctions conferred upon the communications strategist. First, if one fuses the computer sector with its broadcasting and telephony cousins, as this volume recommends, the aforementioned combustible, even the capricious telecommunications industry is arguably the most difficult one in which to design, monitor, measure, and manage strategy. Second, the pace of technological change in telecommunications is reordering change in every other sector of the American and global economies.

To the degree that strategy transforms communications, it surely influences the pace of change permeating the entire economy. While the latter factor is not of immediate concern to telecommunications strategists, it can be inferred that strategic transformation in other sectors of the economy reshape the contours of future demand for communications services. For this reason, the author recommends that planners make a serious effort to forecast the impact of their own strategy on business policies adopted elsewhere. This phenomenon represents not merely a reconfiguration of future telecommunications consumption, but a framework in which to shape strategic management. Above all, we are gradually learning that the most far-sighted, pro-active firms are those which factor this variable into their strategic equations. We start first with a consideration of those common attributes of strategic effectiveness exhibited by successful telecommunications companies.[1]

7.1 Criteria of Strategic Management Effectiveness

A straightforward method of identifying the common attributes of strategic success is to isolate criteria generally accepted as denoting success in the marketplace. There is little question that these criteria include [3]:

1. The methodology applied in this chapter is regarded by me as a contentious treatment, but nevertheless represents a starting point for evaluating strategic efficacy. I operate under the presumption that effective strategic plans result in higher stock valuations over the long-run as investors endorse or repudiate a firm's policies.

1. Sales, measured in aggregate;
2. Profitability, that sum generated by a firm after costs have been subtracted from revenues;
3. Productivity, deduced in monthly, quarterly, yearly increments, and evaluated with respect to organizational and employee output;
4. Customer satisfaction, ultimately manifested in the willingness of consumers to return to the same company for subsequent purchases;
5. Market share, that fraction of the market retained by the firm;
6. Sales growth rate, the percentage a firm's sales are rising, and tied directly to market share;
7. Competitive performance, a collection of measures intended to benchmark the performance of the company to the firm's most aggressive competitors.

These criteria are the ones typically applied by a firm, and embedded in the planning process, to adjudge its strategic effectiveness. Whether viewed individually or collectively, and regardless of the strategic model adopted, these canons afford a reasonable gauge of strategic congruence with the marketplace. Methodology intended to apply these criteria to the marketplace are discussed below, but we pause to observe an important caveat.

7.2 Telecommunications Investment and Strategic Performance

Our objective in weighing the criteria outlined above is to determine whether the strategy imposed by the firm is: (a) maximizing the full potential of the company, measured in sales, profitability, and long-term competitiveness; (b) sustaining long-term growth potential; (c) satisfying the demands of customers so as to attain objectives a and b. Note, however, that these criteria and their methodological evaluation is applied by management itself. The conclusions drawn from this procedure, clearly necessary and essential, are nevertheless judgments formed by strategists—they are not determinations made by the market. For critics of traditional strategic management methods, the ultimate test of strategic efficacy lies in the appraisal of the investor. Do investors think enough of a company (and its accompanying strategy) to invest in it? The answer, argue these observers, lies in a simple assessment of valuation [4].

Market valuation—or capitalization—is computed by multiplying a firm's outstanding shares by their current market (or stock) price. For investors narrowly focused on this criterion to ascertain strategic effectiveness, it is a simple matter to deduce strategic success. A firm's strategy is successful if it results in raising a company's valuation; if its market price falls, the presumption is that strategists have failed in their task. There is good reason to consider this method of evaluating a firm's success. Sales and profitability are clearly linked to an enterprise's value; additionally, this simple method represents a gauge of strategic effectiveness that everyone (inside and outside the firm) can understand. There is nothing arcane or nebulous about the method. There is, too, the larger significance of market valuation: if a firm's stock rises in value, its capacity to raise capital eases, thus resulting in lower costs associated with investment in plant and equipment [5].

Despite the acknowledged simplicity of market valuation as a tool in discerning strategic efficacy, the myopia of this tool limits its usefulness. Strategy is clearly one component in building a firm's market price—it is, however, merely one element. Other factors incite stock valuations: the company's relationship to fluctuations in the business cycle; the movement of funds into and out of capital markets, which can be temporarily unrelated to a firm's capacity to attract customers; the dynamics of new product development outside a firm's own control, but which nevertheless may shift the industry's pool of investment capital; the capriciousness of political, social, or cultural factors which can momentarily influence a stock's market value [5]. In short, a stock's price can be "disconnected" at any given time from a firm's true value—and strategic impact.

For these reasons, applying stock valuation as a measure of strategic success is legitimate only if evaluated over the course of time, relative to the long-run. However, over the space of several years a firm's strategy should result in higher valuations if it was well-conceived and implemented. It is not unreasonable, in this context, to chart a company's stock valuation over the long-term so as to determine its effectiveness. The only exception to this judgment obtains in those instances where a firm explicitly states that its strategic plan is designed to create, attract, or divert market share over a defined duration exceeding thirty-six months.[2]

We therefore have two credible, measurable methods of ascertaining strategic success—a benchmarking of criteria as outlined above, accompanied

2. Contemporary strategic planning has experienced a time-horizon contraction in recent years. Twenty years ago, strategists generally regarded a 5-year time horizon as long-term strategic planning. Short-term strategic planning was generally defined as consisting of a 36-month life-cycle; mid-term planning was widely described as running from years 3 to 5. With product life-cycles shortened by vast improvements in innovation—some high technology products have life-spans today of 60 days or less—communications strategic planning has been reformulated to regard 3-year planning cycles as long-run in nature.

by a serious attempt to weigh the self-interested appraisal of investors. These two approaches to strategic evaluation should be perceived as complementary, but individually inadequate to the task. Given the state of telecommunications strategic management since 1996—in the aftermath of regulatory transformation—it is reasonable now to apply these interlocking methodologies to identify those attributes of firms successful in the competitive landscape.

7.3 Telecommunications Strategic Management Effectiveness: A Post-Act Assessment

We have defined the contours of contemporary telecommunications competition. The empirical evidence affirms:

1. Proliferation of domestic and global competition, with new entrants to market instigated, in part, by advancements in the computer and information technologies industries.
2. An increase in new business formations (alliances, affiliations, partnerships, and intrapreneurships) promulgated by deregulation and privatization, domestic and global.
3. Major improvements in new technologies, impelling prospective substitution of veteran product lines by newly designed alternatives.
4. Business reengineering, the reconfiguration of organizational structures to accommodate technological advances, and concurrently reshaping the future of telecommunications consumption.
5. Customer fastidiousness regarding pricing, service, and quality, and its resultant effect on product lines and the strategies designed to introduce them.

The debate regarding the effect of the Telecommunications Act of 1996 has been provided in Chapter 3. Whether the impact, positive or negative, has been substantial or minimal, it is clear that the Act has had an impact of business formations and strategic management during the past several years. We have witnessed a heightened expansion in the number of mergers and acquisitions, partnerships, affiliations, intrapreneurships, and alliances following adoption of the statute. Whatever else one might deduce about the law's strategic impact, it has clearly set in motion greater interaction between the telephony, broadcasting, and computer industries. This deduction is not conjecture, it is empirical.

In light of these developments, one may begin a serious consideration about strategic success by gauging the stock performance of companies before and after the imposition of deregulation. There are restrictions on this form of analysis, as discussed below, but this approach represents a first step toward the measurement of strategic efficacy. The following case studies are presented in an attempt to clarify strategic performance and evaluation.

7.4 Telecommunications Strategic Management Performance: The Telephony Sector

Thousands of companies, large and small, comprise the telephony segment of the telecommunications industry. Divisions of this sector are generally grouped into long-distance communications, local or residential communications, wireless communications, telecommunications equipment, professional services, data networking, data services, home networking, internetworking, data conversion, networking markets and management, public data and voice networks, and wireless networking. Many observers also include communications information technology as an additional component of the group. The multiplicity of subdivisions within these groups is growing rapidly, as multimedia, internet telephony, video services, video conferencing, and many other technologies are brought to market.

A modest initial attempt to contend with the question of strategic management effectiveness might include a financial survey of publicly held firms whose charge it has been to introduce their products and services in the post-Telecommunications Act of 1996 era. (The reader might wish to examine the websites of those major telephony enterprises featured in Section F of the Appendix to evaluate firms of professional interest. Customized charts, similar to the ones indicated below, can be designed by the reader to track the strategic performance of any public corporation; refer to the economic sites identified in Section F.) Nine case studies are examined (three in each sector of the industry—telephony, computer, and broadcasting—as defined in Chapter 1) in this section.

Three case studies are encased in this portion of the chapter. The author has identified large capitalized firms that possessed the necessary funds to respond to the strategic imperative of deregulation during the past several years. The reader will note that the stock valuation charts, and attached case descriptions, focus on the period beginning in 1994; in this way, we might compare the strategic performance of each company immediately before and after the imposition of deregulation in February of 1996. Three elements combine to form the following strategic evaluation: (a) Company Summary, elaborating the essential organizational activities and mission of the firm; (b) Strategic Profile, specifying past and present strategic emphases; (c) Financial Profile, charting the stock's performance since 1994 (framing a five-year analysis of stock valuation relative to the Dow Jones Industrial Average, and integrating its historical volume of trading); and Strategic Findings, extracting the lessons of the firm's strategic plans relative to the future direction of the industry [6].

7.4.1 Motorola Corporation

Company Summary: A diversified firm, historically involved in radio, two-way paging services, cellular devices, semiconductors, wireless services, satellite communications, and various telecommunications services in both the developed and developing world, Motorola until recently was regarded as the unequivocal leader in most phases of its operations. The corporation has experienced major market penetration by its rivals since the mid-1990s, as implied by its stock valuation since 1994. The company has followed an aggressive path toward diversification within its industry sector, but qualifies as a telephony provider because of wireless services it provides as well as the equipment (business and residential) which supports those services. Its acquisition pattern has followed a vertical integration approach to maximizing market share in the industry.

Strategic Profile: Motorola is a 70-year-old concern which has adhered to a mission statement stressing the need for continuous product improvement through perpetual retraining of its labor force. Motorola annually expends more than $100 million on education and training, and affords its workers major incentives to improve product reliability and durability. Its strategic emphasis focuses on "being first to market" rather than "first to market acceptance," though there have been a few recent exceptions to this direction. Since the mid-1970s Motorola has dramatically improved product quality through innovations linked directly to quality control and the recruitment and retention of high-quality personnel. Motorola controls a fifth of the Iridium satellite project—critical to its efforts in wireless telephony—which has experienced a number of technical problems since its initial planning in 1990. The inability to roll out new products according to its own production scheduling, combined with technical glitches and slow response to emerging competition since deregulation, has led to stagnant or eroding market share for some of its communications services. Significant new telephony programs, however, are projected for launch in international markets shortly after 2000 [7].

Financial Profile: The company's stock valuation since 1994 is charted in the following graph. The pattern of valuations during the past five years reveals the market's disappointment with Motorola's recent technological improve-

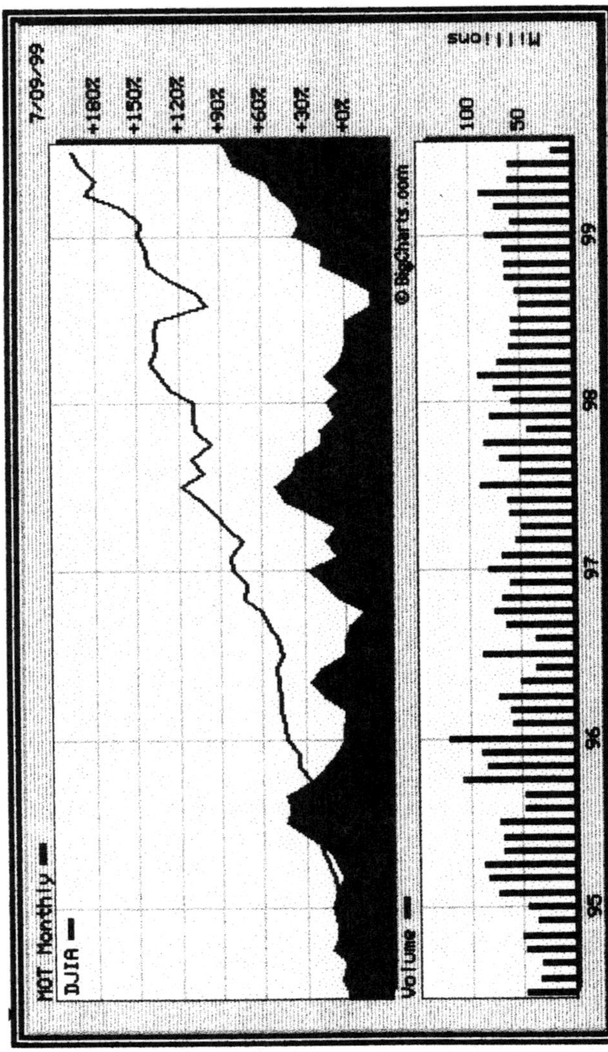

Data source: BigCharts.com.

Figure 7.1 Motorola stock valuation, 1994–1999.

ments, most especially the decline of analog markets in favor of digitized cellular service; the company had responded rather slowly to the transition to digitized wireless equipment and services following 1996. Beginning in late 1998, the company had corrected some of its technical difficulties, recommitted to penetrating wireless markets in both the United States and Asia, and has recovered some of its losses during 1998. It is noteworthy, however, that the company's stock valuation has lagged behind the broader DJIA market gauge since the advent of regulatory reform. Its recent volume in stock trades has been relatively anemic, implying that the firm is not inspiring much investment interest.

Strategic Findings: Motorola has been described since 1980 by most analysts as a well-managed, strategically enlightened firm. Its emphasis on product improvement had strengthened investor confidence until the mid-1990s and led to strong customer loyalty. As noted early in this text, it is cheaper to maintain an existing customer base than to recruit a new one; until recently, Motorola had succeeded mightily in this regard. It is instructive that this firm was slow to respond to the emergence of mergers and acquisitions, partnerships, affiliations, and alliances which have been triggered by deregulation. It is not unreasonable to infer, therefore, that this company probably did not fully anticipate or forecast the repercussions of the Telecommunications Act in terms of its affect on its own strategic priorities. It is probable that Motorola will regain its former market prominence, but the lesson of its recent strategic experience is clear: being anything less than proactive in seeking out new markets or potential cooperative alliances results in passive strategy—and there are great risks associated with this reticence.

7.4.2 SBC Communications

Company Summary: SBC is a local telephone company created as a result of AT&T's break-up during the 1980s (the other divested local telephone companies similarly constructed were Pacific Bell, Bell South, Ameritech, Bell Atlantic, Nynex, and U.S. West). SBC is particularly intriguing because they represent the most aggressive of those firms practicing horizontal integration. SBC engineered the acquisition of Pacific Bell in 1996, extending strategy with a bid for Ameritech in 1998. They have also acquired a number of smaller firms, most noteworthy the absorption of Williams Communications, a national fiber-optic network provider. One can deduce the general stance of this firm with regard to strategy: lock in key acquisitions early during deregulation, building up assets to form the foundation of collateral for comparable future acquisitions. The addition of Ameritech makes SBC the largest local telephony provider in the United States. The firm has not limited this acquisitive philosophy to traditional wireline business, however; it operates wireless services in ten nations. SBC eschews cultivation of elaborate new communications services in favor of solid management of traditional products. It is ambitious in its market planning, but is austere with regard to its palate of services. SBC is more concerned about being "first to market acceptance" as opposed to "first to market." It is a company which avoids incurring the costs associated with new product development, instead preferring to learn from the experience of other providers before incorporating new services. The company has recently been installed as a member of the Dow Jones Industrial Average.

Strategic Profile: SBC is a provider of local telephony services to nearly 40 million customers throughout California, Texas, Arkansas, Connecticut, Kansas, Missouri, Nevada, and Oklahoma. It also serves some seven million wireless consumers in more than fifteen states, complementing voice communications with paging services. SBC operates in ten nations, and is planning to penetrate additional international markets in the near future. Its expansionary strategy was established from the first moment telecommunications deregulation was inaugurated: an offer to acquire Pacific Bell was accepted within weeks following passage of the statute. This was the first major acquisition of a major telecommunications provider by a direct competitor, and prompted the FCC to carefully review the pattern of mergers and acquisitions to follow. Indeed, the Pacific Bell negotiations signaled a simple but aggressive effort to expand market share through competitor absorption rather than innovative product introduction. In this way, SBC has been able to leverage operational costs. After acquiring its competitors, SBC embarked immediately on an internal strategy to reduce administrative costs by stripping personnel redundancies and organizational overlap. It has also carefully cultivated contacts in state and local governments, lobbying effectively to strengthen its image in communities. It has generally been successful in this endeavor.

Financial Profile: SBC's aggressive strategy, measured in market terms, compares most favorably to its peer competitors. Unlike many of the Baby Bells, as well as AT&T, SBC's stock performance has been reasonably congruent with broader market averages since 1998. By avoiding the sometimes excessive costs associated with "first to market" technologies, and adhering strictly to proven market lines while expanding its customer base, SBC has goaded investors into an affirmation of its strategy. SBC is not glamorous, but widely regarded as a dependable provider, and fierce competitor, of standard telephony services.

Strategic Findings: We have in SBC a rugged competitor and acquisitive strategist. This is a firm that will seize upon any opportunity to expand its customer base and market share through horizontal integration. It is also corporation managed in a rather traditional organizational ambiance; centralized decision-making at the top of the its pyramid is principally responsible for its

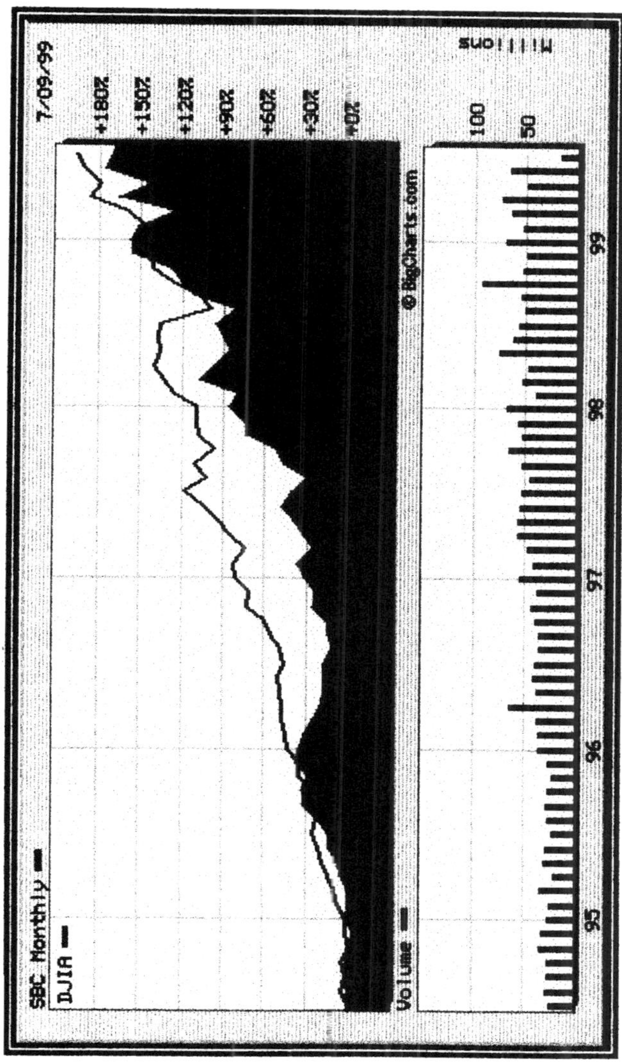

Data Source: Big Charts.com.

Figure 7.2 SBC stock valuation, 1994–1999.

strategic positioning. If the price is justifiable, and a potential competitor can be absorbed through effective leveraging, one can anticipate its strategy. The specter of a firm braced for competition through the application of simple business principles—a path that some dismiss given the risks attendant in a dazzling, dynamic industry—invites serious contemplation regarding its strategic design. Whatever its effect on the competition, SBC in recent years affirms that traditional business strategy has not yet faded from view.

7.4.3 AT&T Corporation

Company Summary: AT&T is the nation's largest market capitalized firm. In the aftermath of divestiture following 1984, the firm groped for an appropriate market strategy, initially emphasizing company spin-offs fused with diversified penetration in a multiplicity of communications markets. During the years immediately following deregulation, the corporation pursued a policy of acquiring telecommunications firms whose assets maximized perceived potential synergies. In essence, the company has embarked on an aggressive acquisition strategy aimed at getting access to as many customers as possible. The company acquired TCI in 1997; this purchase, with its recent acquisition of Media One make AT&T the nation's leading cable operator. We have in this firm, therefore, the largest long-distance provider with a second line to entering the homes of approximately 100 million consumers. No other firm in the industry can make such a claim. An interesting irony, or perhaps paradox, in the strategic development of AT&T concerns the fact that it can now theoretically penetrate the home or business of virtually every American with a greater multiplicity of products than during its most prominent period as a regulated monopoly.

Strategic Profile: AT&T had lacked a clear, cohesive market strategy prior to 1996, although it had committed to experimenting with entering new markets. Its management of strategy since 1996 is salient and unambiguous: it aims to secure the largest conceivable market base in the short-run—based on wired telephony and cable—then introduce value-added services based on bundled packages. This is a costly investment, as the corporation has had to take on extensive debt to accumulate the firms necessary to seize desired market share. One might refer to this strategy as one based on horizontal integration, the obvious motive an attack on as many high-end value users (observers refer to such practices as "cherry picking" the market) as feasible in the short-term. In short, they seek access to as many customers prepared to pay high fees for a variety of services—telephony, cable, Internet, and wireless—as possible. Customer retention will be especially vital for this firm, as interest costs on debt will be significant for years to come. It is clear that AT&T regarded passage of the 1996 statute as a clarion call for aggressive acquisition strategy.

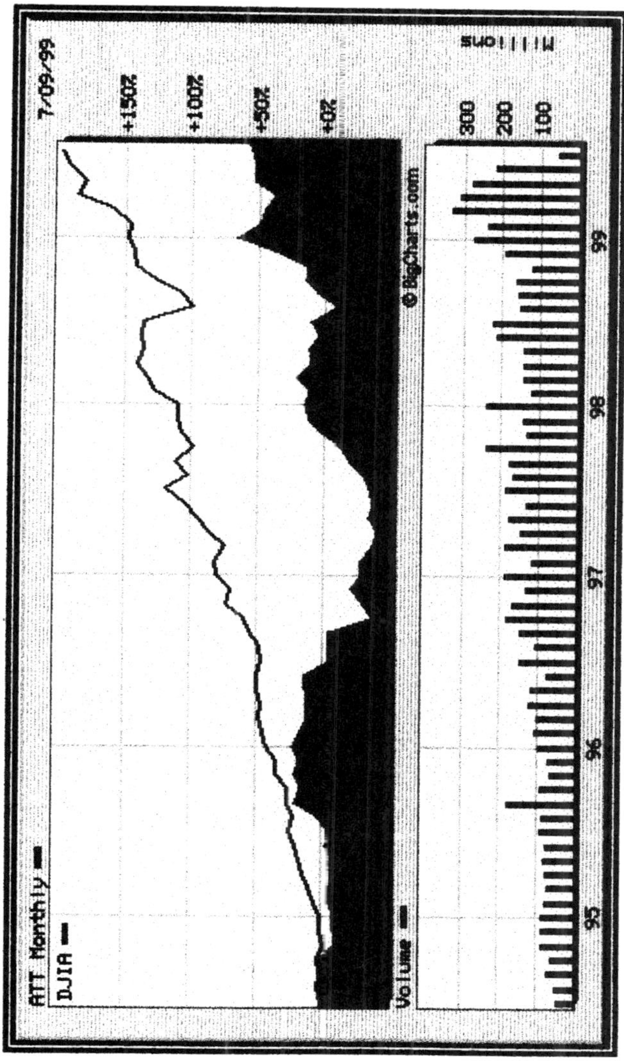

Data Source: BigCharts.com.

Figure 7.3 AT&T stock valuation, 1994–1999.

Financial Profile: Since 1996 the market has adopted a "wait-and-see" attitude with regard to AT&T's recent assertive strategy. One clear concern of investors has been the wisdom of pursuing so expensive a strategy; the return-on-investment (ROI) will have to be substantial to absorb interest costs. If ROI is not significant, the company would be limited in pursuing other strategies. In effect, the company banking its future existence on a single strategy, albeit the most impressive wireline connection to consumers in industry history. The chart below reveals the market assessment of AT&T's strategy, and it reflects an inconclusive reaction. AT&T's stock price lagged broader market averages, both within and outside the telecommunications sector. This is understandable, in part, because AT&T's strategy is clearly focused on the long-term. If its strategic priorities achieve the intended result, profitability will not be enhanced for some years to come, and in a mature corporation such a strategy is not rewarded with investor confidence until such profits are recorded. Nevertheless, the lag in stock profitability suggests that the market questions the wisdom of a one-dimensional strategy that could be rendered ineffective by technological change capable of diverting consumers to wireless substitutes in telephony, broadcasting, and Internet usage.

Strategic Findings: AT&T's long-term strategic focus on the acquisition of customers through traditional wireline access signifies that category of communications providers which rely on horizontal integration to "grow" the business. A cutting-edge debate among telecommunications strategists today concerns strategic positioning and its efficacy. We note a gradual shift in strategic emphasis following deregulation: how does a firm place itself relative to others with regard to the philosophy of accumulating customers. If it is acquisitive, it asserts financial authority to acquire other firms—adapting horizontal integration—to minimize competition while growing market share; if it is interested in securing predictable supplies or reducing the cost of production, it might be inclined to pursue a vertical integration strategy. In the case of AT&T, it is presumed by the company that if it maximizes its available customer base through acquisitions, it can sustain and increase market share. It presumes, indeed, that customer loyalty can be built based on quality and reputation. For AT&T, this strategy remains an untested hypothesis, and will require some years to affirm or reject.

7.5 Telecommunications Strategic Management Performance: The Computer Sector

Some observers might contest the appropriateness of including the computer industry as a sub-component of the communications industry. We have outlined the principal reasons behind this inclusion in Chapters 1 and 2. The computer industry has progressively involved itself in telecommunications during the past decade because it has evolved from an information to communications tool. The multiplicity of features now standard in home computers, for example, has induced households to increase their consumption of telephony and broadcasting services. The elimination of cross-ownership restrictions has invited telephony and broadcasting firms to negotiate alliances and partnerships with firms in computer hardware and software. The proliferation of content providers in the Internet domain has encouraged interaction among all three groups. The list of factors instigating interaction between telephony, broadcasting, and computer firms is lengthy and growing rapidly [8, 9].

There is wide agreement today that an important field of specialization within the telecommunications industry is information technology [8, 9]. Advances in software are driving improvements in communications products and services; they are also speeding organizational changes within other sectors of the economy, thereby encouraging vast prospective demand for sophisticated communications. We might divide information technologies along these lines to gauge its future impact on telephony and broadcasting for both businesses and households:

- Advanced intelligent networks (AINs);
- Data warehousing and mining;
- Geographic information systems;
- Network management hardware and software;
- Object-oriented technology;
- Billing network systems;
- Fault tolerance and clustering;
- Transaction processing software;
- Data networking components;
- Digital signal processing (DSPs);
- Internet telephony;
- Fiber-optic transmission technologies;
- Wireless ASICs;
- Broadband multimedia support systems;
- Web hosting services;
- Extranet support systems; and
- Intranet support systems.

The first half of this list directly impinges on the management of telecommunications firms themselves; the latter half pertains currently or prospectively to the Internet and the growing significance of electronic commerce. The influence of electronic commerce on strategic management is explored in the following chapter, but it is now salient that the computer industry is inspiring much turbulence in the telephony and broadcasting sectors. Indeed, advances in hardware and software—industries untouched by government regulation—are responsible for transforming the method by which services are developed in their regulated counterparts. By extension, this simple logic means the computer sector is transforming strategic management in telephony and broadcasting. The computer is indeed transforming the nature, design, implementation, and evaluation of strategic plans in all industries. We now explore strategic management practices in several leading computer firms, and their role in inciting changes elsewhere in the communications industry.

7.5.1 Intel Corporation

Company Summary: Intel is the premiere microprocessor maker in the world, controlling in excess of 80% of the market. The firm is noted for emphasizing continuous product improvement and rewarding its employees in furthering innovation. During the past two decades the firm has pursued development of microcontrollers, networking products, and flash memories. Its products are regarded as state-of-the-art and sustain targeted market share. Although Intel's earnings have recently been uneven, due largely to the decline in the global economy during the late 1990s, the company nevertheless outperforms its competitors in sales, market share, productivity, and innovation. The company has embarked on a recent strategy to encourage entrepreneurial ventures in every area of the private sector; it believes this approach will assist in sustaining demand for its products in both small businesses and households. Key buyers of Intel chips are Dell Computer and Compaq; the fate of these firms is tied directly to Intel's self-interest. Early in the 1990s Intel had made a number of projections pertaining to productivity. Intel had contended that the economy would experience a rising tide of productivity (labor efficiency) as a result of the full flowering of the computer industry. As noted in Chapter 2, recent empirical evidence has begun to support this lofty claim. During the past three years, the corporation has explored new ways to spur development of electronic commerce, both domestically and internationally. Intel was installed on the Dow Jones Industrial Average in October 1999.

Strategic Profile: As a firm committed to continuously growing its market through product improvement, Intel has few peers. The historic presumption has been that its open organizational style, system of rewards and incentives, and aggressive multinational development have been principally responsible for this growth. These factors may constitute significant variables in its development, but the firm's mission is clearly congruent with the destiny of an economy which pivots on information, knowledge, and organizational intelligence. Twenty years ago Intel could hardly conceive that its product would be as important in driving the communications industry as the indigenous innovations promulgated by the sector itself. It would not be literally accurate to infer that Intel is a communications provider, but its contributions in information technology altered telecommunications strategy. There is little doubt that Intel will play a significant part in advancing Internet, multimedia, e-commerce, and broadband services, and will be pivotal in defining those products in computer software upon which telecommunications depends. The long-term prognosis for Intel is very favorable, provided its strategy is congruent with the future contours of the global economy. If its products are characterized by continuous value-added quality, it seems unlikely that its global competitors will substantially penetrate its overwhelming market share.

Financial Profile: In reviewing Intel's chart, we note that it has experienced exaggerated swings in demand for its product. These movements in stock price reflect the link between the company's product and prevailing economic demand. In other words, this is a firm which must seriously study fluctuations in the business cycle so as to refine its production scheduling and inventory management. Note in the following chart that the stock consistently outperforms the broader market average—a hallmark of a consistently successful firm. However, implicit in these data is volatility in volume and valuation which signify the need to develop strategy consistent with prevailing market demand. How does Intel handle the challenge of strategic fit—ensuring that its strategy exploits market potential without resulting in excessive or unneeded supply of its product. They have adopted remarkably effective forecasting methods to estimate variations in the economy, consumer demand, and competitor behavior. Their methodological framework for the design of strategy, in this regard, has resulted in one of the great high-technology success stories in the post-World War II era. Note, too, that the firm has been even more successful in the years following deregulation, than the era which preceded adoption of the statute. Deregulation of the communications industry has benefited Intel by encouraging dynamic interplay between telephony, broadcasting, and computer and information technology.

Strategic Findings: The evidence suggests that Intel has excelled in three critical endeavors: (a) framing strategy consistent with generating products congruent with market demand; (b) successfully gearing production scheduling to effective forecasts; (c) encouraging consumers and businesses to buy its products by emboldening them—through counsel, technical assistance, and

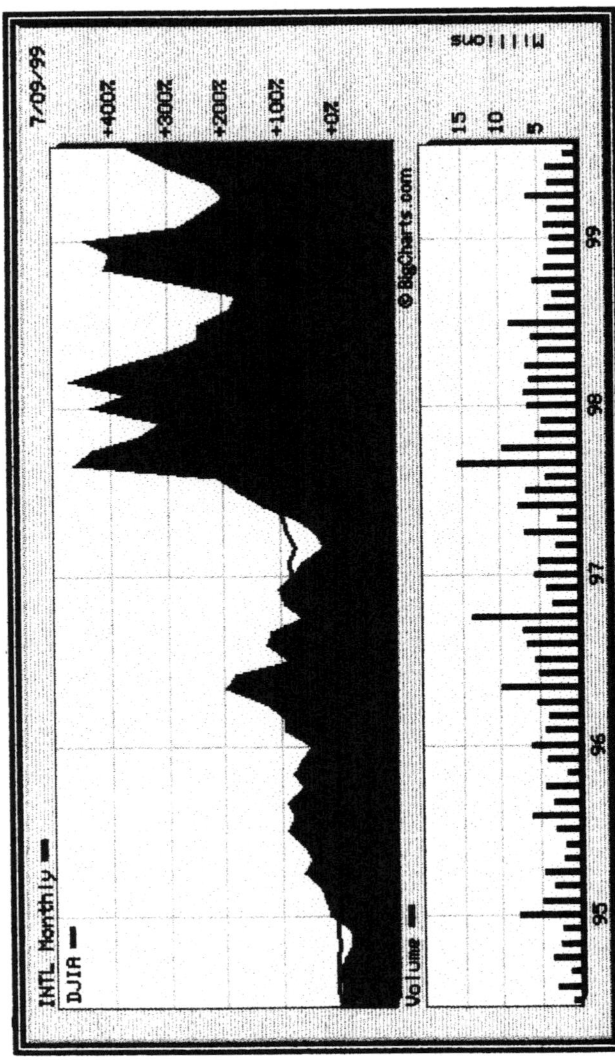

Data Source: BigCharts.com.

Figure 7.4 Intel stock valuation, 1994–1999.

financial support—to pursue entrepreneurial endeavors. This is a proactive, assertive firm which relies chiefly on its internal resources to generate growth rather than pursuing an acquisitive policy. Its strategy is to pursue high value-added products that simply cannot be replicated by its competition at any given moment in the production cycle.

7.5.2 Microsoft Corporation

Company Summary: Microsoft's meteoric success as the global leader in software products has been reinforced recently by a strategy focused on diversification and entry into new markets. With a market capitalization that now makes it the world's wealthiest company, the firm has placed itself in the position of product developer and acquisitive operator. The corporation has pursued new markets in broadcasting, Internet and multimedia services, cable television, and a host of other communications enterprises. It has actively pursued strategic partnerships, affiliations, and alliances with cable television firms, broadcasters, content providers, and a wide array of other firms. It seizes market opportunities in virtually every communications domain, apparently concluding that it cannot pick winners and losers with precision, but can, nevertheless, participate in every growing ancillary market. Equally apparent is the hostility it is now drawing from firms both within and outside its own domain. For this reason, a constant strategic concern involves anti-trust action by the Justice Department (a contingency that lobbyists for other firms pursue with some vigor). Critics charge that Microsoft is less an innovator than copier of competitor products. The fact is that Microsoft is pragmatic, pursuing every opportunity to penetrate new markets. As a result, some observers express concern that the company may lose its strategic focus—that the firm may abandon its original mission as software developer. If an observer makes a serious effort to distill the company's mission today, Microsoft is interested in pursuing any market aimed at creating, transmitting, interpreting, applying, processing, or storing information. This broader strategic focus—beyond that contemplated by any other software manufacturer—makes Microsoft a communications provider in the fullest sense of that term. One should not dismiss Microsoft's potential pursuit of any market within the telephony, broadcasting, or computer sectors. Microsoft has recently been designated a member of the Dow Jones Industrial Average.

Strategic Profile: Microsoft's ventures in Internet content and provision, cable television, satellite communications, and a host of related industries define Microsoft as a modern telecommunications corporation. There is no example of business formation—acquisition, alliance, partnership, affiliation, and intrapreneurship—that it has not exploited during the past decade. Business synergies, creating cross-ownership opportunities in telephony, broadcasting, and cable television—ignited after 1996 have been pursued with vigor. What is remarkable about the proliferation of business alliances the firm has entertained is the fact that it could have pursued many more market opportunities in recent years. The company could have penetrated banking, electronic commerce, publishing, and a wide array of other industries if not intimidated by prospective anti-trust administration. Bill Gates, the firm's CEO, has had to defend his corporate strategy before numerous congressional inquiries in 1997 and 1998.[3] This is a firm whose acquisitive instinct has been tempered, for better or worse, by government prosecution. The merits of the government's anti-trust case are not entertained here, but there is an obvious strategic footnote that all firms communications and related high-technology industries should contemplate: the effect of anti-trust application on the competitive landscape. Should a firm factor such government activity into its strategic formulation? Should a firm renounce potential market opportunities in an effort to placate government regulators? These questions cannot be addressed in the absence of assembling empirical evidence in the years to come. It is instructive that firms of comparable characteristics in other parts of the globe feel no such reticence in the current climate.

Financial Profile: Microsoft's market success has been unrivaled during the past twenty years. Given its number one market capitalization position, its control of in excess of 90% of its market, its global dominance, and tertiary influence on all other industries, it is not unreasonable to contend that this is the most important corporation in the world. As we approach the new millennium, it is conceivable that Microsoft could double in market capitalization every three to four years. Note in Figure 7.5 that Microsoft's stock valuation has greatly exceeded market averages in the DJIA and NASDAQ. Its pattern of trading volume also certifies that it attracts willing investors on a perpetual basis. Its stock price is high but investors continually reward the firm with a plentiful supply of new funds; this is a firm able to leverage its market capitalization to secure funds at minimal costs. In short, Microsoft can pursue its multi-level strategy of acquisition and diversification without fear of enduring inadequate future capital.

Strategic Findings: Microsoft is a multi-layered communications and information-technology company. It is in the most enviable position of all

3. At the time of this writing, the Justice Department's suit against Microsoft has not been resolved. Viewed in terms of its implications for the entire telecommunications industry, the final dispensation of this suit extends beyond Microsoft's relationship with its immediate competitors. An unfavorable judicial outcome for Microsoft is likely to act as a precedent for prospective anti-trust administration in those cases of industry consolidation perceived as restraining potential competition. A favorable outcome for Microsoft may signal major acquisition efforts by this and other well-capitalized computer firms seeking entre to the broadcasting and telephony sectors.

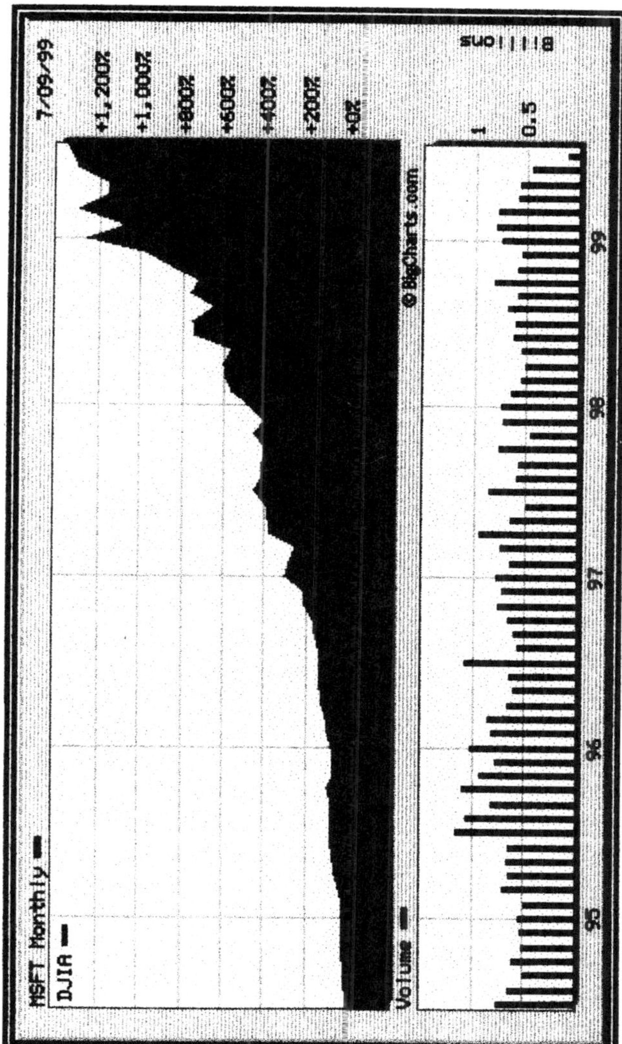

Data Source: BigCharts.com.

Figure 7.5 Microsoft Corporation stock valuation, 1994–1999.

communications entities, save the prospective intervention of government sanctions. Its acquisitive strategy is likely to produce market winners and losers with respect to new product introduction, but the firm has also assured itself that there is no attractive telecommunications market of which it will not be a part. Such is the fortune of the most significant global corporation in 2000. If anti-trust policies intervene to influence strategy, Microsoft can take comfort that the vigor of such enforcement will vary from administration to administration. Microsoft emerges in this era with optimal strategies stressing "first to market" and "first to market acceptance" philosophy depending on the vagaries of each market opportunity.

7.5.3 IBM Corporation

Company Summary: With nearly 300,000 employees worldwide, International Business Machines Corporation is the global leader in computer hardware. It has embarked on providing an ever-expanding array of new networking systems, technologies, software, enterprise services, and electronic commerce platforms. The firm has outperformed most other companies, save Microsoft, in its sector. The corporation has acquired other firms throughout the 1990s, and these acquisitions represent vertical integration strategy. IBM is less concerned about penetrating new communications markets, unlike Microsoft, than developing systems which support both physical and virtual commerce throughout the global economy. Sales have grown exponentially in recent years, reflecting in part growing market demand for IBM's services. The firm went through a difficult organizational transition following the 1990–1991 recession, experiencing falling sales and stock valuations in 1992 and 1993, but recovered dramatically in 1994. This is a company likely to avoid the risks associated with diversification, but poised instead to attack through horizontal or vertical integration any firm which threatens its market share. Computer networking designed for telecommunications and e-commerce providers constitutes a growing segment of its customer base.

Strategic Profile: IBM has apparently projected that its strategy should focus chiefly on facilitating business operations throughout the economy in terms of information technology requirements. IBM's strategy implicitly acknowledges business trends unfolding throughout the 1990s: the need to streamline (reengineer) organizational structures; introduce new cost containment measures; support information technology requirements; adapt software applications in the pursuit of new e-commerce ventures; and dozens of related activities. Its mission—to facilitate business operations across all sectors of the economy—remains in force. The strategy adopted by Microsoft, in pursuing nearly every available market, is a path the company has avoided in favor of

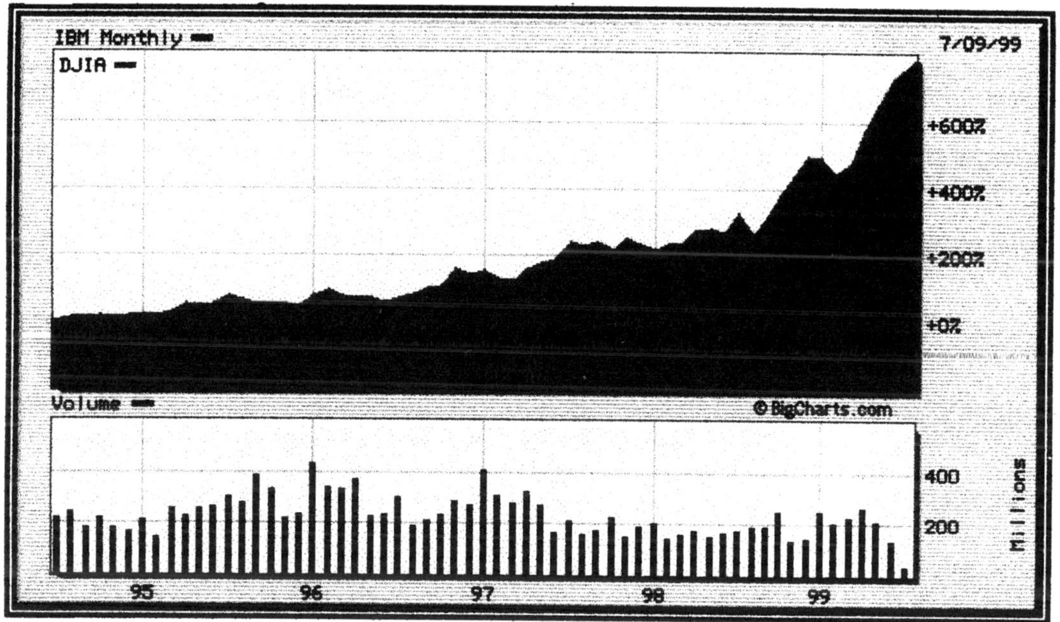

Data Source: BigCharts.com.

Figure 7.6 IBM Corporation stock valuation, 1994–1999.

retention and growth of its traditional customer base. International operations are now given the firm's highest priority.

Financial Profile: After stumbling during the early 1990s, with its stock valuation receding by a factor of one-third, IBM consistently outperformed both competitors and the market averages. The company's strategy was affirmed by both individual and institutional investors, and its sharp improvement in valuation since 1996 reflects the firm's effort to ride the explosion in Internet commerce and global demand for its software products. There have been some concerns recently regarding the quality of services and support associated with post-purchase satisfaction—a traditional hallmark of the firm—but the financial markets remain impressed by the otherwise far-sighted nature of its strategy.

Strategic Findings: IBM is assertive but has avoided the risks associated with Microsoft's acquisitive emphasis on entering every peripheral communications market. As a software giant, IBM has the financial resources to seize market opportunities in any sector of the economy. IBM recognizes, however, that the potential global demand for its existing product line remains immense. The firm evidently perceives that it should remain the innovator in its principal domain, complementing this strategy with the pursuit of partnerships, affiliations, and alliances which encourage firms to buy its products. There is no evidence that this strategy will be fundamentally modified in the immediate future.

7.6 Telecommunications Strategic Management Performance: The Broadcasting Sector

The broadcasting industry has been reconstituted in recent years. It is no longer occupied by competitors exclusively focused on television and radio, or even wireless platforms. The encroachment of the Internet, and its expansive array of new multimedia services, promises industry change on two key fronts: (1) content development, with increased emphasis on customized information and knowledge designed for the unique needs of individuals, households, and businesses; and (2) network delivery, with a striking multiplicity in the number of alternative methods to organize, distribute, archive, and retrieve specialized programming.

If technology facilitates a growing range of alternative providers, this development surely portends transformation of content development. Again, the computer industry has been decisive in spurring this metamorphosis. Innovation in information technologies has led to value-added improvement in the quality of deliverable knowledge; the thirst for such knowledge—in education, health, entertainment, business, and government—further incites consumer demand. The distillation of knowledge—information that can be applied to improving the quality-of-life or material well-being of customers—confers upon the user comparative advantage. Those who possess state-of-the-art information hold the upper hand in all forms of endeavor. This factor ensures that the broadcasting industry will move in the direction of customization, value-added programming for both domestic and global consumers. Content development will reconfigure network delivery in radio, television, satellite, Internet, and other emerging forms of delivery; improvements in network delivery will inevitably enlarge both the number of providers and prospective demand for such services. It is in this environment that we explore the dramatic strategic transformation of the broadcasting sector. Three examples of contemporary strategic management in the industry are explored.

7.6.1 Disney Corporation

Company Summary: In market capitalization, the Walt Disney Corporation is the second leading media firm in the world. The company is involved in television and film production, Internet commerce, the publishing business, amusement parks, transportation, and professional sports. It is at once both a multinational corporation and growing conglomerate. Its ventures in communications, now pronounced with the recent acquisition of ABC/Capital Cities, makes it a broadcaster in the most profound sense of the word. The Disney brand has been cultivated and refined over many decades, and the firm welcomes any project which builds and enhances its global prominence. Disney is both innovative and acquisitive, originating new broadcasting product but also active in seeking out prospective alliances and affiliations which leverage its brand name to greatest effect. The result is that the company has generally exceeded market averages in valuation throughout its history, save recent turbulence in stock trading. The firm maintains a diversified product and service line, but is sensitive to movements in the business cycle, particularly with regard to global fluctuations in consumer income.

Strategic Profile: Strategically, Disney, as a broadcaster, has attempted to penetrate television and Internet industries by developing its own content while simultaneously purchasing networks. It has seized upon its forecasts of changes in consumer preference by diversifying its content; the firm is not reactive to change, but definitely vigorous in its pursuit of new markets. The firm immediately responded to the deregulation of the communications industry, for example, by intervening to purchase ABC Network before other competitors could prepare bids. Disney has little difficulty raising capital in new stock offerings or entering the financial markets for an infusion of new funds. Some competitors in the broadcasting sector therefore view Disney as a predator in the marketplace.

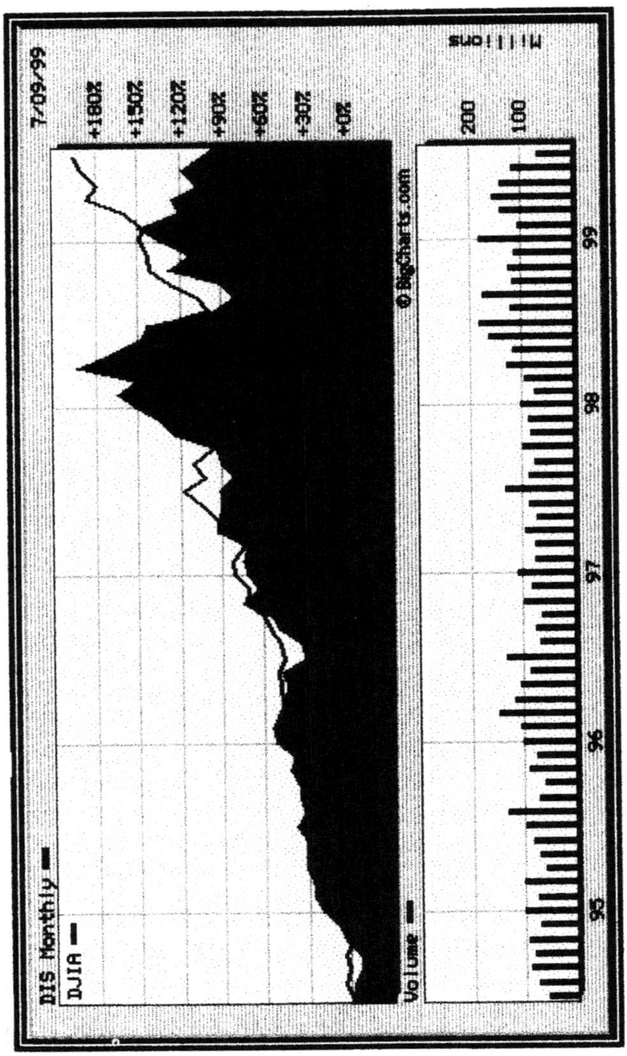

Data Source: BigCharts.com.

Figure 7.7 Disney Corporation stock valuation, 1994–1999.

Financial Profile: During periods of general economic prosperity, Disney excels in outperforming market averages. As a member of the Dow Jones Industrial Average (a select group of thirty of the nation's largest firms), Disney's performance is directly factored into, and responsible for, significant variation in the DJIA. You will note below that Disney's acquisition of ABC in the mid-1990s was well-received on Wall Street, and widely regarded as a harbinger of similar broadcasting acquisitions in the years to come. Such acquisitions provide the network platform for its growing content interests. Recently, the firm has had to digest a series of lawsuits, decline in attendance at its global amusement parks, with concurrent decline in the global economy. This last factor reaffirms the important of the business cycle relative to consumer demand for Disney products. Within the United States, however, Disney's production, broadcasting, and Internet-based businesses have performed well. The decline of Disney's stock valuation in 1998 and 1999 reflect these recent problems, but its long-term strategy remains intact. Most observers regard current financial difficulties as transient, short-term phenomena.

Strategic Findings: Disney's success in the broadcasting sector has been replicated by few rivals. As was the case with Microsoft in its sector of communications, Disney trades on a powerful brand name, diversified product line, and acquisitive tactics. It has succeeded in penetrating every potential market targeted for experimentation, and maintains the financial leverage to do so at reasonable scale economies. In short, it is expansive without being inefficient. Disney's experience in broadcasting signifies an emerging trend in communications, for those fortunate enough to emulate or adapt the appropriate strategy: when content is developed, designate new networks necessary to bring it to market. Ignore no alliance, affiliation, partnership, or acquisition which advances this objective before the competition can react.

7.6.2 America On-Line

Company Summary: This Internet provider is also a broadcaster. AOL has moved from its position as a link to the "information superhighway" to a full-fledged innovator of content. It is the organization, design, and reformulation of content which attracted customers to its service. In an industry of numerous providers formed at the beginning of the World War Web, America On-Line has managed to differentiate its service, build a brand name, and generate a subscriber base approaching twenty million. No company in this sector compares favorably with this performance.

A key element of the company's strategy has been the relentless pursuit of business affiliations and alliances intended to generate value-added content for its subscribers, and minimal cash is diverted to this task. In other words,

AOL has successfully built economies of scale—the cost of producing content has fallen over the long-run—resulting in an expansive array of services that no competitor can match. AOL has complemented this strategy with a sustained emphasis on vertical and horizontal acquisition. The firm has purchased its direct competitors (Compuserve is a recent acquisition), while acquiring network and browser operators (Netscape was acquired in 1999, for example). Continuous improvement in infrastructure management, fused with content development, are the synergies defining the early evolution of this firm.

Strategic Profile: AOL is one of the few major Internet firms presently recording profit—an extraordinary development in light of the fact that companies must necessarily plow all available capital back into network management and content development. In the early stage of a young company operating in a seminal industry, it is unreasonable to anticipate profitability. Yet, AOL's strategy of using other individuals and firms to generate content under its brand name—thus assuring value-added performance at minimal cost—has proved remarkably prescient. We see in this firm the development of a new business model, a model discussed at greater length in the final chapter of this text. This business model establishes a precedent that few firms have begun to understand in the advent of electronic commerce, let alone imitate. The model embraces four interactive tactics:

- Brand name development, substantially cultivated through business alliances forged to exploit complementary attributes of other firms;
- Economies of scale, reducing costs of managing the business concurrent with improved product, and resulting in rising profitability;
- Proprietary technologies, exploiting state-of-the-art hardware and software, differentiating itself from the competition and restraining competitor encroachment on its market share;
- Value-added focus, continuously improving the applicability and customization of services available to consumers.

Whether America On-Line conceived of the criticality of these tactics early in its development can be debated. What cannot be refuted or dismissed is the success which has been realized through these strategic priorities. AOL's growth has not been limited to the United States. The firm has been zealous in its penetration of international markets, as well.

Financial Profile: AOL's financial performance has been extraordinary since its inception in 1990. The firm has grown by a factor of 15,000% in less than a decade, a feat unequaled by its competitors. No company in the telecommunications industry during the past decade has come close to AOL

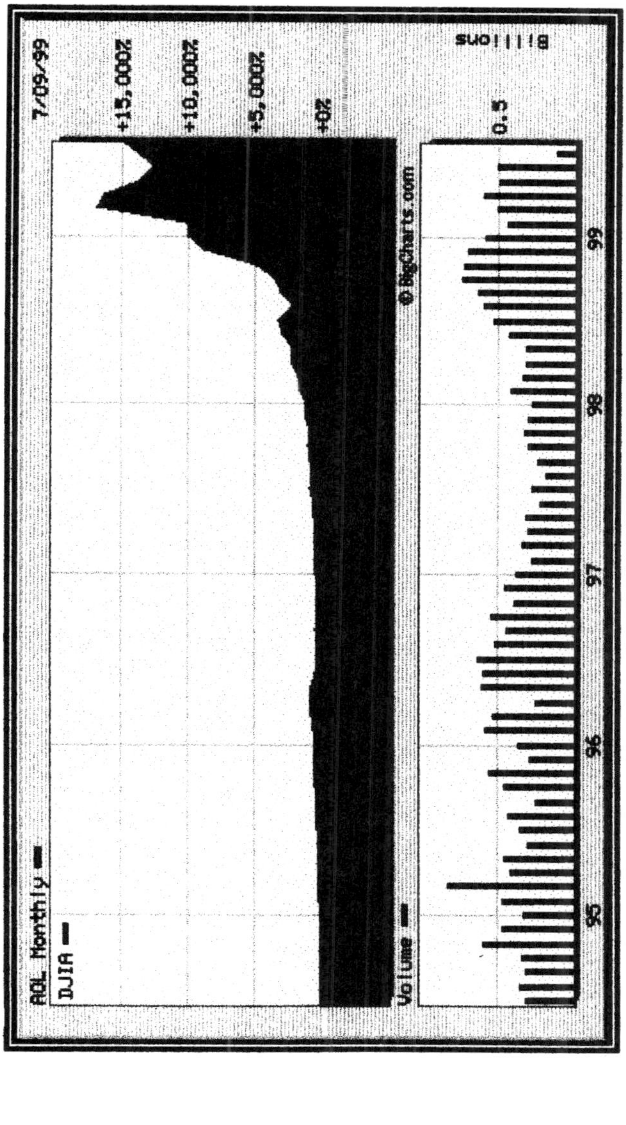

Data Source: BigCharts.com.

Figure 7.8 America On-Line Corporation stock valuation, 1994–1999.

with respect to: (a) generating customers (reach); and (b) sustaining that base (loyalty). It has thus outperformed every market average, every direct and potential competitor, and is thus a broadcaster in the literal sense of that term.

Strategic Findings: AOL literally broadcasts content in every form—data, audio, video—and continually reformulates information and knowledge to expand its customer base. Its financial profile reflects the dynamism of a firm which has not merely stumbled upon, but exploited, the emerging economics of the telecommunications industry. The four tactics defined above, when applied to sustaining customer reach and loyalty, result in a winning financial formula. The business model designed by America On-Line will draw many future imitators in the telecommunications industry.

7.6.3 CBS Corporation

Company Summary: As one of the contingent of seven major network broadcasters (including Disney's ABC, GE's NBC, Fox Entertainment Group, UPN, Turner, and Warner), CBS Corporation has undergone considerable restructuring since 1997. Several factors impelled its current organizational strategy, and there are important lessons for other sectors of the telecommunications industry. Only recently has CBS stock traded as a separate entity, divorced from Westinghouse Electric in the past three years (note the stock's history in Figure 7.9 below).

CBS is one of the original "big three" television broadcasters—NBC and ABC have been its primary competitors in the post-World War II period—and had been consistently successful in delivering product to advertisers, resulting in substantial market share. With the penetration of cable television in the 1980s, network market share eroded substantially; consumers today simply have in the palm of their hand an extraordinary and proliferating array of viewing options. In this environment, the networks find themselves competing not merely with rivals indigenous to their sector, but with emerging multimedia content developers advanced to market via the Internet. Between the encroachment of niche markets delivered via cable television on the one hand, and Internet providers on the other, CBS and its traditional rivals perpetual customer erosion if locked into standard business models. We had referred to this phenomenon in Chapters 2 and 3 as an example of disintermediation: the loss of employment or market share precipitated by advances in technology. Deregulation has spurred part of this development by facilitating acquisition of network infrastructure by emerging cable rivals, a development effectively prevented before the 1996 statute. CBS has responded to this competitive transformation by streamlining its strategy and adding value to its product.

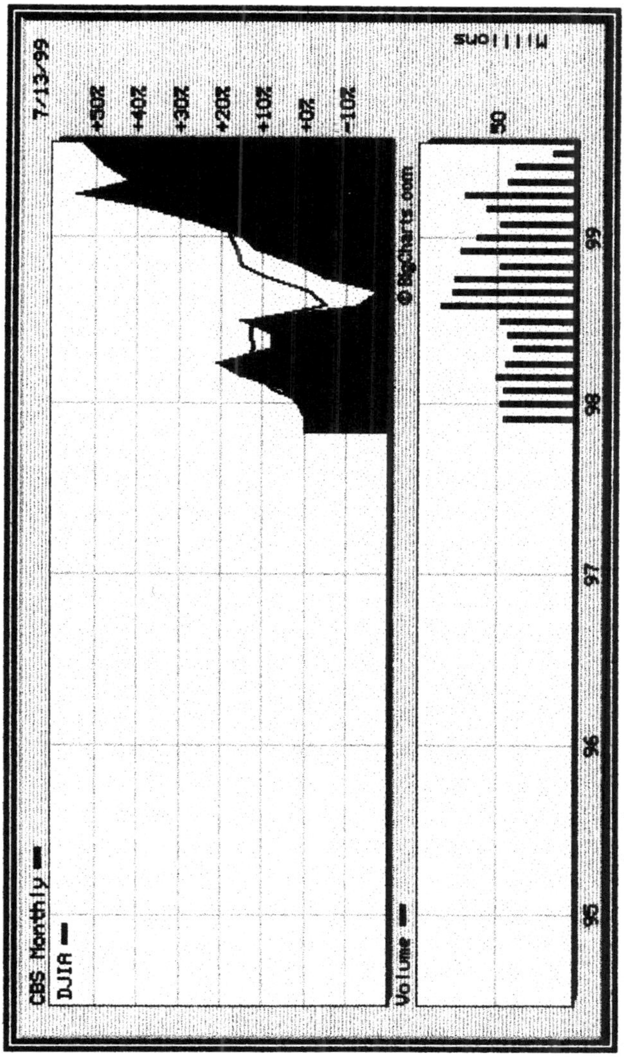

Data Source: BigChart.com.

Figure 7.9 CBS Corporation stock valuation, 1994–1999.

The firm abandoned its industrial assets under its former Westinghouse Electric umbrella, and has now concentrated on the development of content as its singular mission. It has innovated or acquired new product and ancillary companies during the past three years, and has succeeded in enticing investors to buy its stock so as to implement its strategy. CBS now controls the Nashville Network, several cable television outlets, and Infinity Broadcasting (which includes 160 radio stations). It has simultaneously advanced Internet content, including the much-respected MarketWatch.com and SportsLineUSA.com. It is actively pursuing the acquisition of independent motion picture production companies. Recently, CBS has also negotiated alliances with firms across all sectors of the telecommunications industry, among them a recent agreement with AOL as a deliverer of its news content.

Strategic Profile: CBS, evading the pursuit of several potential acquirers following 1996, adopted a strategy simplifying and reinforcing its organizational mission amid competitive tumult. CBS had been a comparatively staid and conservative enterprise before 1990, and its strategic transformation during the past three years has been quite remarkable. It has adopted, in effect, the strategy of its fiercest rivals. CBS proposes to emphasize the development of content for delivery on any telecommunications platform. The term "broad" no longer characterizes the intent of the firm; it will develop programming for a multiplicity of niches and tastes. CBS has deduced that content, not physical infrastructure, is the source of future growth, and it has effected this strategy to regain its position as the number one network since late 1998.

Financial Profile: CBS's stock valuation has excelled in the current environment and certifies market confidence in its reconfigured strategy. CBS stock has outperformed that of its direct rivals, and indeed the broader market averages. After incurring costs of acquisition and new product development in 1997, the stock has moved dramatically upward. The short-term costs associated with strategic redesign have clearly resulted in expansive opportunities across all sectors of the communications industry.

Strategic Findings: We discern in the CBS experience an important lesson for veteran communications firms—that it is possible to fundamentally reshape strategy within a comparatively brief period, and that this is a necessity in volatile, transitional competitive landscapes. Strategic reconfiguration is often defeated in veteran enterprises with long histories. Change disturbs, employees often rebel, and organizational structures are frequently unable to adapt promptly to new technologies, thus assuring success for certain market aspirants. CBS has demonstrated that "you can turn a firm on a dime." Strategic foresight—and the will to incur risks—are necessary preconditions for this eventuality.

7.7 Common Attributes of Effective Telecommunications Strategic Management

Nine case studies have been examined in this chapter, each representative of a company coping effectively with extraordinary strategic risks and uncertainties. We discern in these profiles important lessons about the design, monitoring, and evaluation of successful strategy. Although these cases reflect different strategic experiences in different sectors of the industry, we can reasonably assert that these common attributes of modern strategic management obtain:

- Successful firms in their respective sectors do not react to change, they embrace it and seek new avenues to add value to their product.
- Successful firms in the newly defined competitive landscape have learned to negotiate affiliations, alliances, and partnerships which either improve their product or create new distribution routes to consumers.
- Successful firms exploit the Internet as a new distribution channel to consumers, they do not attempt to supplant this competitive threat through traditional strategic design.
- Successful firms make every effort to reinforce brand image and management, now more important than ever.
- Successful firms have learned to exploit scale economies, critical in strengthening profitability.
- Successful firms develop proprietary technologies or products, where feasible, such that their competitors' efforts at replication are forestalled.
- Successful firms adopt either one of these strategic philosophies: (a) they define themselves as "first-to-market" product developers; or (b) they emphasize a "first-to-market-acceptance" image clearly recognizable to employees, customers, and investors.
- Successful firms "obsolete" their own product before the competition does it for them.
- Successful firms have learned the most important lesson of contemporary telecommunications economics: reach (customer base) × loyalty (customer allegiance) rewards providers and content developers, and lays the foundation for future growth.
- Successful firms exploit organizational synergies to enhance value and growth rates by cooperating with any firm, including competitors, which assures this contingency.

These common strategic attributes hold profound ramifications for the future of the telecommunications industry, especially with regard to defining future business opportunities. These emerging opportunities, and the strategic options necessary to exploit them, are the subject of the following chapter.

References

[1] Peters, T., and R. Waterman, Jr., *In Search of Excellence*, New York: Warner Books, 1982.

[2] Wright, P., M. Kroll, and J. Parnell, *Strategic Management: Concepts and Cases*, Upper Saddle River, NJ: Prentice-Hall, Inc., 1998.

[3] Porter, M., *Competitive Strategy*, New York: The Free Press, 1998, pp. 126–155.

[4] Fahey and Randall, op. cit., pp. 171–194.

[5] Bernstein, P. L., (ed.), *The Portable MBA in Investment*, New York: John Wiley & Sons, Inc., 1995, pp. 111–197.

[6] Data on financial and industrial developments from IndustryStandard.com; CompanySleuth.com; Hoovers.com; and BigCharts.com.

[7] Sweet, W., "Battling for Local Phone Customers," *IEEE Spectrum*, May 1999, pp. 32–37.

[8] Shaw, J., *Telecommunications Deregulation*, Norwood, MA: Artech House, 1998, pp. 38–48.

[9] Heldman, P., R. Heldman, and T. Bystrzycki, *Competitive Telecommunications*, New York: McGraw-Hill Co., 1997.

8

Creating Strategic Options Through Market Planning

Armed with a knowledge of the probable contours of the American and global economies during the coming decade, cognizant of the principal tools of strategic management, and sensitive to the determinants of product innovation, strategists should now focus on expanding the array of options before them. Rarely, if ever, has there been greater latitude in cultivating a multiplicity of new business formations in the telecommunications industry [1]. The object of creating this range of strategic options is growth: in customer base, market share, productivity, new market access, sales, and profitability. Modern strategic management is now directed at assuring continuous growth in the largest sense of that term [2].

Passage of the Telecommunications Act of 1996 has effectively channeled unprecedented capital flow to the communications sector [3]. New capital formations reveal the future intent of organizations and entrepreneurs, as the private sector pursues unfolding opportunities. The shift from regulation to deregulation instigates a reconception of strategic options by introducing new players from industries ancillary to the communications sector. Among strategic concerns promulgated by deregulation are these critical emerging concerns:

- Evolving economics of communications convergence;
- Tactical mergers and acquisitions;
- Proliferation of strategic affiliations, alliances, and partnerships;
- New economics of venture capital;
- Telecommunications Business Opportunity Paradigm (BOP); and
- Tertiary externalities and new product innovation.

8.1 Evolving Economics of Communications Convergence

Although this term is applied in different contexts, confusing the general public as to its meaning, "convergence" may be defined as the interaction and evolving relationship of the television, telephony, and computer industries. The dynamic, continuously changing involvement of these sectors now denote a major restructuring of both domestic and international commerce [3, 12]. The significance of communications convergence lies in the fact that the telecommunications industry is transforming itself through [3, 12]:

1. Expanding merger and acquisition activity, fostering consolidation in both domestic and international markets;
2. Strategic partnering, also referred to as "cooperative management," in which network and content providers increasingly interact to achieve objectives not otherwise attainable;
3. Strategic affiliations, or temporary agreements in which telephony, television, and computer enterprises join in bilateral and multilateral agreements which enhance their value-added capabilities, thus expanding access to new markets.

The early stage of convergence represents the practical recognition that no firm can fully participate in all sectors of the industry-transmission services, content generation, value-added enhancements, and hardware development—in the absence of acquiring, merging, or cooperating with complementary firms. In this regard, cost of accessing new customers falls through economies of scale, though there is no guarantee that a loyal customer base will be obtained. The industry is so vast and expanding at so rapid a rate that no enterprise can successfully accommodate consumer demand while relying exclusively on its own resources. Indeed, to the extent that the industry's innovative products drive operations in other sectors, thus creating a dependence on communications for competitive advantage, it is evident that partnering is about to penetrate every economic endeavor. Today, the largest concerns cannot hope to pursue diversified strategies without complementary support. We are thus at the cutting-edge of new approaches to the design and management of strategic partnering, as outlined below.

8.2 Tactical Mergers and Acquisitions

A merger is a combination of two or more firms, the purpose of which is to establish a singular identity for the newly created enterprise. Mergers are precipitated by a shared vision which maintains that enlarged size results in expanded market share; that expanded market share assigns competitive leverage over rivals; that rising profit margin flows from the transaction; and that greater capital availability will facilitate innovation through enhanced research and development. Mergers are transacted through exchanges of stock or accounting "good-will" techniques, in which the larger partner passively assumes the difference in assets and costs of transaction. Because a merger can be negotiated as a tax free pooling of assets, stockholders can readily realize substantial capital appreciation within a short period of time [4].

An acquisition is defined as the purchase of one firm by another [5]. Multiple factors spur acquisition decisions. An acquisition is prompted by the desire of one firm to engage in vertical or horizontal acquisition. Strategic management aimed at these forms of business integration are viewed with concern by the Anti-Trust Division of the Department of Justice. Vertical integration represents the ownership of networked production or distribution of a product or service [6]. A company buys or merges with those firms that support its delivery of products or services. Vertical integration may be expressed as "forward" or "backward": if the corporation buys a competitor or ancillary enterprise which supports its on-going strategy, it is said to be backward; if a firm acquires a rival or neutral entity solely to secure access to new markets, it is said to be engaged in forward, vertical integration.

Horizontal integration is asserted when a company buys a competitor in order to extend its distribution [7]. Horizontal integration potentially results in a firm's greater influence over market price, since competition can abate. The motive to acquire competitors also frequently involves an attempt to gain access to state-of-the-art knowledge; clearly, competitive advantage is conferred upon those firms successful in coming to market with new product lines. Intellectual capital in the telecommunications industry therefore becomes the lifeblood of new products, in terms of both network infrastructure and content development.

The Justice Department historically views with suspicion the prospective vertical and horizontal integration strategies that are established by the largest firms within an industry. The Sherman Anti-Trust Act, the Clayton Act, and similar anti-trust mechanisms can be applied to block prospective mergers or acquisitions if they are deemed to threaten competitiveness through the manipulation of market share. For firms especially aggressive and innovative, particularly with respect to the pursuit of new markets, great risks are implicit in a strategy focused on mergers and acquisitions if present internal innovation is limited or insignificant.

Large, capital-intensive telecommunications firms must factor this consideration into their long-term strategic plans, in light of the recent investigation of Microsoft by the Justice Department. As was the case with the aviation, trucking, and natural gas industries, we note the emerging wave of merger activity associated with both domestic and global telecommunications firms. Consolidation—or the strategy of tactical mergers and acquisitions—is thus focused on expanding market share, extending industry influence, and exploiting available intellectual capital, while expediting new market access.

Major telecommunications transactions may well be challenged by the Justice Department in the years ahead, with corporations weighing the vicissitudes of market timing and product innovation. As noted, there exists philosophical variation among various presidential administrations as to the usefulness of these actions, and we therefore cannot forecast with precision prospective anti-trust enforcement.

Government action is taken when concerns surface regarding a firm's mounting market share, or barriers to entry are erected by firms successful in exploiting scale economies not attainable by rivals. Despite these implicit suspicions or reservations by regulators, many mergers are deemed to be consistent with the public interest. The Federal Communications Commission had anticipated before passage of regulatory reform that mergers would be the practical, even aspired product of deregulation in the short-run. Removal or relaxation of such restrictions would, it was believed, inevitably energize those firms successful in the market, and induce them to acquire inefficient enterprises. Such a hypothesis has yet to be fully tested in the United States, and has not been evaluated on the international level.

Many mergers and acquisitions, including those cutting across international boundaries, may eventually determine market winners. It can be concluded that some of these transactions will produce losers, as well. Those integrations that do not lead to telecommunications products or services that the consumer desires are destined for market failure, regardless of the form they eventually assume.

The historic justification for mergers, despite anti-trust concerns, lies in the efficacy of economies of scale: price per unit of a product or service drops as the costs of capital, labor, and technology diminishes. The underlying presumption of a benign merger or acquisition is that the firm can reduce the cost of these three factors of production when its size proportionately expands. There are historic examples of a firm's failure to measurably enhance economies of scale through a planned integration. Moreover, through recent advances in software development and inventory management, the cost of producing a product is significantly diminishing; many industries now produce an item in lower quantity while approximating the cost advantages associated with higher levels of output [8]. Competitive advantage associated with scale economies as the key determinant to competitive pricing is thus eroding, thereby impelling boutique enterprises to emerge as major competitors in every sector of the telecommunications industry.

The pattern of consolidation in industries recently undergoing deregulation suggests that horizontal and vertical integration will penetrate the largest corporate structures in communications. Natural gas, aviation, trucking, and banking sectors provide an historical guidepost, the front-end of M/A activity will spur such strategy among major providers and distributors of services. The back-end of M/A strategy will be aimed at the integration of service and content providers. The pattern of corporate consolidation began to unfold shortly after passage of the 1996 Telecommunications Act.

8.3 Proliferation of Strategic Affiliations, Alliances, and Partnerships

Technological innovation intensifies unpredictability and amplifies risks in today's telecommunications markets. If we examine forecasts of the past decade, particularly those involving the advent of wireless communications, interactive television, and electronic commerce, it becomes salient that industry experts have frequently underestimated or exaggerated the waves of change hitting this vital sector. Prominent among underestimates were those projected for mobile communications; these predictions stood in stark contrast to extravagant forecasts for both wireline and wireless interactive television services [9]. In both cases, forecasters had failed to fully anticipate both the value and limitations of emerging technologies and the prospective consumer interest for these tools. Compounding these methodological errors was the challenge of predicting emerging substitutes for veteran product lines.

It is now evident that a firm's nimbleness in designing its competitive strategy, coupled with customer responsiveness to value-added product improvements, dictate future competitive opportunities. In light of today's market realities, to rely exclusively on mergers and acquisitions to expand market opportunities impoverishes a firm's range of strategic options. Deregulation has, in part, encouraged a number of firms to seek speedy entry into new markets via supportive alliances: agreements which join service providers and content developers to exploit market trends. These agreements endure as long as the interests of both parties are served. Speedy response to changes in consumer preference is the vital ingredient which moves firms far-sighted enough to engineer such agreements.

Well-designed strategic alliances are flexible, easy to dissolve. This advantage may present a dilemma, however: exceptional long-term opportunities can be forsaken at the discretion of a partner. In markets characterized by rapid, profound technological innovation, this option remains an important tactic in a firm's strategic arsenal. Not surprisingly, given the dexterity of strategic alliances, we find mounting evidence that larger firms are prepared to partner with firms of any size, provided attractive synergies result. Thus, an unprecedented development has occurred in the history of strategic management: company size is no longer an obstacle in facilitating agreements among diverse organizational structures.

For communications firms which must invest substantial capital to enter new markets or product lines, and whose assets prevent merger or acquisition activity, strategic partnerships represent an attractive alternative. Partnerships designate "strategic fit": the chance to forge a symbiotic relationship in which complementary and supportive expertise and resources extend competitive advantage. Such relationships can be formed between any service or content provider in the computer, television, or telephony industries.

Notable strategic partnerships have recently included Microsoft and NBC broadcasting content; Bell Atlantic, Nynex, and Pacific Telesis in the formation of Tele-TV; Motorola and Nextel in the creation of a specialized mobile radio network; NetTV and Intel in the development of integrated television/personal computer systems; WRQ and GTE in the delivery of wireless data; OMNI and LitelNet in the formation of a new electronic software distribution network; and CNBC and Street.com in producing financial analysis for stock investors [9]. Both the frequency and diversity of partnerships are proliferating rapidly. We identify a central theme in these agreements—building a lasting collaboration while leveraging or minimizing financial risk. Strategic partnerships reflect a far-sighted attempt to gauge consumer demand. They also represent an admission that no firm can create and maintain market share and base in the absence of such an agreement. Partnerships pivot on the presumption that a long-term commitment is required to develop new markets.

Strategic alliances serve the interests of two or more firms pursuing an alternative to the cumbersome and time-consuming machinations of mergers or acquisitions. These agreements can be expedited in a fraction of the time required to negotiate mergers or acquisitions. While traditional market consolidations are attractive options, the time necessary to consummate them may put a firm at a competitive disadvantage. Alliances do not inspire intervention by the Federal Trade Commission or the Anti-Trust Division of the Department of Justice, another enticing consideration during an era in which anti-trust enforcement is increasingly prominent.

It is now evident that flexibility, adaptability, and customer responsiveness define prospective market success. To rely exclusively or even chiefly on mergers and acquisitions to pursue market opportunities impoverishes a firm's range of strategic options. Deregulation has now spurred many firms to speed product to market through alliances: agreements, often temporary in nature, that typically join service providers with content developers to seize upon emerging market trends. Either party may sever the agreement the moment conditions warrant. The great advantage of a strategic alliance—leveraged risk coupled with minimized financial exposure—represents a potential pitfall. Long-run profitability in growing markets can be forsaken at the whim of a partner. In markets characterized by extraordinary technological innovation, this option signifies a valuable strategic weapon in a company's arsenal.

The strategic alliance serves the objectives of two or more firms pursuing practical alternatives to traditional merger and acquisition strategy. Strategic partnerships, in contrast, require greater time to negotiate and implement than do strategic alliances. Despite this intrinsic disadvantage, they are clearly an expedient alternative to merger and acquisition strategy, given the fact that they can be accomplished in far less time. Partnerships, like alliances, ordinarily do not serve to invite anti-trust investigation. The strategic motive underlying partnerships rarely inspire government attorneys to propose prosecution.

Strategic partnerships evolve when equipment providers seek channels of distribution via service providers or in situations where software providers can stimulate the productivity of either equipment or service providers [10]. Exceptional software innovation now drives the service and content sectors. The great organizational value of computer innovation—improved productivity—translates into lower pricing, resulting in a competitor's capture of greater market share [11]. It has become something of a bromide in recent years, but firms in every sector have had difficulty raising prices in the current competitive landscape. This phenomenon can be traced to the impact of computer technology fused with communications equipment (see Section D of the Appendix for a treatment of productivity analysis during the 1990s). It is therefore reasonable to project that a substantial number of future telecommunications partnerships will result from evolving relationships between telecommunications providers and the most advanced, incisive software developers.

8.4 The New Economics of Venture Capital

Communications business start-ups require venture capital in great quantity. This source of funds has been forthcoming during the 1990s. Venture—or risk—capital is a principal funding source for firms whose projections of revenue growth typically exceed 25% annually with the initial stage of development [3]. Often, venture capitalists seek returns of up to 50% annually within the first five years of creation, with the caveat that profits are plowed back in the company.

The venture capital industry has expanded significantly since the passage of favorable federal legislation in 1978, and then again in 1998.[1] Further reductions in capital gains tax rates have been projected by Congress in the immediate future. Favorable tax treatment for these investments is highly encouraging for the future of telecommunications development, although recent studies suggest that a disproportionate fraction of such funds have been directed at e-commerce start-ups.

Several forms of venture capital define the channeling of funds to the communications and its ancillary information-technology sectors. In sequential order these venture funds can be categorized as:

1. Seed capital, directed at financing research and development, securing equipment, and financing initial payroll;
2. Working capital, focused on finalizing the initial phase of development while defining and implementing strategic planning, marketing strategy, sales objectives, and essential overhead; and
3. Acquisition capital, often used to secure expertise of another firm so as to promulgate product to market.

Studies indicated over and over again that working capital remains the critical ingredient for successful start-ups. Effective Initial Public Offering (IPO) strategy is tied to a demonstration of effective use of working capital. The vicissitudes of the IPO market, which may be characterized as favorable, unfavorable, or stagnant, outline future success or failure. Extraordinary IPO markets, periods in which investors are aggressive in their risks, include the late 1970s, mid-1980s, and post mid-1990s. Decidedly unfavorable eras included much of the 1970s and late 1980s. We note that favorable IPO environments are strongly correlated with reduced capital gains tax rates; the fact that Congress is presently contemplating reducing rates augers well for venture funds directed to the communications industry. Among the factors critical to luring funds, particularly in the most exotic of emerging technologies, are:

- Accuracy and persuasiveness of sales and profitability projections;
- Volatility and unpredictability of technological innovation, and the probability of product substitutes; and
- The capacity of firms to simulate or copy proprietary technologies.

1. As of this writing, two new capital gains tax bills have been advanced in Congress, lowering prospective rates to as little as 10% long-term holdings.

Venture capitalists are extremely reticent to commit major sums to product development before they secure credible estimates relative to the variables listed above. We do know, on the basis of empirical evidence, and particularly with respect to the raging bull stock market of the late 1990s, that entrepreneurs are finding it very difficult to attract funds in those areas in which products or services cannot be clearly differentiated. In other words, proprietary technology—products patented or otherwise uniquely conceived—is prized above all in the current competitive environment.

8.5 Telecommunications Business Opportunity Paradigm (BOP)

Technological innovation, fused with growing demand, precipitates the development of new communications products. Strategic planners cannot assume that a sound product will find its market niche merely on the basis of product attributes. Several concerns confront telecommunications firms when contemplating market response to their product:

1. We have learned empirically that when a product or service attracts 10% or more of a market, competition unfolds. Product adoption stimulates the availability product alternatives.
2. Tertiary forces are unleashed after a product succeeds in securing major new market share. Tertiary forces are the product of externalities—unintended consequences associated with the interaction of producers and consumers (this subject is explored below).
3. The emergence of successful, young firms invites established enterprises to absorb such start-ups.

These three dynamic forces illuminate the realities of a predatory marketplace. Entrepreneurs and veteran enterprises must construct strategy which anticipates the effect of their own products on prospective customer behavior. Success in forecasting these trends allows firms to project future consumer and business demand for their products, and simultaneously reveals important information about likely competitor behavior. Among the incisive strategic questions a firm must contemplate in this environment are:

1. Should our newly introduced product succeed in capturing market share, how will our competition respond? Is the product designed

such that it cannot be immediately replicated? When can we anticipate invasive competitive response?

2. Will opportunities emerge from the success of our product, and should we divert resources to pursue these secondary market alternatives, as well?
3. When and how will the predatory instincts of our competitors be activated? Will the competition usurp our market through competitive pricing, manipulation of distribution channels, or other strategies?

Abundant data confirm that these three interactive strategic concerns often confront the successful communications firm, emerging or veteran. These matters are dealt with effectively when they are framed within a strategic paradigm. Figure 8.1 defines a Business Opportunity Paradigm, designed to describe the typical developing market impact of a new product. The dynamics of new product development define future customer, competitor, and industry interaction and responsiveness.

8.6 Telecommunications Business Opportunity Paradigm (BOP)

One may infer from the model outlined below that the most successful firms are beginning to seriously forecast the long-term market consequences associated with their own actions, though this is a precarious exercise in so volatile an

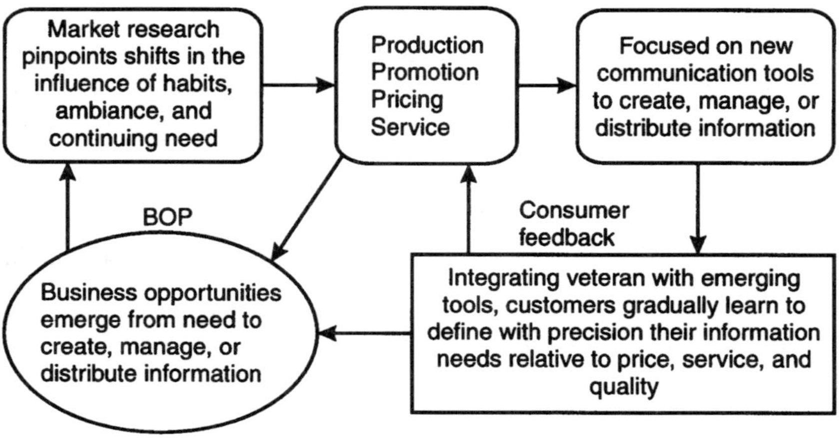

Figure 8.1 Telecommunications business opportunity paradigm.

industry. Strategists now contemplate the long-run strategic implications of new product development and its influence in driving the future of the firm. A simple axiom applies for telecommunications providers: there is no such thing as a telephone, cable, television, or computer company, per se. Instead today's telecommunications company is a creator, manager, or distributor of information and should therefore be willing to pursue any avenue that advances this mission. Traditional boundaries separating these functions are systematically disappearing, redesigning strategic management as the new millennium unfolds.

8.7 Tertiary Externalities and New Product Innovation

The BOP model defines the importance tertiary of economic analysis as a forecasting tool for strategic planners. Tertiary analysis—the study of externalities associated with new product development—endeavors to estimate the type, size, and role of new products unfolding in response to the release of new technologies. A firm typically brings its product to market and rarely evaluates the long-term significance of current strategic plans and their role in reshaping products and markets. American business people, for example, traditionally market a product and exploit its market value until its life-cycle has been terminated. Support industries are often necessary to distribute, process, market, or maintain that product line. Every major technology which has met with market success—defined as broad consumer acceptance—has adhered to this pattern of development in the twentieth century.

We therefore identify the linkage between primary, secondary, and tertiary effects. Emerging technology has a primary impact on its market and precipitates a pattern of household or business consumption which follows. A secondary consequence of that adoption is typically the maintenance of that product once purchased. Thereafter, third-, fourth-, and fifth-ordered tertiary consequences of that product are unleashed. These predictable patterns signify important strategic implications. Among these important developments are:

- The limited influence of the originating firm on the tertiary products or services it provokes.
- A pattern of "creative destruction" results, in which business reconfigures its strategy in response to product innovation.

In both instances, no public or private entity can contain the opportunities or dysfunctions which result from new product introduction. If a firm succeeds in seducing a large customer base, there is no fool-proof method of predicting the long-term tertiary impacts of new telecommunications products. However, tertiary effects should not be dismissed or ignored in strategic management. Techniques are evolving to improve strategic design so as to exploit emerging business opportunities.

8.8 Creating Strategic Options: The Emerging Significance of Relational Marketing

The pattern of success evidenced by successful telecommunications firms in recent years reveals a series of common strategic attributes, many of which are discussed in chapter seven. We have learned, in addition, as elaborated in the final chapter of this volume, that creative strategies are closely tied to customer responsiveness. Sensitivity to customer needs represents more than a commitment to satisfy consumers, or even customize products; it hinges, ultimately, on exploiting customer knowledge and predisposition to define future production and organize resources; to manage inventory, even design strategy. Such an effort is based, finally, on relational marketing: establishing a permanent relationship with the customer, such that customer loyalty is reinforced and perpetuated to enhance strategic advantage.[2]

We are now witnessing the very early stage of what might be described as "customer advantage"—that circumstance in which the most attractive customers are cultivated so as to direct the least profitable consumers to our uneducated rivals. Simply, customers can now be leveraged to maximize profitability of one firm at the expense of industry rivals. In the same way, global competitors can divert "undesirable" customers to those firms whose profits—and thus strategic advantage—are proportionately reduced. This strategic insight is counter-intuitive: ordinarily, a firm solicits any and all prospective customers. Every customer is thought to be valuable. This emerging logic holds that such strategy effectively diminishes the firm's market position. Transitional strategic design is elaborated further in the final chapter.

2. Seminal research on relational marketing has been conducted by Martha Rogers and Don Peppers since 1994. Refer to website listings under marketing-specific functions in Section F of the Appendix to evaluate their latest findings in the telephony and ancillary communications sectors.

References

[1] Shaw, J. K., "Future Scenarios for the Telecom Industry: A Ten-Year Forecast," *New Telecom Quarterly*, December 1996, pp. 25–26.

[2] Pappalardo, D., "The State of Convergence: Study Reiterates Importance of Killer Apps," *Business Week*, (Special Insert), November 25, 1996.

[3] Argenti, P., *The Portable MBA Desk Reference*, New York: John Wiley & Sons, Inc. 1994.

[4] Thatcher, B., and R. McNamara, "How Merger Mania Has Redefined the Communications Landscape," *Telecommunications*, October 1997, pp. 42–44.

[5] Toffler, A., Testimony before United States Congress, Subcommittee on Commerce and Telecommunications, February 12, 1996.

[6] Lim, K., ed., "The Top Ten Myths of Digital Convergence," *Cybermedia 2001*, March 1997, V1.2, pp. 1–3.

[7] Wood, W., "Can Telcos Survive," *Telephony*, March 4, 1996, pp. 22–29.

[8] Pettis, C., *TechnoBrands*, New York: American Management Association, 1996, pp. 165–200.

[9] Bygrave, W. D., *The Portable MBA in Entrepreneurship*, New York: John Wiley & Sons, Inc., 1994, pp. 184–185.

[10] Mills, M., "An Interactive Dream Unfulfilled," *Washington Post National Weekly Edition*, July 10–16, 1995, p. 20.

[11] Shaw, J. K., "Wireless Communications and Technology Substitution: What S-Curves Reveal About Pending Cellular Competition," *New Telecom Quarterly*, Summer 1996

[12] Egan, G., *Adding Value*, San Francisco: Jossey-Bass, 1997.

9

Global Strategic Management in Telecommunications

The management of strategy in multinational telecommunications corporations is no less daunting than that confronting the most energetic domestic firms. Deregulation, privatization, technological innovation, creative business alliances, and international treaties have transformed the global strategic environment. One might reasonably point to two critical factors permeating every aspect of international strategic management at the beginning of the new millennium: the first, key features of the World Trade Organization (WTO) agreement of 1997, defining future intellectual property rights; the second, the emergence of new patterns in telecommunications trade, affecting export and import activity in both developed and developing nations [1]. These issues are explored in this chapter, and a proposed model of strategy is posited to exploit these developments.

In reviewing this material, the reader may wish to consult Section C of the Appendix, which consists of summary features of the 1997 World Trade Organization Agreement, including subsequent amendments adjoined in 1998 and 1999, and export/import telecommunications data, generated by the United States government, since 1995. These two statistical snapshots provide the strategist with a contemporary perspective on emerging opportunities for both small and large enterprises aspiring to form multinational agreements [2–4]. Note that these data are regularly evaluated by strategic planning departments in major multinational corporations.

9.1 Strategic Management: International Economic Context

In contrast to the economic growth enjoyed by the United States during the second half of the 1990s (discussed in Chapter 2), the global economy during the same era has been anemic (see Sections D and E of the Appendix for a review of these data). Most European economies stagnated during this period, as growth in profitability, stock valuations, employment opportunities, and tax base remained comparatively static. In Asia, economic growth began to abate in 1996, and was followed by severely contracted economic growth; indeed, most Asian economies lapsed into depression by 1997 as first Japan, then Indonesia, South Korea, Thailand, Malaysia, Singapore, Hong Kong, and other nations followed suit. The declining economies experienced first a collapse in their financial markets, then stagnant demand for their exports. Similar conditions prevailed in various Latin American and African nations. The years stretching from 1997 to 2000 might generally be referred to as the worst three-year global economic performance in the post-World War II period. Paradoxically, the American economy performed splendidly at this time, despite a decline in demand for American exports of all kinds.

For telecommunications strategists, an important distinction must be drawn between merchandise and services as components of exports. The export figures outlined in the Appendix differentiate merchandise (physical products) from services (intangible items, including management consulting, engineering advice, and all other forms of services). The data suggest that, in percentage terms, demand for American telecommunications products diminished beginning in 1996; concurrently, the demand for technical and managerial assistance generally increased in telecommunications, despite constrained economic growth elsewhere. This significant finding suggests that, as the world economy gradually recovers after 2000, the demand for telecommunications services (telephony, broadcasting, and computer/information technologies), will invariably increase and probably profoundly. For American firms, this inference bodes well for new market development; it also constitutes an opportunity for all global enterprises with access to state-of-the-art technologies. The fact that the demand for telecommunications services improved during the midst of global economic diminution implies that management consulting in the industry is likely to be pivotal in the emerging international economy. Developing nations will depend greatly on communications products and services, but especially consulting services in the design, deployment, and management of cutting-edge telecommunications equipment.

We note in these data the growing importance of what might be referred to as *international convergence*, defined as the interaction of the multinational broadcasting, telephony, and computer industries. Strategists should regard the evolving relationship of these international organizations, and the governments that regulate them, as signifying critical trends in management cooperation and collaboration. This development represents an extraordinary challenge, as common standards of technical, financial, political, regulatory, and cultural interaction must be accepted [5].

International telecommunications convergence has been expedited by regulatory reform in the United States. Although many nations have been hesitant to introduce competition to state-owned or regulated enterprises, to risk delayed regulatory relaxation would only have the effect of strengthening the competitive positions of deregulated industries in Great Britain and the United States. Speed to market with new telecommunications services becomes the predominant objective as communications deployment transcends national boundaries [6]. In forecasting international convergence, we affirm the following dynamics:

- First, the proliferation of mergers and acquisitions is likely to precede international strategic alliances and partnerships, given the need by major competitors to secure immediate market share.
- Second, fundamental regulatory reform, or deregulation, is likely to be applied only after deregulation is perceived as effective in the United States.
- Third, necessary preconditions for the earmarking of international venture capital include political stability, predictable regulatory administration, reduced capital gains tax rates, and the embrace of electronic commerce.
- Fourth, the economic value of strategic partnering will not be realized until reciprocity is enacted by all nations through negotiated trade agreements, such as that adopted in 1997. International strategic alliances, critical in establishing the synergies of cooperative agreements, will accelerate and likely follow the pattern of development instigated in the United States.

The growth of the American postindustrial economy—or knowledge-based service economy—hinges on the vital role telecommunications plays in reconfiguring transactions between producers and consumers. Telecommunications—and the resultant influence it has on productivity and competitive advantage—will provoke economic growth in the global arena. Telecommunications development will seemingly inspire comparable growth in the global economy. It may be argued that American telecommunications strategic effectiveness is connected to that of other post-industrial economies (most notably, Japan, Canada, and Europe). Communications progress in these nations stimulates economic development in industrial and agrarian economies elsewhere in the world. Deregulation, if effective, will promote similar efforts around the globe. Deregulation begets convergence, and convergence instigates economic growth through mergers, acquisitions, cooperative alliances, affiliations, partnerships, and even cross-national intrapreneurial ventures.

9.2 Trade Organizations and Strategic Management

Global trade in the post-World War II era has been guided by the interactive influence of two negotiating entities: General Agreements on Tariffs on Trade or GATT, and the World Trade Organization (WTO). GATT, established in 1948, has had its principal influence on the relaxation of tariffs. Economists have long held that lower tariffs stimulate international trade, facilitating economic growth while concurrently introducing greater competition. The WTO, created in 1967, has served to promote international property rights (copyrights, patents, trademarks, and other forms of intellectual property).

In recent years GATT has receded as an influential facilitator of trade negotiations. The two hundred nations comprising the global economy represent an extraordinarily cumbersome negotiating vehicle. With countries in various stages of development, and national economic objectives mutually incompatible in many cases, GATT has not succeeded in forming a common basis of intellectual property protection—precisely the underpinning of a knowledge-based service sector. In this vacuum, trade negotiations—and the protection of intellectual property—have been promulgated at WTO meetings. Only in the 1990s has WTO been a decisive trade mechanism in furthering the interests of telecommunications product and service providers.

A momentous development in late 1996 was the WTO initiative to eliminate all tariffs on information technology imports. This proposal was adopted and implemented in most nations in December 1996 and calls for phased-in elimination of all tariffs by the end of 1999. If fully administered in the manner intended, an essentially free-market economy will prevail for all information technology products and services in virtually all nations. This agreement effectively establishes precedent for the elimination of duties on all high-technology products, including those specified in the Appendix, Section C. This cutting-edge trade agreement could stimulate substantial economic growth in all nations. Post-industrial economies (the United States, European Community, Canada, and Japan) would theoretically benefit enormously in the short-run, as exports of equipment and services would rise throughout the three intersecting telecommunications industries.

Strategists should now factor into their competitive analysis the influence of WTO's recent trade negotiations. It is one thing to advance intellectual property rights, but the protection of these rights depends solely on multinational cooperation in the adjudication and enforcement of those privileges. American and European firms, in particular, must carefully monitor such adjudication, especially in the Far East where such laws have been unevenly administered.

9.3 Strategy and International Telecommunications Regulatory Procedures

The advent of deregulation in the British and American markets has created a precedent for regulatory relaxation. There exists substantial variation in European, Asian, Latin American, East European, and African regulatory procedures. The variation in governmental authority is a function of political, legal, and business traditions and philosophical perspectives. The fear of the "competitive wedge," inherent competitive advantages exploited by those nations which permit their telecommunications providers to enter into global agreements, has prompted one analyst to observe that "an inexorable, worldwide trend toward privatization and competition" will extend into the foreseeable future [7, 8]. To yield competitive advantage places domestic telecommunications enterprises at great competitive risk, with rivals potentially seizing upon their customer base.

We note variation in regulatory practices throughout Europe. In Germany, the first step toward German deregulation was taken in late 1996 with the privatization of Deutsche Telekom. This firm began offering stock to domestic and international buyers, and did so with great notoriety and success at that time. Competition is now intensifying in the German telecommunications market. Alliances and partnerships throughout the region immediately followed [9]. Serious negotiations have now taken place among primary and secondary communications providers throughout the region, including France, Italy, Belgium, and Germany itself; these negotiations are aimed at consolidating major firms in an effort to exploit scale economies and leveraging capital and technology [10].

British Telecommunications remains the most influential private telecommunications corporation in Europe, particularly in terms of its aggressive strategy in aligning itself with innovative American firms. The British, under the guidance of the Thatcher administration, dismantled a number of domestic monopolies during the 1980s. In the shipbuilding, energy, transportation, and communications industries the British embraced competition so as to mitigate inflation while encouraging productivity. This strategy, accompanied by tax reduction, propelled deregulation in the European Union. Under terms of the "Europe 1992" agreement, competition in the telecommunications sector was to commence in 1998. In response to this agreement, the Germans, French, Italians, and other EU affiliates were compelled to assume similar strategies.

Governments which adhere to standard regulatory policies contend that monopolies or oligopolies will define market price at the expense of consumers. Regulatory sanctions become the only appropriate vehicle in a capital-intensive industry with the common foundation of interacting networks. Even if empirical evidence should prove the contrary, these governments claim that the introduction of competition could not expedited to mitigate consumer disadvantage. This dogma is gradually changing as technology evolves and reconfigures the competitive landscape.

Whether the EU will allow substantial imports of global telecommunications goods and services in the immediate future remains a vital concern of other nations [11]. Whether European nations, individually or collectively, design a fortress in which telecommunications imports are effectively limited is not yet clear. The next several years will be decisive in forging cooperative agreements between the telecommunications monopolies of Germany, Italy, France, and competitors located outside the region. Generally, the Italians and French are most resistant to dramatic regulatory transformation, while the Germans, Belgians, and Dutch are now pursuing reform in earnestness [12].

Eastern European nations—most especially Slovakia and the Czech Republic, Hungary, and Poland—appear more responsive than their Western counterparts to extensive telecommunications competition. In the wake of the collapse of the Soviet Union, Eastern Europeans are eager to incorporate economic change with their political revolutions. American firms, particularly those companies hoping to seize immediate market share, are pursuing opportunities throughout the whole of Europe. A key stumbling block to such agreements are common legal standards, particularly with regard to intellectual property rights and anti-trust administration; these are prerequisites to international mergers, acquisitions, and other business formations. Among these concerns are mutually acceptable standards of stock issuance and equity interest, hiring and personnel criteria, and similar items where government sanctions impinge on corporate administration.

Americans and Europeans share a common competitive context, as much of their telecommunications markets are already saturated—growth in the number of domestic consumers is limited—although the appetite for emerging services is growing. A population explosion in both Europe and the United States unfolded after World War II, and it is just now having its blast effect on the communications industry. However, annual population growth in most European nations is presently less than 1%. In some nations, the aggregate population may eventually fall in the immediate future. Population in the United States expands at the rate of about 1% annually.

Future global growth in demand of telecommunications products and services will be a function of two forces:

1. Growth in new services supplanting veteran ones; and
2. Aggregation of consumer markets.

For Europeans, it is vital that they pursue both avenues so as to introduce growth trajectories. In terms of developing new technologies and associated product lines, the merger/partnership/alliance route appears inviting because it is a ready technique to bring new products to market, results in favorable technology transfer, and strengthens industries by encouraging rising productivity [9].

European merger and acquisition strategic patterns are likely to resemble those now unfolding in the United States. In those cases where mergers can be consummated without provoking regulatory and judicial bodies, Americans and British firms will pursue immediate access to prospective customers. This strategy effectively deflects regulatory sanctions, preventing such consolidation, that might later be installed by the European community. Should the EU alternatively liberalize trade constraints and permit this form of investment in their region, these firms would retain a competitive advantage over their rivals [13].

The engine of economic growth in Asia during most of the post-World War II period has been Japan, the world's second largest economy. The Japanese established a biregulatory environment to facilitate telecommunications services following the end of World War II. One element of this regulatory strategy was the creation of the government-sponsored Nippon Telephone and Telegraph Corporation, guiding the domestic communications market. The second component is Kokusai Denshin Denwa Company, whose legal monopoly position steered international telecommunications development.

The bimodal Japanese regulatory strategy has forestalled much telecommunications innovation, and constrained the nation's international competitive position. This conclusion has been tacitly conceded by the government during the late 1990s, and Japan has now dedicated itself publicly to the systematic elimination of such restrictions [14].

The decline of the Japanese economy during the past decade has put its domestic communications industry at international competitive disadvantage. NTT maintains a very close relationship with public officials who are reluctant to divest this megamonopoly [15]. Multinational corporations are thus effectively discouraged from entering this enormous market, although there has been some penetration by rivals in the domestic market since 1998. Nevertheless, "red tape" continues to restrain innovation in the domestic communications market, and consumers effectively pay as much as three times the prices Americans bear for comparable services and products.

Substantial skepticism remains, despite recent regulatory liberalization, that the two incumbent monopolies may mitigate the effects of competition by securing advantageous concessions from senior politicians, particularly with respect to domestic content provisions (rules requiring that a certain fraction of equipment or services can only be provided by domestic incumbents). This contingency is now a matter of conjecture, but it holds profound strategic ramifications. The Japanese have limited their own ability to export telecommunications products to both Europe and the United States by resisting reciprocal trade arrangements. In endeavoring to insulate their domestic industries, including communications, the Japanese may have laid the groundwork for diminished future export revenues. Simultaneously, the government may have deprived itself of the benefits of competition: lower prices, superior quality, and more vigorous competition that might have been realized in its domestic sector. Some observers have inferred that the two most promising global communications markets within the next thirty years are China and India. With enormous customer markets now primed for telecommunications demand—China and India are the two most populous countries—multinational corporations are developing strategic plans for these markets. The regulatory structure of China, controlled by local authorities, has gradually removed restrictions [9]. The central government has recently encouraged joint partnerships, alliances, and other financial vehicles to spur telecommunications activity. A contingent of venture capitalists have seized upon the Chinese market, and joint projects are now unfolding.

India is primed for enormous growth. Domestic communications providers have penetrated less than 1% of the market [16]. Traditionally, the Indian government has strictly managed the communications industry with rigorous regulatory administration. This government-owned monopoly is a rigid one, but recent attempts to encourage private capital investment—both internal and external—have established precedent for international joint ventures. Indeed, the government has now embarked on a major effort to entice European and American firms to form joint alliances with several government agencies so as to facilitate the creation of state-of-the-art systems. Indian entrepreneurs have been successful recently, for example, in building wireless cable systems throughout the nation; an infusion of capital from the United States has been a key ingredient to this development.

Most Asian nations appear to fall into one of two parallel categories with respect to their strategic designs for communications: those nations committed to building state-of-the-art telecommunications infrastructures, contingent on international capital investment, and especially dependent on venture capital; and in another vein, those countries dedicated to traditional monopoly structures, seemingly resistant to the mobilization of outside capital and privatization. In Hong Kong, Singapore, South Korea, Malaysia, and Taiwan telecommunications systems are comparatively advanced, even cutting-edge in some cases. In Indonesia, Burma, and North Korea little or no outside capital investment exists. Indonesians encourage limited foreign ownership, ordinarily confined to 20%, of domestic enterprises [9]. Should the Indonesian government conclude that economic growth is efficiently encouraged by the introduction of such private investment, especially in the aftermath of depressionary conditions that have been manifested since 1997, restrictive policies may be lifted. The rapidity with which international investment penetrates these emerging economies will hinge on the licensing approval and regulatory processes. Foreign firms and venture capitalists may be reluctant to invest substantial monies if they perceive clumsy or time-consuming regulatory procedures. Legal and regulatory reform is thus a necessary precondition for the infusion of new capital to these nations.

There is wide variance in the administration of government regulation throughout Latin America. The most alluring markets are clearly Brazil and Mexico, owing to their large consumer base. At this time, Chile is the most attractive market from the point-of-view of international entrepreneurs. Chile embarked on deregulation of most major industries during the 1980s. Domestic and international capital has continued to infuse the nation with advanced wireline and wireless systems now under construction. Chile is approaching a highly competitive telecommunications market structure, noted for a rising level of industry rivals. The significance of this experiment concerns its effect on other nations on the continent. Strategists will carefully monitor emerging telecommunications developments in Chile so as to gauge comparable opportunities in Latin America.

The Brazilian economy, a prospective customer base of nearly 160,000,000, is the most attractive in all of South America. Despite continuing problems of financial instability throughout the 1990s, the Brazilian economy promises long-term economic expansion. The Brazilian government introduced privatization in its communications industry during the mid-1990s. The Brazilians are particularly interested in triggering growth of value-added services, pivotal to development of areas of the economy [19]. The government has resisted attempts to sell majority interest in its state-owned communications enterprise. By December 1996, however, the Communications Ministry of Brazil emphasized to investors, particularly multinational corporations, that it was unlikely to fully deregulate domestic telecommunications for at least three to five years. Recent indications by the government have moved many strategists to forecast that multinationals will be permitted majority ownership in these operations shortly after 2001. The reluctance to experiment with immediate deregulation as practiced in Chile has retarded development of cutting-edge infrastructure.

Mexico holds a consumer base of roughly 100 million. Communications development is tied to the relaxation of government sanctions, particularly as they pertain to privatization of TelMex, a publicly-sanctioned monopoly. A complex system of shared ownership, including equity held by employees, configures TelMex, and reflects a nationalism antithetical to multinational corporate investment. Nonetheless, privatization has systematically penetrated the industry, and led to innovations in communications productivity. Value-added services have evolved proportionate to the level of foreign investment [20]. One gauge of the effectiveness of privatization in Mexican telecommunications has been the extraordinary increase in TelMex stock valuations during the past decade. Profitability has exceeded that of telecommunications firms in many nations.

In Latin America, ranging from Costa Rica and El Salvador in the North to Uruguay, Paraguay, Bolivia, Peru, Argentina, and Venezuela in the South, the communications sectors in Mexico, Brazil, and Chile are regarded as pivotal. Government resistance to deregulatory experiments will only subside as improvements in employment, government revenue, and productivity ensue.

In Australia deregulation has been systematically implemented in communications, simultaneous to experiments in the United States and Britain. Significantly, the Australian economy has outperformed all other Asian nations during the recent financial crisis. The Australians have opted for a gradual relaxation of government restrictions, and forged reform of the state-sponsored communications monopoly. Private investment, particularly ventures funds directed at emerging technologies, is vigorously pursued [7]. The Australians especially desirous of exploiting advanced wireless technologies, including MMDS, PCS, and other cellular communications, owing to their geographic terrain.

Russia and the juxtaposed republics that align this region require vast capital from foreign firms. In many sectors of the economy, reduction, elimination, or modification of government regulations have already spurred private investment; however, many such firms, in the aftermath of several devaluations since 1997, have pulled back capital previously promised. Telecommunications infrastructure has not progressed in recent years, therefore. Complex legal problems have further slowed the growth of this vital sector for the country. Compounding these dilemmas are the obvious political uncertainties associated with investment projected to build networks. Foreign firms presently have little incentive or motivation to build systems necessary to sustain business infrastructure. Truly, Russia's inability to compete effectively in world markets has been partly a function of inadequate communications and information technology infrastructure.

In Africa government regulatory control remains the norm. This pattern of government edict is changing in some nations, but there is little incentive for foreign firms to invest in countries whose political systems are unstable and in which leadership reconfigures with great frequency. Entrepreneurial inexperience discourages foreign investment on the continent. One recent positive development has been manifested in the area of wireless communications infrastructure; because capital costs of installing wireless facilities have fallen relative to wireline facilities, the continent may be poised to adopt telecommunications infrastructure at a rate faster than previously anticipated. Paradoxically, certain regions of Africa may deploy cutting-edge technology exceeding that presently utilized in more developed regions, owing to differentials in capital construction.

Regulatory policy in the Middle East may generally be described as restrictive. In Israel, Saudi Arabia, Iran, Iraq, and most other Mideast nations, there exists substantial government control over the build-out of telecommunications systems. Monopoly control is applied in a manner consistent with Western European policy. Deregulation varies from rigid regulatory control to political influence over the means of information transmission. It should be noted, however, that cutting-edge software development is especially pronounced in Israel [21]. Recent technical advances introduced by Israeli firms in specialized mobile radio and Internet telephony have been particularly impressive.

There has been a marked increase in international mergers and acquisitions since 1994 [22]. The multiplicity of government regulations and anti-trust sanctions that dictate successful mergers in the United States are further promoted by comparable cross-national agreements. Whereas domestic agreements can be negotiated between buyers and sellers with some intuitive sense of probable government reaction, global communications agreements must be shepherded through a maze of regulations and legal interpretation adjusted for regional political machinations. Nevertheless, it is anticipated that, for reasons later identified, international telecommunications mergers and acquisitions will proliferate within the next several years, particularly in the developed nations

9.4 Strategy and International Mergers and Acquisitions

Merger strategy in Europe has never been characterized as dynamic or aggressive. Europeans refrain from the vigorous pursuit of acquisitions, traditionally a hallmark of American multinational corporate policy. American firms regard mergers and acquisitions as frequently a necessity in gaining entry to new market or critical in securing market share. In the background is the continuous need to recruit high-quality talent embedded in other firms. Competitive advantage is pursued at all costs. Europeans frequently look upon mergers critically; for them, mergers involve an uncivil strategic design [23, 24]. The surge in American mergers, both domestic and international, cajoles European firms to discard their traditional antipathy toward such strategies.

In anticipation of EU regulatory relaxation set for 1998, mergers were launched by British Telecommunications and Cable & Wireless PLC, Bertelsmann, and CLT [25]. British Telecommunications' merger with MCI spurred similar discussions between European corporations and their competitors in other regions. Merger and acquisition activity is now underway in response to strategic opportunities. To delay market penetration is to allow competitors to seize emerging markets. American deregulation has thus prompted Europeans to preempt further gains by U.S. multinational corporations. Mergers consistent with economic convergence designate the first phase of global communications collaboration.

Multinational corporate mergers must satisfy four critical regulatory considerations: universal access, price manipulation, licensing, and interconnection [26]. Government authorities in many nations have been influenced by the continuing American commitment to service an entire national customer base. Problems of interconnection and efficient networking remain continuing concerns, as does price stability. Licensing, particularly in terms of electromagnetic spectrum controlled by wireless firms, remains a pivotal issue. The key to consummating these agreements and surmounting government regulatory reticence is embedded in that nation's cultural and political environment. The political ambiance of the regulatory agency in question must therefore be factored into the competitive strategy of the firm.

9.5 Global Strategic Alliances and Partnerships

Empirical evidence provided in the preceding chapter provided a working definition of strategic alliances and partnerships. These are agreements that join service and content providers in telecommunications, and are designed to leverage the assets of complementary firms. International partnerships are negotiated with the concrete understanding that the agreement endures only as long as the interests of both parties are served.

Unlike mergers or acquisitions, the partnership is dedicated to preserving the independence of both parties while encouraging the attainment of a task or objective. Partnerships promote organizational flexibility and adaptability in foreign markets. Partnerships represent an effective tool in minimizing strategic risk. The alternative risk is that success in a partnered enterprise does not assure long-term growth and opportunity; either party may venture in a new direction with a vested proprietary interest. Contracts which fuse common interests while neutralizing organizational differences are critical to efficient collaboration. The best of strategic partnerships are noted for their adroitness. The speed with which they are negotiated and introduced to market should be the gauge by which their success is measured, as competitive advantage is conferred upon the firm which acts quickly in response to market changes.

The terms mergers, acquisitions, alliances, and partnerships are applied with increasing frequency in telecommunications usage, and the multiplicity of contexts in which these expressions are used has led to substantial confusion. The distinction between partnerships and alliances lies in the motive supporting the agreement and the time frame projected for that collaboration. An alliance may be defined as a long-term commitment between two or more enterprises that find that the merger/acquisition route may not be feasible or desirable. Productive relationships can be formed through this endeavor without necessitating legal and regulatory delays prompted by foreign governments.

Alliances, if formed to the long-term satisfaction of both or multiple partners, also signify an efficient method of integrating two or more partners in a venture that could not otherwise be financed. Costs of capital, labor, and technology can be leveraged proportionate to the number of partners advancing the alliance. In the United States, alliances formed in interactive television in 1994 and 1995, for example, have typically resulted in three or more partners consummating an agreement so as to reduce considerable financial risks. The same development has been observed in broadcasting, telephony, and various collaborative software forums.

9.6 Global Venture Capital and Telecommunications Strategy

International venture capitalists furnish a vital supply of financial reserves for emerging telecommunications companies. Venture capitalists have already been fueling industry start-ups in telephony, wireless communications, and information technology; this growth is linked to trade liberalization commencing in 1997. It is instructive to note that these industries have been growing in the midst of a global economic slowdown extending through the same period. International risk capital is often controlled by fund managers who assess market potential based on political stability, regulatory restrictions, and rising consumer consumption. These fund managers search for firms that are projected to grow in excess of 25% during their first three years of growth. There is variation in the dispensation of start-up funds based on geographic territory and density of consumer markets.

Global venture capital markets are defined as "visible" or "invisible" [29]. The visible venture capital sector consists of between one and two thousand firms worldwide. A majority of these firms reside in the United States. An estimated $85 billion is now directed by these companies, much of which is channeled to high-technology and communications start-ups. The standard method of obtaining international venture capital, as is the case in the American market, lies in designing a structured business plan stressing differentiated products aimed at identifiable market needs or niches. In regions where political risk remains intense, these capitalists will invariably require higher rates of return. New ventures borne in nations other than the United States retain a higher probability of procuring capital if an American partner can be lured into collaborating on a business plan.

The "invisible" venture capital market consists of an estimated three million persons whose net worth exceeds $1 million, and who embrace the adventure of venture capital investment [30]. These persons live in many nations, act as independent entrepreneurs, and generate an annual investment pool of $30 billion. This group is often labeled "business angels." Business angels provide a foundation of capital essential for international ventures. Referrals to business angels are sometimes directed by venture capital firms, but the development of the Technology Capital Network (TCN) is rapidly becoming a critical link between entrepreneurs and venture capitalists [30]. TCN, a forum for businesspeople and investors, acts as a conduit for global venture capital in the high-technology sector.

9.7 Strategy and the International Business Opportunity Paradigm

A Business Opportunity Paradigm (BOP) for telecommunications was introduced in the preceding chapter. The pattern of empirical evidence now suggests that new technologies, when achieving 10% market penetration, inevitably invite imitators. These imitators form the first layer of competition sparked by each successive innovation in telecommunications. Established telecommunications firms aim to absorb such innovators early in their stage of growth. The interaction of veteran and entrepreneurial companies outline an IBOP which specifies emerging opportunities throughout the competitive landscape. In Figure 9.1 we note the interaction of consumer behavior, product design, strategy, and consumer feedback which isolates emerging business opportunities.

Tertiary ramifications are the products of telecommunications innovation. The BOP for domestic telecommunications signifies three developments: (1) constant innovation, subsequently provoking product substitution; (2) established firms act to preempt small firms by exploiting scale economies; and (3) ancillary firms or industries will emerge to sustain or support the originating technology. The synthesis of these three themes defines a model which allows strategists to forecast communications development.

The Business Opportunity described in the last chapter set forth essential competitive questions to which contemporary strategists must respond.

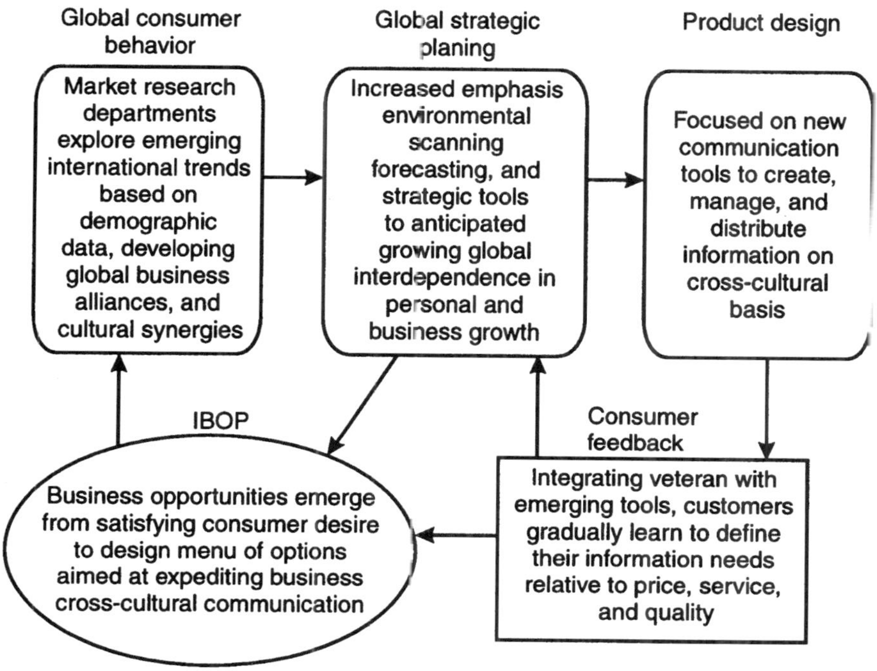

Figure 9.1 International telecommunications business opportunity paradigm.

Adjusted for firms managing strategy in the international environment, these questions are:

1. If we introduce this product, how will domestic and global competitors respond?
2. In bringing this product to market, have we prepared adequately for supporting sales, distribution, and maintenance infrastructure?
3. Are we vulnerable to predatory instincts of larger firms? Can our product and strategy be quickly imitated, simulated, or copied?

The model indicates that many economic opportunities emerging from telecommunications innovation are embedded in the tertiary consequences prompted by that creativity. Communications firms are increasingly compelled to evaluate and forecast unfolding business and consumer transformation in order to specify the firm's future strategic path. The IBOP implies that in the next century a global telecommunications provider should design a strategy emphasizing its role as creator, manager, and distributor of information. This flexible mission statement enables the global firm to seize upon opportunities otherwise unrealized in a narrowly defined strategic profile.

References

[1] Tapscott, D. (ed.), *Blueprint to the Digital Economy*, New York: McGraw-Hill Co., 1998, pp. 355–370.

[2] Porter, M., *Competitive Analysis*, New York: Free Press, 1998, pp. 275–299.

[3] Adams, W., and J. Brock, *The Structure of American Industry*, Upper Saddle River, NJ: Prentice-Hall Inc., 1995.

[4] Shepherd, W., *Industrial Economics*, Upper Saddle River, NJ: Prentice-Hall Inc., 1997.

[5] Elliott III, T. L., and D. Torkko, "World-Class Outsourcing Strategies," *Telecommunications*, April 1996, pp. 47–49.

[6] "Global Alliances Must Fight for Growth," *Reuters News* (London Newsroom), Nov. 3, 1996.

[7] Taoka, G. M., and D. R. Beeman, *International Business: Environments, Institutions and Operations*, New York: Harper Collins Publishers, 1991.

[8] Hudson, H., *Global Connections: International Telecommunications Infrastructure and Policy*, New York: Van Nostrand Press, 1997.

[9] Kennedy, C. H., and M. V. Pastor, *An Introduction to International Telecommunications Law*, Norwood, MA: Artech House, 1996, p. xvii.

[10] "Investments in Engineering Telecommunications," *PR Newswire*, Nov. 19, 1996.

[11] "Getting a Line on Europe's Telecom Free-for-All," *Business Week*, June 19, 1997.

[12] "Europe: The Age of Mergers," *Business Week*, April 15, 1996.

[13] "Tearing Up Today's Organization Chart," *Business Week*, Nov. 18, 1996.

[14] "The New Asian Manager," *Business Week*, Sept. 2, 1997.

[15] "A Free Phone Market in Japan," *Business Week* (*International Edition*), March 11, 1996.

[16] Appleyard, D. R., and A. J. Field, Jr., *International Economics*, Homewood, IL: Irwin Press, 1997, pp. 152–153.

[17] Schiller, B., *The Economy Today*, New York: McGraw-Hill Co., 1999.

[18] "Chile: The Strangest Telecom War Yet," *Business Week*, Jan. 15, 1997.

[19] "Wireless on a Global Scale," *Wireless*, Oct. 1998, Vol. 5: No. 9., pp. 18–20.

[20] "A $15 Trillion Private Piggy Bank," *Business Week*, June 24, 1998.

[21] "Israel . . . Or Go Forth to the U.S. and Prosper," *Business Week*, Feb. 26, 1996.

[22] "Telecommunications: The Developing Leap," *Wall Street Journal*, Feb. 11, 1998, p. R15.

[23] "Global Competition: The Second Act," *Tele.com*, Nov. 15, 1998, pp. 11–33.

[24] "Deregulation and a Segmented European Market," *Telecommunications*, July 1996, pp. 64–66.

[25] "How Merger Mania Has Redefined the Communications Landscape," *Telecommunications,* Aug. 1996.

[26] "Europe: The Age of Mergers," *Business Week,* April 15, 1997.

[27] "British Telecommunications/MCI Announce Definitive Merger," *PR Newswire* Nov. 3, 1996.

[28] "Answer the Call of the Future," *PR Newswire,* Oct. 22, 1996.

[29] "World Telecommunications Is Undergoing a Revolution," *Reuters News,* Nov. 3, 1998

[30] Bygrave, W. D., *The Portable MBA in Entrepreneurship,* New York: John Wiley & Sons 1994, pp. 172–194.

10

Emerging Issues in Strategic Management in Telecommunications

Today's strategist now confronts a baffling array of challenges embedded within methodological content and context. Superimposed over the daunting goal of designing competitive strategy is a watershed era of economic, social, political, and technological volatility. The economy is reconfiguring to accommodate the technological innovations introduced by the communications industry. There is little dispute among strategic planners that this is indeed the environment of contemporary management. This chapter discusses those developments unfolding as decisive in differentiating competitive performance, and the problems strategists are likely to encounter during the coming decade. The material presented here reflects a blend of strategic theory and practice, admittedly combined with a good deal of conjecture.

10.1 Strategic Management Methodology: Context vs. Content

The author has conceded that, during the year of research preceding this writing, he concluded that it was impossible to elaborate on the components of strategic design without projecting an interpretation of economic data. The interpretation of these statistics provides context for the strategist. While these data are culled from primary sources, and the most credible gauge we presently have for analysis of the competitive landscape, the author's assertions are intended to be contentious and provocative.

Increasingly, a firm's capacity to construct effective strategic plans is contingent on economic forecasting. Hence, some consideration of economic analysis, particularly with respect to the coming decade, is essential to define strategic context. A recapitulation of the forecasts specified in Chapters 2 and 3 of this manuscript, based on empirical evidence described in the Appendix (Sections C, D, and E) suggests the following:

- The American economy is likely to grow substantially between 2000 and 2010, based on positive demographics coinciding with rising productivity. The world economy is now poised for significant economic growth, beginning in 2000 and extending through the same period. The engine of growth will be technological, with people only now learning to exploit the intrinsic value of information. This positive economic environment reflects the growing significance of communications technologies, but also translates into mounting prospective demand for the technologies themselves. Information cannot be depleted; unlike all other resources, it increases proportionately, indeed geometrically, to its use.
- The economy is reconfiguring to reward those—individuals and firms—who learn to exploit and leverage information and communications technologies. Competitive advantage is conferred upon those who speed products to market without sacrificing customer responsiveness or sensitivity.
- Mergers, acquisitions, affiliations, alliances, partnerships, intrapreneurships will proliferate and cross national boundaries. No firm can control its own fate, except in those instances of proprietary technologies. More importantly, no firm retains full control over infrastructure, distribution, or content. Public monopolies are about to end in the form to which we were accustomed, and will be replaced by a mixture of competitive, oligopolistic, and monopolistic competitive industrial market structures.
- The telephony, broadcasting, and computer industries will continue to intersect, with principal strategic synergies driven by software development. The computer sector will come to predominate in influence over its telephony and broadcasting cousins.
- New business formations in other nations will intensify in frequency, complexity, and reconfiguration. The wave of deregulation and privatization typified in the United States after 1996 is likely to be replicated elsewhere, with cross-national collaboration strengthening as a result of implementation of key provisions of the 1997 World Trade Organization agreement.

- Demand for American telecommunications equipment and services (exports) will rise substantially after 2000, as the global economy recovers and it becomes increasingly apparent that competitive advantage is conferred upon those able to exploit cutting-edge technologies.
- Productivity will rise significantly after 2000, resulting in higher wages and profitability, and will instigate continuing demand for communications services. Inadequate skilled labor in the communications industry will simultaneously serve as a stimulus to higher wages within this industry sector; for communications firms, a critical managerial issue will become the recruitment, retention, and continuous retraining of skilled personnel.
- Regional trade (multinational) agreements will combine with bilateral (country-to-country) arrangements to forge dramatic demand for communications products in both developed and developing nations. Command-control nations, economies in which government outlines the goals and means of production, will not benefit from these improvements in technological innovation. The empirical evidence suggests that those nations committed to relaxing trade restrictions or sanctions (tariffs, quotas, regulations) will profit most by the coming innovations introduced by both domestic and global telecommunications enterprises. Liberalized trade will define the future of telecommunications imports and exports.
- The demand exerted by nations for American communications services of all kinds will rise faster than the demand for physical products, resulting in a boom for telecommunications management consulting. A substantial fraction of telecommunications equipment will be produced in other nations; American communications services will be provided almost exclusively by Americans themselves. All American and European telecommunications firms should prime themselves for emerging demand for these value-added services.

These are the broad contextual developments about to assert their influence. Their blast effect will be felt very shortly.

With respect to content—strategic design—we find a distilled array of tools at our disposal, but our capacity to exploit them is severely constrained by shortened product life-cycles. These tools include:

- Classic strategic management;
- Systems (or process) strategy;

- Competitive analysis; and
- Co-opetitive analysis.

Strategists operating in different competitive environments may find one or more applications for any of these methods. The dilemma in applying any of these techniques lies in the time required to successfully implement them. Volatile market environments, fused with the rapidity of technological change, diminish the immediate value of any of these approaches. The strategist is thus left with an awkward choice: it may be necessary to create a nimble hybrid of these methodologies, uniquely suitable and appropriate for the firm in which he or she works. The ability of a strategist to forecast changes in the competitive landscape gains heightened value in this environment.

We discern common attributes in the most successful communications firms currently designing strategy. Successful firms, and the strategists which guide them, speed product to market; listen and respond to customers; benchmark their performance against the best global—not domestic—standards; elect and adhere to one of two product tactics: first-to-market or first-to-market-acceptance; embrace continuous improvement in product design, distribution, and post-purchase satisfaction; are now planning or introducing customized product menus; are learning the precepts of tertiary analysis and their role in projecting future demand; and are determining the suitability of relational marketing as a component of strategic management. The effectiveness of a firm's strategic management can be legitimately monitored, over the long-run, by consulting stock valuations—the ultimate expression of investor confidence in those plans.

10.2 Electronic Commerce and Strategic Management

Few will quarrel with the commercial application of networked information and knowledge in terms of its role in reconfiguring the domestic and international economies. Those sufficiently prescient to anticipate the exponential growth of electronic commerce have already profited from the transition. The data inserted in the Appendix (Sections B and C) reveal what is increasingly obvious in this regard. What is less apparent is the importance of e-commerce in reshaping relationships between employers and employees; between consumers and producers, between suppliers, distributers, wholesalers, retailers; between governments and citizens. We have already asserted empirically the unfolding phenomenon known as disintermediation, in which individuals and firms are already being removed from the distribution chain.

Viewed strictly in competitive terms, we are beginning to divine essential strategic principles governing effective commercial use of the Internet:

- Most firms will be compelled to cultivate narrow market niches in this competitive environment.
- High-quality value-added services will be essential to establish customer reach and loyalty.
- Raw information or data offers little value to consumers in this environment: it must be distilled in a form which can be applied to efficient decision-making. Strategy in firms will be redirected to attain this objective.
- Aggregation of a customer base becomes the greatest strategic—or competitive advantage—in networked environments.
- Firms gravitating to Internet-based services must educate customers as to the intrinsic value of the information they provide, and must persuade them that their firm is superior in delivering information and knowledge.
- Target-marketing, illuminating precisely the information which improves consumer decision-making (in professional or personal endeavors) will confer competitive advantage to the firms which learn to exploit this tactic.

10.3 Customer-Led Customization

The emerging significance of one-to-one marketing is outlined in Figure 10.1. Formerly, telecommunications providers developed a new product line and then sought to create a market for it. Providers must now verify the peculiar requirements of individual customers and then innovate services which satisfy those multifaceted needs. Market research becomes a highly complex task in this environment. The industry may well be reshaped—mass marketing will give way to customized market research, and success or failure will pivot on the ability to meet the dynamic and fastidious requirements of consumer tastes. Consumers will dictate the terms and distribution of the services they require, dramatically reconfiguring organizations. Competitive advantage will fall to those firms which are first off the mark in customizing their products. "Total marketing" therefore emerges as a model equipped to forecast changing consumer tastes, and establishes a framework for strategic management in this environment.

Assuming this is a credible paradigm for the delivery of communications services, then we may conclude that service and content providers will treat customers as the principal focus of strategy. An increasing fraction of a firm's earnings will be redirected to marketing and strategic planning departments, where the firm is closer and most sensitive to customer behavior. Customers, in a customer-led customization environment, dictate what is to be produced. A firm which chooses to ignore this development will be superceded and supplanted by companies prepared to serve this emerging market.

10.4 Strategy and the Emergence of Continuous Markets

Telecommunications continuous markets are ones in which demand for a variety of services is perpetually manifested through communications technology. Primary, secondary, and tertiary markets for telecommunications in health care, education, transportation, environment, entertainment, agriculture, engineering, science, construction, and manufacturing are created and reconfigured when consumers learn to exploit information and knowledge to perform daily undertakings. In this environment, there is little difference between work and learning; work is learning, and is tied directly to the production of value-added services necessary to maintain competitive advantage. For communications providers, a strategy stressing learning, in the largest sense of that term, may be vital in securing new markets and sustaining market share.

Learning is the pursuit and acquisition of knowledge. Whether one dichotomizes or synthesizes a vision of the telephony, cable, or broadcasting markets, the recurring need of the individual to assimilate information and knowledge may become the core of future demand for communications services. To tap the continuous learning market, one must understand the prospective market response to the nature of learning itself. Two developments are reconfiguring knowledge-based, post-industrial societies. Education involves the transmission of knowledge through formal instruction and results in a body of knowledge whose economic value does not erode. Language, history, mathematics, and the sciences, for example, represent bodies of knowledge not subject to obsolescence. Training, however, is practical instruction focused on accomplishing specific professional or avocational goals. Innovation spurs training, and requires periodic updating. Education and training must be updated and expanded on a regular basis, as the useful half-life of knowledge is contracting. There is little dispute that markets for education and training will expand proportionate to the level of knowledge and skill compensated by the private sector.

Communications technology increasingly directs economic change. It is already common for professionals to return to school at two or three junctures in their lives. These institutions are very likely to focus on customized learning, geared to professional objectives. Telecommunications providers, whether focused on service or content, will inevitably respond to this emerging demand. Learning, personal and vocational, represents the most attractive long-term market for this industry, and strategists are just now awakening to this market reality.

10.5 Strategy and Developments in Relational Marketing

Using the customer as a tool to project, define, and launch new product line is the heart of relational marketing. Euphemisms for relational marketing include one-to-one marketing or particle marketing. The undergirding strategic implications of relational interaction with consumers, discussed in Chapter 2, impel a firm to eschew unprofitable customers in favor of profitable ones. Relational strategists presume that customers whose demand for services is limited—or unprofitable—can be deflected to unsuspecting firms. Because companies traditionally invest heavily in recruiting all available customers, the cost of recruitment per customer is higher than that incurred by firms focused exclusively on acquiring the most profitable consumers.

The incisive insight of relational strategists, however, concerns a shift in customer strategy. Historically, a firm creates or improves a product, then finds a market for it. This has been the typical pattern of new product development in telecommunications and high-technology. Relational strategy holds that customers should be managed, not products. In short, strategic effectiveness and profitability result from managing a "lifetime customer." Because the cost of retaining a veteran customer is cheaper than recruiting a new one, the firm that achieves customer loyalty among its most profitable consumers is likely to be the most successful. There exists variance by product and service, but it is generally accepted in the telecommunications industry that solicitation of new customers exceeds the cost of retention by a factor of five to eight times.

If one adopts this line of reasoning, the economics of customer loyalty will come to dominate strategic design in the most successful communications firms. Hence, we return to the theme that strategists might consider managing customers rather than products. We would determine what tactics would be necessary to manage a customer's demand over the lifetime of that consumption. Some analysts have noted that in telecommunications, the top quarter of the customer base is responsible for as much as 100% of the firm's profitability. If this statistic holds in the immediate future, we will witness a reinvestment of profits in market research aimed at retaining a firm's best customers. Customer prospecting will thus give way to reinvestment in the firm's most loyal and dedicated consumers.

10.6 Envisioning Change in Telecommunications Strategic Management

We have noted through empirical verification the determinants of industry change, and thus the emerging requirements of strategic management. These determinants are both contextual and content-based. The array of regulatory, technological, economic, and demographic variables which comprise these two categories are growing increasingly difficult to forecast with precision. These difficulties will be further compounded by the intensification of global interaction in the years ahead. In the face of these challenges, we require new methods to project the pattern of determinants destined to define industry change. We have at our disposal a set of methods designed to anticipate such change, thus aiding in effective design and implementation of strategy.

10.6.1 Historical Analogy and Analysis of Industry Life-cycle

Historical analogy is an economic methodology which asserts that all industries endure a series of evolutionary phases, in which a strategist may define the competitive environment of a firm. Understanding those forces governing a firm's competitive landscape enables one to position a firm and estimate probable strategic interactions with industry rivals. The most compelling feature of the contemporary communications environment, as noted in Chapter 2, is the prospective unfolding of competition associated with deregulation. A mature industry, steeped in regulatory compliance and incumbent control of market share, is now headed for intense competition and technological convergence.

We have some understanding of the dynamics of deregulation and the role it plays in reordering industry structure while innovating product line. In aviation, trucking, petroleum, natural gas, and financial institutions, among others—industries all subjected to deregulation since 1978—we can outline a similar pattern of competitive development. These data suggest a standard paradigm of performance, as indicated in Figure 10.1.

Some observers contend that market confusion and chaos will inevitably follow implementation of the Telecommunications Act of 1996. It is their contention, however, that it is just as inevitable that market stability will be restored after ten to fifteen years of experimentation. A highly capital-intensive industry whose progress was guided by government regulation, they believe, must educate itself in market dynamics never before experienced. The absence of competitive knowledge and experience means that industrial giants must learn during a period of chaotic experimentation. Competitive exposure and the assumption of risk by new entrants establish thereafter a framework of growing maturity and stability.

We note in this model the instability associated with short-term innovation and new product line in telecommunications. The proliferation of new entrants, accompanied by uncertainty and risk in infrastructure maintenance, translate into an industry shakeout before 2005. The model holds that long-term stability will manifest itself before 2008. It is during this period that risk diminishes as new infrastructure providers finally stabilize the economic environment. Market dynamism eventually begets consolidation, stability, and product maintenance. This model must be adjusted for each sector of the industry. Computer and broadcasting market structures, largely unburdened by regulatory litigation, are characteristically far more dynamic in the early stages of deregulation. The telephony sector, presently embroiled a series of lawsuits designated to forestall regulatory sanctions, has effectively slowed the process of competitive metamorphosis.

10.6.2 Applying Scenarios to Strategic Management

Scenario-writing, applied to strategic management, itemizes prospective avenues for industry development and company positioning. A scenario is model-built on a potential determinant or catalyst for change. These determinants may be economic, social, regulatory, political, or technological in nature. Every scenario carries a set of assumptions about how industry rivals interact in response to these determinants. Scenarios are aimed at one or more of these objectives:

1. Describing and thus categorizing future competition;
2. Prescribing appropriate strategy in response to industry developments;
3. Projecting future outcomes based on continuing changes in the marketplace.

Scenario-writers bring both intuitive insight as well as empirical analysis to their models. As a result, scenarios vary considerably in terms of model-building and drawn inferences. The essential concern here is to bring order and clarity to strategic management when quantitative methodology cannot be applied effectively. Scenarists have designed a multitude of theoretical constructs to their design of models and have borrowed extensively from similar methodology in the public sector.

10.6.3 Industrial Economics and the Rule of Three

Another method of projecting the future of the communications industry involves a key lesson culled from industrial economics and the stage of consolidation known as the "Rule of Three." The Rule of Three states that oligopolical forces—market structures characteristically consisting of three (or sometimes four) primary providers—emerge in capital-intensive industries.

Such a contingency is likely in most areas of telecommunications. In such a competitive landscape, three firms would generally acquire 80% or more market share, would offer a full array of bundled services, and concentrate on industry volume (scale economies) to sustain their strategic positions.

The Rule of Three asserts that the exclusive avenue open to smaller firms hoping to sustain a competitive position would involve strategic planning aimed at exploiting a market niche. This theory of industrial evolution, if appropriate to prediction of telecommunications markets, implies that the telephony, broadcasting, and computer sectors would inevitably consolidate after uniform technical standards and infrastructure are fully adopted across national boundaries. Again, if credible, this theory, adjusted for the peculiarities of communications markets in the aftermath of deregulation, suggests that the period extending from 2008 to 2012, will be decisive in restructuring international competition within the industry.

10.7 Strategists and the Knowledge-Based Economy

Scarcely any observer now contests the inevitability of a knowledge-based, customized economy evolved from the current standardized market system. Among other compelling features of the newly shaped economy are these as recently reported by the International Engineering Consortium:

- After the year 2000, in excess of two-thirds of new American jobs will be assumed by highly educated, well-trained workers dependent on the generation, distribution, or interpretation of knowledge.
- Today, in excess of 60% of Americans can be classified as knowledge-workers.
- Presently, in excess of 50% of OECD (Organization for Economic Cooperation and Development) nations may be regarded as knowledge-based, post-industrial societies.

If organizations are to become knowledge-based structures, what does this development portend for the management of strategy in telecommunications firms? We have already reviewed the implications of knowledge-based institutions for prospective consumer demand: communications firms are very likely to pursue strategies focused on bundled packages of high value-added services and products. If this contingency holds, then those firms which speed services to market while simultaneously facilitating the facile manipulation of information are likely to achieve greatest success. Knowledge-based institutions managed with high-quality human capital (workers who distill information for effective decision-making) will generate a voracious appetite for communications products.

If information cannot be depleted, then the same can be said for knowledge. Strategists must now prepare for a spectacular and exponential demand for their firm's services: we are at the earliest and most primitive stage of a globalized appetite for the industry's most enlightened products. No firm should hesitate in pursuing any strategic affiliation, alliance, or partnership—with rival or ally—in seizing upon such opportunities.

Section A: Telecommunications Act of 1996 (Abridged Version): Strategic Management Interpretation

APPENDIX
(Section A)

Section A of the Appendix includes excerpts from the 1996 Telecommunications Act of 1996. This modified version of the Act includes only those elements having a direct bearing on the future of strategic management within the telecommunications industry.

Excerpts pertain only to those sections of the statute which inspire direct strategic management ramifications. Those passages of the Act which pertain to emerging competitive trends, new market opportunities, or the reconfiguration of existing business enterprises are italicized in the body of the text.

The full text of the Telecommunications Act of 1996 can be downloaded through the Federal Communications Commission archives (website: fcc.gov/act). This measure was passed by Congress and signed into law by the president in February, 1996.

When distilled to its core business applications, what emerges is the following spectrum of immediate economic and regulatory consequences, as indicated in the right-hand margin:

(1) **RELAXES, GRADUALLY PHASES OUT, OR ELIMINATES FCC REGULATORY POWERS**

(2) **IDENTIFIES AREAS OF PRESENT AND/OR FUTURE REGULATORY CONTENTIOUSNESS**

(3) **LEADS TO INDUSTRY CONSOLIDATION AND MERGER/ACQUISITION ACTIVITY**

(4) **INSTIGATES OR SUSTAINS COMPETITION, WITH PROBABLE PRICE REDUCTION IN SERVICES AND/OR EQUIPMENT**

(5) **SPURS PROBABLE TECHNOLOGICAL INNOVATION OR CREATION OF NEW BUSINESS FORMATIONS**

(6) **PROMOTES ECONOMIC CONVERGENCE THROUGH PROSPECTIVE INTEGRATION OF TELEPHONY, CABLE, OR COMPUTER INDUSTRIES**

(7) **DIMINISHES STATE REGULATORY POWER AND INFLUENCE THROUGH FEDERAL INTERVENTION**

TELECOMMUNICATIONS ACT OF 1996 (Abridged Version): STRATEGIC MANAGEMENT INTERPRETATION

TITLE I--TELECOMMUNICATION SERVICES
SUBTITLE A--TELECOMMUNICATIONS SERVICES
SEC. 101. ESTABLISHMENT OF PART II OF TITLE II.

(a) AMENDMENT- Title II is amended by inserting after section 229 (47 U.S.C. 229) the following new part:

'PART II--DEVELOPMENT OF COMPETITIVE MARKETS
'SEC. 251. INTERCONNECTION.
'(a) GENERAL DUTY OF TELECOMMUNICATIONS CARRIERS- Each telecommunications carrier has the duty--
'(1) to interconnect directly or indirectly with the facilities and equipment of other telecommunications carriers; and
'(2) not to install network features, functions, or capabilities that do not comply with the guidelines and standards established pursuant to section 255 or 256.
'(b) OBLIGATIONS OF ALL LOCAL EXCHANGE CARRIERS- Each local exchange carrier has the following duties:
'(1) RESALE- The duty not to prohibit, and not to impose unreasonable or discriminatory conditions or limitations on, the resale of its telecommunications services.
'(2) NUMBER PORTABILITY- The duty to provide, to the extent technically feasible, number portability in accordance with requirements prescribed by the Commission.
'(3) DIALING PARITY- The duty to provide dialing parity to competing providers of telephone exchange service and telephone toll service, and the duty to permit all such providers to have nondiscriminatory access to telephone numbers, operator services, directory assistance, and directory listing, with no unreasonable dialing delays.
'(4) ACCESS TO RIGHTS-OF-WAY- The duty to afford access to the poles, ducts, conduits, and rights-of-way of such carrier to competing providers of telecommunications services on rates, terms, and conditions that are consistent with section 224.
'(5) RECIPROCAL COMPENSATION- The duty to establish reciprocal compensation arrangements for the transport and termination of telecommunications.
'(c) ADDITIONAL OBLIGATIONS OF INCUMBENT LOCAL EXCHANGE CARRIERS- In addition to the duties contained in subsection (b), each incumbent local exchange carrier has the following duties:
'(1) DUTY TO NEGOTIATE- The duty to negotiate in good faith in accordance with section 252 the particular terms and conditions of agreements to fulfill the duties described in paragraphs (1) through (5) of subsection (b) and this subsection. The requesting telecommunications carrier also has the duty to negotiate in good faith the terms and conditions of such agreements.
'(2) INTERCONNECTION- The duty to provide, for the facilities and equipment of any requesting telecommunications carrier, interconnection with the local exchange carrier's network--
'(A) for the transmission and routing of telephone exchange service and exchange access;
'(B) at any technically feasible point within the carrier's network;

`(C) that is at least equal in quality to that provided by the local exchange carrier to itself or to any subsidiary, affiliate, or any other party to which the carrier provides interconnection; and
`(D) on rates, terms, and conditions that are just, reasonable, and nondiscriminatory, in accordance with the terms and conditions of the agreement and the requirements of this section and section 252.
`(3) UNBUNDLED ACCESS- The duty to provide, to any requesting telecommunications carrier for the provision of a telecommunications service, nondiscriminatory access to network elements on an unbundled basis at any technically feasible point on rates, terms, and conditions that are just, reasonable, and nondiscriminatory in accordance with the terms and conditions of the agreement and the requirements of this section and section 252. An incumbent local exchange carrier shall provide such unbundled network elements in a manner that allows requesting carriers to combine such elements in order to provide such telecommunications service.
`(4) RESALE- The duty--
`(A) to offer for resale at wholesale rates any telecommunications service that the carrier provides at retail to subscribers who are not telecommunications carriers; and
`(B) not to prohibit, and not to impose unreasonable or discriminatory conditions or limitations on, the resale of such telecommunications service, except that a State commission may, consistent with regulations prescribed by the Commission under this section, prohibit a reseller that obtains at wholesale rates a telecommunications service that is available at retail only to a category of subscribers from offering such service to a different category of subscribers.
`(5) NOTICE OF CHANGES- The duty to provide reasonable public notice of changes in the information necessary for the transmission and routing of services using that local exchange carrier's facilities or networks, as well as of any other changes that would affect the interoperability of those facilities and networks.
`(6) COLLOCATION- The duty to provide, on rates, terms, and conditions that are just, reasonable, and nondiscriminatory, for physical collocation of equipment necessary for interconnection or access to unbundled network elements at the premises of the local exchange carrier, except that the carrier may provide for virtual collocation if the local exchange carrier demonstrates to the State commission that physical collocation is not practical for technical reasons or because of space limitations.
`(d) IMPLEMENTATION-
`(1) IN GENERAL- Within 6 months after the date of enactment of the Telecommunications Act of 1996, the Commission shall complete all actions necessary to establish regulations to implement the requirements of this section.
`(2) ACCESS STANDARDS- In determining what network elements should be made available for purposes of subsection (c)(3), the Commission shall consider, at a minimum, whether--
`(A) access to such network elements as are proprietary in nature is necessary; and
`(B) the failure to provide access to such network elements would impair the ability of the telecommunications carrier seeking access to provide the services that it seeks to offer.
`(3) PRESERVATION OF STATE ACCESS REGULATIONS- In prescribing and enforcing regulations to implement the requirements of this section, the Commission shall not preclude the enforcement of any regulation, order, or policy of a State commission that--
`(A) establishes access and interconnection obligations of local exchange carriers;
`(B) is consistent with the requirements of this section; and

`(C) does not substantially prevent implementation of the requirements of this section and the purposes of this part.
`(e) NUMBERING ADMINISTRATION-
`(1) COMMISSION AUTHORITY AND JURISDICTION- The Commission shall create or designate one or more impartial entities to administer telecommunications numbering and to make such numbers available on an equitable basis. The Commission shall have exclusive jurisdiction over those portions of the North American Numbering Plan that pertain to the United States. Nothing in this paragraph shall preclude the Commission from delegating to State commissions or other entities all or any portion of such jurisdiction.
`(2) COSTS- The cost of establishing telecommunications numbering administration arrangements and number portability shall be borne by all telecommunications carriers on a competitively neutral basis as determined by the Commission.
`(f) EXEMPTIONS, SUSPENSIONS, AND MODIFICATIONS-
`(1) EXEMPTION FOR CERTAIN RURAL TELEPHONE COMPANIES-
`(A) EXEMPTION- Subsection (c) of this section shall not apply to a rural telephone company until (i) such company has received a bona fide request for interconnection, services, or network elements, and (ii) the State commission determines (under subparagraph (B)) that such request is not unduly economically burdensome, is technically feasible, and is consistent with section 254 (other than subsections (b)(7) and (c)(1)(D) thereof).
`(B) STATE TERMINATION OF EXEMPTION AND IMPLEMENTATION SCHEDULE- The party making a bona fide request of a rural telephone company for interconnection, services, or network elements shall submit a notice of its request to the State commission. The State commission shall conduct an inquiry for the purpose of determining whether to terminate the exemption under subparagraph (A). Within 120 days after the State commission receives notice of the request, the State commission shall terminate the exemption if the request is not unduly economically burdensome, is technically feasible, and is consistent with section 254 (other than subsections (b)(7) and (c)(1)(D) thereof). Upon termination of the exemption, a State commission shall establish an implementation schedule for compliance with the request that is consistent in time and manner with Commission regulations.
`(C) LIMITATION ON EXEMPTION- The exemption provided by this paragraph shall not apply with respect to a request under subsection (c) from a cable operator providing video programming, and seeking to provide any telecommunications service, in the area in which the rural telephone company provides video programming. The limitation contained in this subparagraph shall not apply to a rural telephone company that is providing video programming on the date of enactment of the Telecommunications Act of 1996.
`(2) SUSPENSIONS AND MODIFICATIONS FOR RURAL CARRIERS- A local exchange carrier with fewer than 2 percent of the Nation s subscriber lines installed in the aggregate nationwide may petition a State commission for a suspension or modification of the application of a requirement or requirements of subsection (b) or (c) to telephone exchange service facilities specified in such petition. The State commission shall grant such petition to the extent that, and for such duration as, the State commission determines that such suspension or modification--
`(A) is necessary--
`(i) to avoid a significant adverse economic impact on users of telecommunications services generally;
`(ii) to avoid imposing a requirement that is unduly economically burdensome; or
`(iii) to avoid imposing a requirement that is technically infeasible; and

`(B) is consistent with the public interest, convenience, and necessity. The State commission shall act upon any petition filed under this paragraph within 180 days after receiving such petition. Pending such action, the State commission may suspend enforcement of the requirement or requirements to which the petition applies with respect to the petitioning carrier or carriers.
`(g) CONTINUED ENFORCEMENT OF EXCHANGE ACCESS AND INTERCONNECTION REQUIREMENTS- On and after the date of enactment of the Telecommunications Act of 1996, each local exchange carrier, to the extent that it provides wireline services, shall provide exchange access, information access, and exchange services for such access to interexchange carriers and information service providers in accordance with the same equal access and nondiscriminatory interconnection restrictions and obligations (including receipt of compensation) that apply to such carrier on the date immediately preceding the date of enactment of the Telecommunications Act of 1996 under any court order, consent decree, or regulation, order, or policy of the Commission, until such restrictions and obligations are explicitly superseded by regulations prescribed by the Commission after such date of enactment. During the period beginning on such date of enactment and until such restrictions and obligations are so superseded, such restrictions and obligations shall be enforceable in the same manner as regulations of the Commission.
`(h) DEFINITION OF INCUMBENT LOCAL EXCHANGE CARRIER-
`(1) DEFINITION- For purposes of this section, the term `incumbent local exchange carrier' means, with respect to an area, the local exchange carrier that--
`(A) on the date of enactment of the Telecommunications Act of 1996, provided telephone exchange service in such area; and
`(B)(i) on such date of enactment, was deemed to be a member of the exchange carrier association pursuant to section 69.601(b) of the Commission's regulations (47 C.F.R. 69.601(b)); or
`(ii) is a person or entity that, on or after such date of enactment, became a successor or assign of a member described in clause (i).
`(2) TREATMENT OF COMPARABLE CARRIERS AS INCUMBENTS- The Commission may, by rule, provide for the treatment of a local exchange carrier (or class or category thereof) as an incumbent local exchange carrier for purposes of this section if--
`(A) such carrier occupies a position in the market for telephone exchange service within an area that is comparable to the position occupied by a carrier described in paragraph (1);
`(B) such carrier has substantially replaced an incumbent local exchange carrier described in paragraph (1); and
`(C) such treatment is consistent with the public interest, convenience, and necessity and the purposes of this section.
`(i) SAVINGS PROVISION- Nothing in this section shall be construed to limit or otherwise affect the Commission's authority under section 201.

`SEC. 252. PROCEDURES FOR NEGOTIATION, ARBITRATION, AND APPROVAL OF AGREEMENTS.

`(a) AGREEMENTS ARRIVED AT THROUGH NEGOTIATION-
`(1) VOLUNTARY NEGOTIATIONS- Upon receiving a request for interconnection, services, or network elements pursuant to section 251, an incumbent local exchange carrier may negotiate and enter into a binding agreement with the requesting telecommunications carrier or carriers without regard to the standards set forth in subsections (b) and (c) of section 251.

The agreement shall include a detailed schedule of itemized charges for interconnection and each service or network element included in the agreement. The agreement, including any interconnection agreement negotiated before the date of enactment of the Telecommunications Act of 1996, shall be submitted to the State commission under subsection (e) of this section.

`(2) MEDIATION- Any party negotiating an agreement under this section may, at any point in the negotiation, ask a State commission to participate in the negotiation and to mediate any differences arising in the course of the negotiation.

`(b) AGREEMENTS ARRIVED AT THROUGH COMPULSORY ARBITRATION-

`(1) ARBITRATION- During the period from the 135th to the 160th day (inclusive) after the date on which an incumbent local exchange carrier receives a request for negotiation under this section, the carrier or any other party to the negotiation may petition a State commission to arbitrate any open issues.

`(2) DUTY OF PETITIONER-

`(A) A party that petitions a State commission under paragraph (1) shall, at the same time as it submits the petition, provide the State commission all relevant documentation concerning--

`(i) the unresolved issues;

`(ii) the position of each of the parties with respect to those issues; and

`(iii) any other issue discussed and resolved by the parties.

`(B) A party petitioning a State commission under paragraph (1) shall provide a copy of the petition and any documentation to the other party or parties not later than the day on which the State commission receives the petition.

`(3) OPPORTUNITY TO RESPOND- A non-petitioning party to a negotiation under this section may respond to the other party's petition and provide such additional information as it wishes within 25 days after the State commission receives the petition.

`(4) ACTION BY STATE COMMISSION-

`(A) The State commission shall limit its consideration of any petition under paragraph (1) (and any response thereto) to the issues set forth in the petition and in the response, if any, filed under paragraph (3).

`(B) The State commission may require the petitioning party and the responding party to provide such information as may be necessary for the State commission to reach a decision on the unresolved issues. If any party refuses or fails unreasonably to respond on a timely basis to any reasonable request from the State commission, then the State commission may proceed on the basis of the best information available to it from whatever source derived.

`(C) The State commission shall resolve each issue set forth in the petition and the response, if any, by imposing appropriate conditions as required to implement subsection (c) upon the parties to the agreement, and shall conclude the resolution of any unresolved issues not later than 9 months after the date on which the local exchange carrier received the request under this section.

`(5) REFUSAL TO NEGOTIATE- The refusal of any other party to the negotiation to participate further in the negotiations, to cooperate with the State commission in carrying out its function as an arbitrator, or to continue to negotiate in good faith in the presence, or with the assistance, of the State commission shall be considered a failure to negotiate in good faith.

`(c) STANDARDS FOR ARBITRATION- In resolving by arbitration under subsection (b) any open issues and imposing conditions upon the parties to the agreement, a State commission shall--

`(1) ensure that such resolution and conditions meet the requirements of section 251,

including the regulations prescribed by the Commission pursuant to section 251;
`(2) establish any rates for interconnection, services, or network elements according to subsection (d); and
`(3) provide a schedule for implementation of the terms and conditions by the parties to the agreement.
`(d) PRICING STANDARDS-
`(1) INTERCONNECTION AND NETWORK ELEMENT CHARGES- Determinations by a State commission of the just and reasonable rate for the interconnection of facilities and equipment for purposes of subsection (c)(2) of section 251, and the just and reasonable rate for network elements for purposes of subsection (c)(3) of such section--
`(A) shall be--
`(i) based on the cost (determined without reference to a rate-of-return or other rate-based proceeding) of providing the interconnection or network element (whichever is applicable), and
`(ii) nondiscriminatory, and
`(B) may include a reasonable profit.
`(2) CHARGES FOR TRANSPORT AND TERMINATION OF TRAFFIC-
`(A) IN GENERAL- For the purposes of compliance by an incumbent local exchange carrier with section 251(b)(5), a State commission shall not consider the terms and conditions for reciprocal compensation to be just and reasonable unless--
`(i) such terms and conditions provide for the mutual and reciprocal recovery by each carrier of costs associated with the transport and termination on each carrier's network facilities of calls that originate on the network facilities of the other carrier; and
`(ii) such terms and conditions determine such costs on the basis of a reasonable approximation of the additional costs of terminating such calls.
`(B) RULES OF CONSTRUCTION- This paragraph shall not be construed--
`(i) to preclude arrangements that afford the mutual recovery of costs through the offsetting of reciprocal obligations, including arrangements that waive mutual recovery (such as bill-and-keep arrangements); or
`(ii) to authorize the Commission or any State commission to engage in any rate regulation proceeding to establish with particularity the additional costs of transporting or terminating calls, or to require carriers to maintain records with respect to the additional costs of such calls.
`(3) WHOLESALE PRICES FOR TELECOMMUNICATIONS SERVICES- For the purposes of section 251(c)(4), a State commission shall determine wholesale rates on the basis of retail rates charged to subscribers for the telecommunications service requested, excluding the portion thereof attributable to any marketing, billing, collection, and other costs that will be avoided by the local exchange carrier.
`(e) APPROVAL BY STATE COMMISSION-
`(1) APPROVAL REQUIRED- Any interconnection agreement adopted by negotiation or arbitration shall be submitted for approval to the State commission. A State commission to which an agreement is submitted shall approve or reject the agreement, with written findings as to any deficiencies.
`(2) GROUNDS FOR REJECTION- The State commission may only reject--
`(A) an agreement (or any portion thereof) adopted by negotiation under subsection (a) if it finds that--
`(i) the agreement (or portion thereof) discriminates against a telecommunications carrier not a

party to the agreement; or
`(ii) the implementation of such agreement or portion is not consistent with the public interest, convenience, and necessity; or
`(B) an agreement (or any portion thereof) adopted by arbitration under subsection (b) if it finds that the agreement does not meet the requirements of section 251, including the regulations prescribed by the Commission pursuant to section 251, or the standards set forth in subsection (d) of this section.
`(3) PRESERVATION OF AUTHORITY- Notwithstanding paragraph (2), but subject to section 253, nothing in this section shall prohibit a State commission from establishing or enforcing other requirements of State law in its review of an agreement, including requiring compliance with intrastate telecommunications service quality standards or requirements.
`(4) SCHEDULE FOR DECISION- If the State commission does not act to approve or reject the agreement within 90 days after submission by the parties of an agreement adopted by negotiation under subsection (a), or within 30 days after submission by the parties of an agreement adopted by arbitration under subsection (b), the agreement shall be deemed approved. No State court shall have jurisdiction to review the action of a State commission in approving or rejecting an agreement under this section.
`(5) COMMISSION TO ACT IF STATE WILL NOT ACT- If a State commission fails to act to carry out its responsibility under this section in any proceeding or other matter under this section, then the Commission shall issue an order preempting the State commission's jurisdiction of that proceeding or matter within 90 days after being notified (or taking notice) of such failure, and shall assume the responsibility of the State commission under this section with respect to the proceeding or matter and act for the State commission.
`(6) REVIEW OF STATE COMMISSION ACTIONS- In a case in which a State fails to act as described in paragraph (5), the proceeding by the Commission under such paragraph and any judicial review of the Commission's actions shall be the exclusive remedies for a State commission's failure to act. In any case in which a State commission makes a determination under this section, any party aggrieved by such determination may bring an action in an appropriate Federal district court to determine whether the agreement or statement meets the requirements of section 251 and this section.
`(f) STATEMENTS OF GENERALLY AVAILABLE TERMS-
`(1) IN GENERAL- A Bell operating company may prepare and file with a State commission a statement of the terms and conditions that such company generally offers within that State to comply with the requirements of section 251 and the regulations thereunder and the standards applicable under this section.
`(2) STATE COMMISSION REVIEW- A State commission may not approve such statement unless such statement complies with subsection (d) of this section and section 251 and the regulations thereunder. Except as provided in section 253, nothing in this section shall prohibit a State commission from establishing or enforcing other requirements of State law in its review of such statement, including requiring compliance with intrastate telecommunications service quality standards or requirements.
`(3) SCHEDULE FOR REVIEW- The State commission to which a statement is submitted shall, not later than 60 days after the date of such submission--
`(A) complete the review of such statement under paragraph (2) (including any reconsideration thereof), unless the submitting carrier agrees to an extension of the period for such review; or
`(B) permit such statement to take effect.

`(4) AUTHORITY TO CONTINUE REVIEW- Paragraph (3) shall not preclude the State commission from continuing to review a statement that has been permitted to take effect under subparagraph (B) of such paragraph or from approving or disapproving such statement under paragraph (2).
`(5) DUTY TO NEGOTIATE NOT AFFECTED- The submission or approval of a statement under this subsection shall not relieve a Bell operating company of its duty to negotiate the terms and conditions of an agreement under section 251.
`(g) CONSOLIDATION OF STATE PROCEEDINGS- Where not inconsistent with the requirements of this Act, a State commission may, to the extent practical, consolidate proceedings under sections 214(e), 251(f), 253, and this section in order to reduce administrative burdens on telecommunications carriers, other parties to the proceedings, and the State commission in carrying out its responsibilities under this Act.
`(h) FILING REQUIRED- A State commission shall make a copy of each agreement approved under subsection (e) and each statement approved under subsection (f) available for public inspection and copying within 10 days after the agreement or statement is approved. The State commission may charge a reasonable and nondiscriminatory fee to the parties to the agreement or to the party filing the statement to cover the costs of approving and filing such agreement or statement.
`(i) AVAILABILITY TO OTHER TELECOMMUNICATIONS CARRIERS- A local exchange carrier shall make available any interconnection, service, or network element provided under an agreement approved under this section to which it is a party to any other requesting telecommunications carrier upon the same terms and conditions as those provided in the agreement.
`(j) DEFINITION OF INCUMBENT LOCAL EXCHANGE CARRIER- For purposes of this section, the term `incumbent local exchange carrier' has the meaning provided in section 251(h).

`SEC. 253. REMOVAL OF BARRIERS TO ENTRY.
`(a) IN GENERAL- No State or local statute or regulation, or other State or local legal requirement, may prohibit or have the effect of prohibiting the ability of any entity to provide any interstate or intrastate telecommunications service.
`(b) STATE REGULATORY AUTHORITY- Nothing in this section shall affect the ability of a State to impose, on a competitively neutral basis and consistent with section 254, requirements necessary to preserve and advance universal service, protect the public safety and welfare, ensure the continued quality of telecommunications services, and safeguard the rights of consumers.
`(c) STATE AND LOCAL GOVERNMENT AUTHORITY- Nothing in this section affects the authority of a State or local government to manage the public rights-of-way or to require fair and reasonable compensation from telecommunications providers, on a competitively neutral and nondiscriminatory basis, for use of public rights-of-way on a nondiscriminatory basis, if the compensation required is publicly disclosed by such government.
`(d) PREEMPTION- If, after notice and an opportunity for public comment, the Commission determines that a State or local government has permitted or imposed any statute, regulation, or legal requirement that violates subsection (a) or (b), the Commission shall preempt the enforcement of such statute, regulation, or legal requirement to the extent necessary to correct such violation or inconsistency.
`(e) COMMERCIAL MOBILE SERVICE PROVIDERS- Nothing in this section shall affect the

application of section 332(c)(3) to commercial mobile service providers.
`(f) RURAL MARKETS- It shall not be a violation of this section for a State to require a telecommunications carrier that seeks to provide telephone exchange service or exchange access in a service area served by a rural telephone company to meet the requirements in section 214(e)(1) for designation as an eligible telecommunications carrier for that area before being permitted to provide such service. This subsection shall not apply--
`(1) to a service area served by a rural telephone company that has obtained an exemption, suspension, or modification of section 251(c)(4) that effectively prevents a competitor from meeting the requirements of section 214(e)(1); and
`(2) to a provider of commercial mobile services.

`SEC. 256. COORDINATION FOR INTERCONNECTIVITY.
`(a) PURPOSE- It is the purpose of this section--
`(1) to promote nondiscriminatory accessibility by the broadest number of users and vendors of communications products and services to public telecommunications networks used to provide telecommunications service through--
`(A) coordinated public telecommunications network planning and design by telecommunications carriers and other providers of telecommunications service; and
`(B) public telecommunications network interconnectivity, and interconnectivity of devices with such networks used to provide telecommunications service; and
`(2) to ensure the ability of users and information providers to seamlessly and transparently transmit and receive information between and across telecommunications networks.
`(b) COMMISSION FUNCTIONS- In carrying out the purposes of this section, the Commission--
`(1) shall establish procedures for Commission oversight of coordinated network planning by telecommunications carriers and other providers of telecommunications service for the effective and efficient interconnection of public telecommunications networks used to provide telecommunications service; and
`(2) may participate, in a manner consistent with its authority and practice prior to the date of enactment of this section, in the development by appropriate industry standards-setting organizations of public telecommunications network interconnectivity standards that promote access to--
`(A) public telecommunications networks used to provide telecommunications service;
`(B) network capabilities and services by individuals with disabilities; and
`(C) information services by subscribers of rural telephone companies.
`(c) COMMISSION'S AUTHORITY- Nothing in this section shall be construed as expanding or limiting any authority that the Commission may have under law in effect before the date of enactment of the Telecommunications Act of 1996.
`(d) DEFINITION- As used in this section, the term `public telecommunications network interconnectivity' means the ability of two or more public telecommunications networks used to provide telecommunications service to communicate and exchange information without

degeneration, and to interact in concert with one another.

`SEC. 257. MARKET ENTRY BARRIERS PROCEEDING.
`(a) ELIMINATION OF BARRIERS- Within 15 months after the date of enactment of the Telecommunications Act of 1996, the Commission shall complete a proceeding for the purpose of identifying and eliminating, by regulations pursuant to its authority under this Act (other than this section), market entry barriers for entrepreneurs and other small businesses in the provision and ownership of telecommunications services and information services, or in the provision of parts or services to providers of telecommunications services and information services.
`(b) NATIONAL POLICY- In carrying out subsection (a), the Commission shall seek to promote the policies and purposes of this Act favoring diversity of media voices, vigorous economic competition, technological advancement, and promotion of the public interest, convenience, and necessity.
`(c) PERIODIC REVIEW- Every 3 years following the completion of the proceeding required by subsection (a), the Commission shall review and report to Congress on--
`(1) any regulations prescribed to eliminate barriers within its jurisdiction that are identified under subsection (a) and that can be prescribed consistent with the public interest, convenience, and necessity; and
`(2) the statutory barriers identified under subsection (a) that the Commission recommends be eliminated, consistent with the public interest, convenience, and necessity.

`SEC. 259. INFRASTRUCTURE SHARING.
`(a) REGULATIONS REQUIRED- The Commission shall prescribe, within one year after the date of enactment of the Telecommunications Act of 1996, regulations that require incumbent local exchange carriers (as defined in section 251(h)) to make available to any qualifying carrier such public switched network infrastructure, technology, information, and telecommunications facilities and functions as may be requested by such qualifying carrier for the purpose of enabling such qualifying carrier to provide telecommunications services, or to provide access to information services, in the service area in which such qualifying carrier has requested and obtained designation as an eligible telecommunications carrier under section 214(e).
`(b) TERMS AND CONDITIONS OF REGULATIONS- The regulations prescribed by the Commission pursuant to this section shall--
`(1) not require a local exchange carrier to which this section applies to take any action that is economically unreasonable or that is contrary to the public interest;
`(2) permit, but shall not require, the joint ownership or operation of public switched network infrastructure and services by or among such local exchange carrier and a qualifying carrier;
`(3) ensure that such local exchange carrier will not be treated by the Commission or any State as a common carrier for hire or as offering common carrier services with respect to any infrastructure, technology, information, facilities, or functions made available to a qualifying carrier in accordance with regulations issued pursuant to this section;
`(4) ensure that such local exchange carrier makes such infrastructure, technology, information, facilities, or functions available to a qualifying carrier on just and reasonable terms and conditions that permit such qualifying carrier to fully benefit from the economies of

scale and scope of such local exchange carrier, as determined in accordance with guidelines prescribed by the Commission in regulations issued pursuant to this section;
`(5) establish conditions that promote cooperation between local exchange carriers to which this section applies and qualifying carriers;
`(6) not require a local exchange carrier to which this section applies to engage in any infrastructure sharing agreement for any services or access which are to be provided or offered to consumers by the qualifying carrier in such local exchange carrier's telephone exchange area; and
`(7) require that such local exchange carrier file with the Commission or State for public inspection, any tariffs, contracts, or other arrangements showing the rates, terms, and conditions under which such carrier is making available public switched network infrastructure and functions under this section.
`(c) INFORMATION CONCERNING DEPLOYMENT OF NEW SERVICES AND EQUIPMENT- A local exchange carrier to which this section applies that has entered into an infrastructure sharing agreement under this section shall provide to each party to such agreement timely information on the planned deployment of telecommunications services and equipment, including any software or upgrades of software integral to the use or operation of such telecommunications equipment.
`(d) DEFINITION- For purposes of this section, the term `qualifying carrier' means a telecommunications carrier that--
`(1) lacks economies of scale or scope, as determined in accordance with regulations prescribed by the Commission pursuant to this section; and
`(2) offers telephone exchange service, exchange access, and any other service that is included in universal service, to all consumers without preference throughout the service area for which such carrier has been designated as an eligible telecommunications carrier under section 214(e).

`SEC. 261. EFFECT ON OTHER REQUIREMENTS.
`(a) COMMISSION REGULATIONS- Nothing in this part shall be construed to prohibit the Commission from enforcing regulations prescribed prior to the date of enactment of the Telecommunications Act of 1996 in fulfilling the requirements of this part, to the extent that such regulations are not inconsistent with the provisions of this part.
`(b) EXISTING STATE REGULATIONS- Nothing in this part shall be construed to prohibit any State commission from enforcing regulations prescribed prior to the date of enactment of the Telecommunications Act of 1996, or from prescribing regulations after such date of enactment, in fulfilling the requirements of this part, if such regulations are not inconsistent with the provisions of this part.
`(c) ADDITIONAL STATE REQUIREMENTS- Nothing in this part precludes a State from imposing requirements on a telecommunications carrier for intrastate services that are necessary to further competition in the provision of telephone exchange service or exchange access, as long as the State's requirements are not inconsistent with this part or the Commission's regulations to implement this part.'.
(b) DESIGNATION OF PART I- Title II of the Act is further amended by inserting before the heading of section 201 the following new heading:
`PART I--COMMON CARRIER REGULATION'.

(c) STYLISTIC CONSISTENCY- The Act is amended so that--
(1) the designation and heading of each title of the Act shall be in the form and typeface of the designation and heading of this title of this Act; and
(2) the designation and heading of each part of each title of the Act shall be in the form and typeface of the designation and heading of part I of title II of the Act, as amended by subsection (a).

SEC. 102. ELIGIBLE TELECOMMUNICATIONS CARRIERS.
(a) IN GENERAL- Section 214 (47 U.S.C. 214) is amended by adding at the end thereof the following new subsection:
`(e) PROVISION OF UNIVERSAL SERVICE-
`(1) ELIGIBLE TELECOMMUNICATIONS CARRIERS- A common carrier designated as an eligible telecommunications carrier under paragraph (2) or (3) shall be eligible to receive universal service support in accordance with section 254 and shall, throughout the service area for which the designation is received--
`(A) offer the services that are supported by Federal universal service support mechanisms under section 254(c), either using its own facilities or a combination of its own facilities and resale of another carrier's services (including the services offered by another eligible telecommunications carrier); and
`(B) advertise the availability of such services and the charges therefor using media of general distribution.
`(2) DESIGNATION OF ELIGIBLE TELECOMMUNICATIONS CARRIERS- A State commission shall upon its own motion or upon request designate a common carrier that meets the requirements of paragraph (1) as an eligible telecommunications carrier for a service area designated by the State commission. Upon request and consistent with the public interest, convenience, and necessity, the State commission may, in the case of an area served by a rural telephone company, and shall, in the case of all other areas, designate more than one common carrier as an eligible telecommunications carrier for a service area designated by the State commission, so long as each additional requesting carrier meets the requirements of paragraph (1). Before designating an additional eligible telecommunications carrier for an area served by a rural telephone company, the State commission shall find that the designation is in the public interest.
`(3) DESIGNATION OF ELIGIBLE TELECOMMUNICATIONS CARRIERS FOR UNSERVED AREAS- If no common carrier will provide the services that are supported by Federal universal service support mechanisms under section 254(c) to an unserved community or any portion thereof that requests such service, the Commission, with respect to interstate services, or a State commission, with respect to intrastate services, shall determine which common carrier or carriers are best able to provide such service to the requesting unserved community or portion thereof and shall order such carrier or carriers to provide such service for that unserved community or portion thereof. Any carrier or carriers ordered to provide such service under this paragraph shall meet the requirements of paragraph (1) and shall be designated as an eligible telecommunications carrier for that community or portion thereof.

SEC. 103. EXEMPT TELECOMMUNICATIONS COMPANIES.
The Public Utility Holding Company Act of 1935 (15 U.S.C. 79 and following) is amended by redesignating sections 34 and 35 as sections 35 and 36, respectively, and by inserting the following new section after section 33:

`SEC. 34. EXEMPT TELECOMMUNICATIONS COMPANIES.

`(a) DEFINITIONS- For purposes of this section--

`(1) EXEMPT TELECOMMUNICATIONS COMPANY- The term `exempt telecommunications company' means any person determined by the Federal Communications Commission to be engaged directly or indirectly, wherever located, through one or more affiliates (as defined in section 2(a)(11)(B)), and exclusively in the business of providing---

`(A) telecommunications services;

`(B) information services;

`(C) other services or products subject to the jurisdiction of the Federal Communications Commission; or

`(D) products or services that are related or incidental to the provision of a product or service described in subparagraph (A), (B), or (C). No person shall be deemed to be an exempt telecommunications company under this section unless such person has applied to the Federal Communications Commission for a determination under this paragraph. A person applying in good faith for such a determination shall be deemed an exempt telecommunications company under this section, with all of the exemptions provided by this section, until the Federal Communications Commission makes such determination. The Federal Communications Commission shall make such determination within 60 days of its receipt of any such application filed after the enactment of this section and shall notify the Commission whenever a determination is made under this paragraph that any person is an exempt telecommunications company. Not later than 12 months after the date of enactment of this section, the Federal Communications Commission shall promulgate rules implementing the provisions of this paragraph which shall be applicable to applications filed under this paragraph after the effective date of such rules.

`(2) OTHER TERMS- For purposes of this section, the terms `telecommunications services' and `information services' shall have the same meanings as provided in the Communications Act of 1934.

`(b) STATE CONSENT FOR SALE OF EXISTING RATE-BASED FACILITIES- If a rate or charge for the sale of electric energy or natural gas (other than any portion of a rate or charge which represents recovery of the cost of a wholesale rate or charge) for, or in connection with, assets of a public utility company that is an associate company or affiliate of a registered holding company was in effect under the laws of any State as of December 19, 1995, the public utility company owning such assets may not sell such assets to an exempt telecommunications company that is an associate company or affiliate unless State commissions having jurisdiction over such public utility company approve such sale. Nothing in this subsection shall preempt the otherwise applicable authority of any State to approve or disapprove the sale of such assets. The approval of the Commission under this Act shall not be required for the sale of assets as provided in this subsection.

`(c) OWNERSHIP OF ETCS BY EXEMPT HOLDING COMPANIES- Notwithstanding any provision of this Act, a holding company that is exempt under section 3 of this Act shall be permitted, without condition or limitation under this Act, to acquire and maintain an interest in the business of one or more exempt telecommunications companies.

`(d) OWNERSHIP OF ETCS BY REGISTERED HOLDING COMPANIES- Notwithstanding any provision of this Act, a registered holding company shall be permitted (without the need to apply for, or receive, approval from the Commission, and otherwise without condition under this Act) to acquire and hold the securities, or an interest in the business, of one or more exempt telecommunications companies.

SEC. 104. NONDISCRIMINATION PRINCIPLE.
Section 1 (47 U.S.C. 151) is amended by inserting after `to all the people of the United States' the following: `, without discrimination on the basis of race, color, religion, national origin, or sex,'.

SUBTITLE B--SPECIAL PROVISIONS CONCERNING BELL OPERATING COMPANIES
SEC. 151. BELL OPERATING COMPANY PROVISIONS.

(a) ESTABLISHMENT OF PART III OF TITLE II- Title II is amended by adding at the end of part II (as added by section 101) the following new part:

`PART III--SPECIAL PROVISIONS CONCERNING BELL OPERATING COMPANIES
`SEC. 271. BELL OPERATING COMPANY ENTRY INTO INTERLATA SERVICES.
`(a) GENERAL LIMITATION- Neither a Bell operating company, nor any affiliate of a Bell operating company, may provide interLATA services except as provided in this section.
`(b) INTERLATA SERVICES TO WHICH THIS SECTION APPLIES-
`(1) IN-REGION SERVICES- A Bell operating company, or any affiliate of that Bell operating company, may provide interLATA services originating in any of its in-region States (as defined in subsection (i)) if the Commission approves the application of such company for such State under subsection (d)(3).
`(2) OUT-OF-REGION SERVICES- A Bell operating company, or any affiliate of that Bell operating company, may provide interLATA services originating outside its in-region States after the date of enactment of the Telecommunications Act of 1996, subject to subsection (j).
`(3) INCIDENTAL INTERLATA SERVICES- A Bell operating company, or any affiliate of a Bell operating company, may provide incidental interLATA services (as defined in subsection (g)) originating in any State after the date of enactment of the Telecommunications Act of 1996.
`(4) TERMINATION- Nothing in this section prohibits a Bell operating company or any of its affiliates from providing termination for interLATA services, subject to subsection (j).
`(c) REQUIREMENTS FOR PROVIDING CERTAIN IN-REGION INTERLATA SERVICES-
`(1) AGREEMENT OR STATEMENT- A Bell operating company meets the requirements of this paragraph if it meets the requirements of subparagraph (A) or subparagraph (B) of this paragraph for each State for which the authorization is sought.
`(A) PRESENCE OF A FACILITIES-BASED COMPETITOR- A Bell operating company meets the requirements of this subparagraph if it has entered into one or more binding agreements that have been approved under section 252 specifying the terms and conditions under which the Bell operating company is providing access and interconnection to its

network facilities for the network facilities of one or more unaffiliated competing providers of telephone exchange service (as defined in section 3(47)(A), but excluding exchange access) to residential and business subscribers. For the purpose of this subparagraph, such telephone exchange service may be offered by such competing providers either exclusively over their own telephone exchange service facilities or predominantly over their own telephone exchange service facilities in combination with the resale of the telecommunications services of another carrier. For the purpose of this subparagraph, services provided pursuant to subpart K of part 22 of the Commission's regulations (47 C.F.R. 22.901 et seq.) shall not be considered to be telephone exchange services.
`(B) FAILURE TO REQUEST ACCESS- A Bell operating company meets the requirements of this subparagraph if, after 10 months after the date of enactment of the Telecommunications Act of 1996, no such provider has requested the access and interconnection described in subparagraph (A) before the date which is 3 months before the date the company makes its application under subsection (d)(1), and a statement of the terms and conditions that the company generally offers to provide such access and interconnection has been approved or permitted to take effect by the State commission under section 252(f). For purposes of this subparagraph, a Bell operating company shall be considered not to have received any request for access and interconnection if the State commission of such State certifies that the only provider or providers making such a request have (i) failed to negotiate in good faith as required by section 252, or (ii) violated the terms of an agreement approved under section 252 by the provider's failure to comply, within a reasonable period of time, with the implementation schedule contained in such agreement.
`(2) SPECIFIC INTERCONNECTION REQUIREMENTS-
`(A) AGREEMENT REQUIRED- A Bell operating company meets the requirements of this paragraph if, within the State for which the authorization is sought--
`(i)(I) such company is providing access and interconnection pursuant to one or more agreements described in paragraph (1)(A), or
`(II) such company is generally offering access and interconnection pursuant to a statement described in paragraph (1)(B), and
`(ii) such access and interconnection meets the requirements of subparagraph (B) of this paragraph.
`(B) COMPETITIVE CHECKLIST- Access or interconnection provided or generally offered by a Bell operating company to other telecommunications carriers meets the requirements of this subparagraph if such access and interconnection includes each of the following:
`(i) Interconnection in accordance with the requirements of sections 251(c)(2) and 252(d)(1).
`(ii) Nondiscriminatory access to network elements in accordance with the requirements of sections 251(c)(3) and 252(d)(1).
`(iii) Nondiscriminatory access to the poles, ducts, conduits, and rights-of-way owned or controlled by the Bell operating company at just and reasonable rates in accordance with the requirements of section 224.
`(iv) Local loop transmission from the central office to the customer's premises, unbundled from local switching or other services.
`(v) Local transport from the trunk side of a wireline local exchange carrier switch unbundled from switching or other services.
`(vi) Local switching unbundled from transport, local loop transmission, or other services.
`(vii) Nondiscriminatory access to--
`(I) 911 and E911 services;

`(II) directory assistance services to allow the other carrier's customers to obtain telephone numbers; and
`(III) operator call completion services.
`(viii) White pages directory listings for customers of the other carrier's telephone exchange service.
`(ix) Until the date by which telecommunications numbering administration guidelines, plan, or rules are established, nondiscriminatory access to telephone numbers for assignment to the other carrier's telephone exchange service customers. After that date, compliance with such guidelines, plan, or rules.
`(x) Nondiscriminatory access to databases and associated signaling necessary for call routing and completion.
`(xi) Until the date by which the Commission issues regulations pursuant to section 251 to require number portability, interim telecommunications number portability through remote call forwarding, direct inward dialing trunks, or other comparable arrangements, with as little impairment of functioning, quality, reliability, and convenience as possible. After that date, full compliance with such regulations.
`(xii) Nondiscriminatory access to such services or information as are necessary to allow the requesting carrier to implement local dialing parity in accordance with the requirements of section 251(b)(3).
`(xiii) Reciprocal compensation arrangements in accordance with the requirements of section 252(d)(2).
`(xiv) Telecommunications services are available for resale in accordance with the requirements of sections 251(c)(4) and 252(d)(3).
`(d) ADMINISTRATIVE PROVISIONS-
`(1) APPLICATION TO COMMISSION- On and after the date of enactment of the Telecommunications Act of 1996, a Bell operating company or its affiliate may apply to the Commission for authorization to provide interLATA services originating in any in-region State. The application shall identify each State for which the authorization is sought.
`(2) CONSULTATION-
`(A) CONSULTATION WITH THE ATTORNEY GENERAL- The Commission shall notify the Attorney General promptly of any application under paragraph (1). Before making any determination under this subsection, the Commission shall consult with the Attorney General, and if the Attorney General submits any comments in writing, such comments shall be included in the record of the Commission's decision. In consulting with and submitting comments to the Commission under this paragraph, the Attorney General shall provide to the Commission an evaluation of the application using any standard the Attorney General considers appropriate. The Commission shall give substantial weight to the Attorney General's evaluation, but such evaluation shall not have any preclusive effect on any Commission decision under paragraph (3).
`(B) CONSULTATION WITH STATE COMMISSIONS- Before making any determination under this subsection, the Commission shall consult with the State commission of any State that is the subject of the application in order to verify the compliance of the Bell operating company with the requirements of subsection (c).
`(3) DETERMINATION- Not later than 90 days after receiving an application under paragraph (1), the Commission shall issue a written determination approving or denying the authorization requested in the application for each State. The Commission shall not approve the authorization requested in an application submitted under paragraph (1) unless it finds

that--
`(A) the petitioning Bell operating company has met the requirements of subsection (c)(1) and--
`(i) with respect to access and interconnection provided pursuant to subsection (c)(1)(A), has fully implemented the competitive checklist in subsection (c)(2)(B); or
`(ii) with respect to access and interconnection generally offered pursuant to a statement under subsection (c)(1)(B), such statement offers all of the items included in the competitive checklist in subsection (c)(2)(B);
`(B) the requested authorization will be carried out in accordance with the requirements of section 272; and
`(C) the requested authorization is consistent with the public interest, convenience, and necessity. The Commission shall state the basis for its approval or denial of the application.
`(4) LIMITATION ON COMMISSION- The Commission may not, by rule or otherwise, limit or extend the terms used in the competitive checklist set forth in subsection (c)(2)(B).
`(5) PUBLICATION- Not later than 10 days after issuing a determination under paragraph (3), the Commission shall publish in the Federal Register a brief description of the determination.
`(6) ENFORCEMENT OF CONDITIONS-
`(A) COMMISSION AUTHORITY- If at any time after the approval of an application under paragraph (3), the Commission determines that a Bell operating company has ceased to meet any of the conditions required for such approval, the Commission may, after notice and opportunity for a hearing--
`(i) issue an order to such company to correct the deficiency;
`(ii) impose a penalty on such company pursuant to title V; or
`(iii) suspend or revoke such approval.
`(B) RECEIPT AND REVIEW OF COMPLAINTS- The Commission shall establish procedures for the review of complaints concerning failures by Bell operating companies to meet conditions required for approval under paragraph (3). Unless the parties otherwise agree, the Commission shall act on such complaint within 90 days.
`(e) LIMITATIONS-
`(1) JOINT MARKETING OF LOCAL AND LONG DISTANCE SERVICES- a Bell operating company is authorized pursuant to subsection (d) to provide interLATA services in an in-region State, or until 36 months have passed since the date of enactment of the Telecommunications Act of 1996, whichever is earlier, a telecommunications carrier that serves greater than 5 percent of the Nation's presubscribed access lines may not jointly market in such State telephone exchange service obtained from such company pursuant to section 251(c)(4) with interLATA services offered by that telecommunications carrier.
`(2) INTRALATA TOLL DIALING PARITY-
`(A) PROVISION REQUIRED- A Bell operating company granted authority to provide interLATA services under subsection (d) shall provide intraLATA toll dialing parity throughout that State coincident with its exercise of that authority.
`(B) LIMITATION- Except for single-LATA States and States that have issued an order by December 19, 1995, requiring a Bell operating company to implement intraLATA toll dialing parity, a State may not require a Bell operating company to implement intraLATA toll dialing parity in that State before a Bell operating company has been granted authority under this section to provide interLATA services originating in that State or before 3 years after the date of enactment of the Telecommunications Act of 1996, whichever is earlier. Nothing in this

subparagraph precludes a State from issuing an order requiring intraLATA toll dialing parity in that State prior to either such date so long as such order does not take effect until after the earlier of either such dates.

`(f) EXCEPTION FOR PREVIOUSLY AUTHORIZED ACTIVITIES- Neither subsection (a) nor section 273 shall prohibit a Bell operating company or affiliate from engaging, at any time after the date of enactment of the Telecommunications Act of 1996, in any activity to the extent authorized by, and subject to the terms and conditions contained in, an order entered by the United States District Court for the District of Columbia pursuant to section VII or VIII(C) of the AT&T Consent Decree if such order was entered on or before such date of enactment, to the extent such order is not reversed or vacated on appeal. Nothing in this subsection shall be construed to limit, or to impose terms or conditions on, an activity in which a Bell operating company is otherwise authorized to engage under any other provision of this section.

`(g) DEFINITION OF INCIDENTAL INTERLATA SERVICES- For purposes of this section, the term `incidental interLATA services' means the interLATA provision by a Bell operating company or its affiliate--

`(1)(A) of audio programming, video programming, or other programming services to subscribers to such services of such company or affiliate;

`(B) of the capability for interaction by such subscribers to select or respond to such audio programming, video programming, or other programming services;

`(C) to distributors of audio programming or video programming that such company or affiliate owns or controls, or is licensed by the copyright owner of such programming (or by an assignee of such owner) to distribute; or

`(D) of alarm monitoring services;

`(2) of two-way interactive video services or Internet services over dedicated facilities to or for elementary and secondary schools as defined in section 254(h)(5);

`(3) of commercial mobile services in accordance with section 332(c) of this Act and with the regulations prescribed by the Commission pursuant to paragraph (8) of such section;

`(4) of a service that permits a customer that is located in one LATA to retrieve stored information from, or file information for storage in, information storage facilities of such company that are located in another LATA;

`(5) of signaling information used in connection with the provision of telephone exchange services or exchange access by a local exchange carrier; or

`(6) of network control signaling information to, and receipt of such signaling information from, common carriers offering interLATA services at any location within the area in which such Bell operating company provides telephone exchange services or exchange access.

`(h) LIMITATIONS- The provisions of subsection (g) are intended to be narrowly construed. The interLATA services provided under subparagraph (A), (B), or (C) of subsection (g)(1) are limited to those interLATA transmissions incidental to the provision by a Bell operating company or its affiliate of video, audio, and other programming services that the company or its affiliate is engaged in providing to the public. The Commission shall ensure that the provision of services authorized under subsection (g) by a Bell operating company or its affiliate will not adversely affect telephone exchange service ratepayers or competition in any telecommunications market.

`(i) ADDITIONAL DEFINITIONS- As used in this section--

`(1) IN-REGION STATE- The term `in-region State' means a State in which a Bell operating company or any of its affiliates was authorized to provide wireline telephone exchange service

pursuant to the reorganization plan approved under the AT&T Consent Decree, as in effect on the day before the date of enactment of the Telecommunications Act of 1996.
`(2) AUDIO PROGRAMMING SERVICES- The term `audio programming services' means programming provided by, or generally considered to be comparable to programming provided by, a radio broadcast station.
`(3) VIDEO PROGRAMMING SERVICES; OTHER PROGRAMMING SERVICES- The terms `video programming service' and `other programming services' have the same meanings as such terms have under section 602 of this Act.
`(j) CERTAIN SERVICE APPLICATIONS TREATED AS IN-REGION SERVICE APPLICATIONS- For purposes of this section, a Bell operating company application to provide 800 service, private line service, or their equivalents that--
`(1) terminate in an in-region State of that Bell operating company, and
`(2) allow the called party to determine the interLATA carrier, shall be considered an in-region service subject to the requirements of subsection (b)(1).

`SEC. 272. SEPARATE AFFILIATE; SAFEGUARDS.
`(a) SEPARATE AFFILIATE REQUIRED FOR COMPETITIVE ACTIVITIES-
`(1) IN GENERAL- A Bell operating company (including any affiliate) which is a local exchange carrier that is subject to the requirements of section 251(c) may not provide any service described in paragraph (2) unless it provides that service through one or more affiliates that--
`(A) are separate from any operating company entity that is subject to the requirements of section 251(c); and
`(B) meet the requirements of subsection (b).
`(2) SERVICES FOR WHICH A SEPARATE AFFILIATE IS REQUIRED- The services for which a separate affiliate is required by paragraph (1) are:
`(A) Manufacturing activities (as defined in section 273(h)).
`(B) Origination of interLATA telecommunications services, other than--
`(i) incidental interLATA services described in paragraphs (1), (2), (3), (5), and (6) of section 271(g);
`(ii) out-of-region services described in section 271(b)(2); or
`(iii) previously authorized activities described in section 271(f).
`(C) InterLATA information services, other than electronic publishing (as defined in section 274(h)) and alarm monitoring services (as defined in section 275(e)).
`(b) STRUCTURAL AND TRANSACTIONAL REQUIREMENTS- The separate affiliate required by this section--
`(1) shall operate independently from the Bell operating company;
`(2) shall maintain books, records, and accounts in the manner prescribed by the Commission which shall be separate from the books, records, and accounts maintained by the Bell operating company of which it is an affiliate;
`(3) shall have separate officers, directors, and employees from the Bell operating company of which it is an affiliate;
`(4) may not obtain credit under any arrangement that would permit a creditor, upon default, to have recourse to the assets of the Bell operating company; and
`(5) shall conduct all transactions with the Bell operating company of which it is an affiliate on an arm's length basis with any such transactions reduced to writing and available for public inspection.

`(c) NONDISCRIMINATION SAFEGUARDS- In its dealings with its affiliate described in subsection (a), a Bell operating company--
`(1) may not discriminate between that company or affiliate and any other entity in the provision or procurement of goods, services, facilities, and information, or in the establishment of standards; and
`(2) shall account for all transactions with an affiliate described in subsection (a) in accordance with accounting principles designated or approved by the Commission.
`(d) BIENNIAL AUDIT-
`(1) GENERAL REQUIREMENT- A company required to operate a separate affiliate under this section shall obtain and pay for a joint Federal/State audit every 2 years conducted by an independent auditor to determine whether such company has complied with this section and the regulations promulgated under this section, and particularly whether such company has complied with the separate accounting requirements under subsection (b).
`(2) RESULTS SUBMITTED TO COMMISSION; STATE COMMISSIONS- The auditor described in paragraph (1) shall submit the results of the audit to the Commission and to the State commission of each State in which the company audited provides service, which shall make such results available for public inspection. Any party may submit comments on the final audit report.
`(3) ACCESS TO DOCUMENTS- For purposes of conducting audits and reviews under this subsection--
`(A) the independent auditor, the Commission, and the State commission shall have access to the financial accounts and records of each company and of its affiliates necessary to verify transactions conducted with that company that are relevant to the specific activities permitted under this section and that are necessary for the regulation of rates;
`(B) the Commission and the State commission shall have access to the working papers and supporting materials of any auditor who performs an audit under this section; and
`(C) the State commission shall implement appropriate procedures to ensure the protection of any proprietary information submitted to it under this section.
`(e) FULFILLMENT OF CERTAIN REQUESTS- A Bell operating company and an affiliate that is subject to the requirements of section 251(c)--
`(1) shall fulfill any requests from an unaffiliated entity for telephone exchange service and exchange access within a period no longer than the period in which it provides such telephone exchange service and exchange access to itself or to its affiliates;
`(2) shall not provide any facilities, services, or information concerning its provision of exchange access to the affiliate described in subsection (a) unless such facilities, services, or information are made available to other providers of interLATA services in that market on the same terms and conditions;
`(3) shall charge the affiliate described in subsection (a), or impute to itself (if using the access for its provision of its own services), an amount for access to its telephone exchange service and exchange access that is no less than the amount charged to any unaffiliated interexchange carriers for such service; and
`(4) may provide any interLATA or intraLATA facilities or services to its interLATA affiliate if such services or facilities are made available to all carriers at the same rates and on the same terms and conditions, and so long as the costs are appropriately allocated.
`(f) SUNSET-
`(1) MANUFACTURING AND LONG DISTANCE- The provisions of this section (other than subsection (e)) shall cease to apply with respect to the manufacturing activities or the

interLATA telecommunications services of a Bell operating company 3 years after the date such Bell operating company or any Bell operating company affiliate is authorized to provide interLATA telecommunications services under section 271(d), unless the Commission extends such 3-year period by rule or order.
`(2) INTERLATA INFORMATION SERVICES- The provisions of this section (other than subsection (e)) shall cease to apply with respect to the interLATA information services of a Bell operating company 4 years after the date of enactment of the Telecommunications Act of 1996, unless the Commission extends such 4-year period by rule or order.
`(3) PRESERVATION OF EXISTING AUTHORITY- Nothing in this subsection shall be construed to limit the authority of the Commission under any other section of this Act to prescribe safeguards consistent with the public interest, convenience, and necessity.
`(g) JOINT MARKETING-
`(1) AFFILIATE SALES OF TELEPHONE EXCHANGE SERVICES- A Bell operating company affiliate required by this section may not market or sell telephone exchange services provided by the Bell operating company unless that company permits other entities offering the same or similar service to market and sell its telephone exchange services.
`(2) BELL OPERATING COMPANY SALES OF AFFILIATE SERVICES- A Bell operating company may not market or sell interLATA service provided by an affiliate required by this section within any of its in-region States until such company is authorized to provide interLATA services in such State under section 271(d).
`(3) RULE OF CONSTRUCTION- The joint marketing and sale of services permitted under this subsection shall not be considered to violate the nondiscrimination provisions of subsection (c).
`(h) TRANSITION- With respect to any activity in which a Bell operating company is engaged on the date of enactment of the Telecommunications Act of 1996, such company shall have one year from such date of enactment to comply with the requirements of this section.

`SEC. 273. MANUFACTURING BY BELL OPERATING COMPANIES.
`(a) AUTHORIZATION- A Bell operating company may manufacture and provide telecommunications equipment, and manufacture customer premises equipment, if the Commission authorizes that Bell operating company or any Bell operating company affiliate to provide interLATA services under section 271(d), subject to the requirements of this section and the regulations prescribed thereunder, except that neither a Bell operating company nor any of its affiliates may engage in such manufacturing in conjunction with a Bell operating company not so affiliated or any of its affiliates.
`(b) COLLABORATION; RESEARCH AND ROYALTY AGREEMENTS-
`(1) COLLABORATION- Subsection (a) shall not prohibit a Bell operating company from engaging in close collaboration with any manufacturer of customer premises equipment or telecommunications equipment during the design and development of hardware, software, or combinations thereof related to such equipment.
`(2) CERTAIN RESEARCH ARRANGEMENTS; ROYALTY AGREEMENTS- Subsection (a) shall not prohibit a Bell operating company from--
`(A) engaging in research activities related to manufacturing, and
`(B) entering into royalty agreements with manufacturers of telecommunications equipment.
`(c) INFORMATION REQUIREMENTS-
`(1) INFORMATION ON PROTOCOLS AND TECHNICAL REQUIREMENTS- Each Bell operating company shall, in accordance with regulations prescribed by the

Commission, maintain and file with the Commission full and complete information with respect to the protocols and technical requirements for connection with and use of its telephone exchange service facilities. Each such company shall report promptly to the Commission any material changes or planned changes to such protocols and requirements, and the schedule for implementation of such changes or planned changes.
`(2) DISCLOSURE OF INFORMATION- A Bell operating company shall not disclose any information required to be filed under paragraph (1) unless that information has been filed promptly, as required by regulation by the Commission.
`(3) ACCESS BY COMPETITORS TO INFORMATION- The Commission may prescribe such additional regulations under this subsection as may be necessary to ensure that manufacturers have access to the information with respect to the protocols and technical requirements for connection with and use of telephone exchange service facilities that a Bell operating company makes available to any manufacturing affiliate or any unaffiliated manufacturer.
`(4) PLANNING INFORMATION- Each Bell operating company shall provide, to interconnecting carriers providing telephone exchange service, timely information on the planned deployment of telecommunications equipment.

`(5) ALTERNATE DISPUTE RESOLUTION- Within 90 days after the date of enactment of the Telecommunications Act of 1996, the Commission shall prescribe a dispute resolution process to be utilized in the event that a dispute resolution process is not agreed upon by all the parties when establishing and publishing any industry-wide standard or industry-wide generic requirement for telecommunications equipment or customer premises equipment, pursuant to paragraph (4)(A)(v). The Commission shall not establish itself as a party to the dispute resolution process. Such dispute resolution process shall permit any funding party to resolve a dispute with the entity conducting the activity that significantly affects such funding party's interests, in an open, nondiscriminatory, and unbiased fashion, within 30 days after the filing of such dispute. Such disputes may be filed within 15 days after the date the funding party receives a response to its comments from the entity conducting the activity. The Commission shall establish penalties to be assessed for delays caused by referral of frivolous disputes to the dispute resolution process.
`(6) SUNSET- The requirements of paragraphs (3) and (4) shall terminate for the particular relevant activity when the Commission determines that there are alternative sources of industry-wide standards, industry-wide generic requirements, or product certification for a particular class of telecommunications equipment or customer premises equipment available in the United States. Alternative sources shall be deemed to exist when such sources provide commercially viable alternatives that are providing such services to customers. The Commission shall act on any application for such a determination within 90 days after receipt of such application, and shall receive public comment on such application.
`(7) ADMINISTRATION AND ENFORCEMENT AUTHORITY- For the purposes of administering this subsection and the regulations prescribed thereunder, the Commission shall have the same remedial authority as the Commission has in administering and enforcing the provisions of this title with respect to any common carrier subject to this Act.
`(8) DEFINITIONS- For purposes of this subsection:
`(A) The term `affiliate' shall have the same meaning as in section 3 of this Act, except that, for purposes of paragraph (1)(B)--
`(i) an aggregate voting equity interest in Bell Communications Research, Inc., of at least 5

percent of its total voting equity, owned directly or indirectly by more than 1 otherwise unaffiliated Bell operating company, shall constitute an affiliate relationship; and

`(ii) a voting equity interest in Bell Communications Research, Inc., by any otherwise unaffiliated Bell operating company of less than 1 percent of Bell Communications Research's total voting equity shall not be considered to be an equity interest under this paragraph.

`(B) The term `generic requirement' means a description of acceptable product attributes for use by local exchange carriers in establishing product specifications for the purchase of telecommunications equipment, customer premises equipment, and software integral thereto.

`(C) The term `industry-wide' means activities funded by or performed on behalf of local exchange carriers for use in providing wireline telephone exchange service whose combined total of deployed access lines in the United States constitutes at least 30 percent of all access lines deployed by telecommunications carriers in the United States as of the date of enactment of the Telecommunications Act of 1996.

`(D) The term `certification' means any technical process whereby a party determines whether a product, for use by more than one local exchange carrier, conforms with the specified requirements pertaining to such product.

`(E) The term `accredited standards development organization' means an entity composed of industry members which has been accredited by an institution vested with the responsibility for standards accreditation by the industry.

`(e) BELL OPERATING COMPANY EQUIPMENT PROCUREMENT AND SALES-

`(1) NONDISCRIMINATION STANDARDS FOR MANUFACTURING- In the procurement or awarding of supply contracts for telecommunications equipment, a Bell operating company, or any entity acting on its behalf, for the duration of the requirement for a separate subsidiary including manufacturing under this Act--

`(A) shall consider such equipment, produced or supplied by unrelated persons; and

`(B) may not discriminate in favor of equipment produced or supplied by an affiliate or related person.

`(2) PROCUREMENT STANDARDS- Each Bell operating company or any entity acting on its behalf shall make procurement decisions and award all supply contracts for equipment, services, and software on the basis of an objective assessment of price, quality, delivery, and other commercial factors.

`(3) NETWORK PLANNING AND DESIGN- A Bell operating company shall, to the extent consistent with the antitrust laws, engage in joint network planning and design with local exchange carriers operating in the same area of interest. No participant in such planning shall be allowed to delay the introduction of new technology or the deployment of facilities to provide telecommunications services, and agreement with such other carriers shall not be required as a prerequisite for such introduction or deployment.

`(4) SALES RESTRICTIONS- Neither a Bell operating company engaged in manufacturing nor a manufacturing affiliate of such a company shall restrict sales to any local exchange carrier of telecommunications equipment, including software integral to the operation of such equipment and related upgrades.

`(5) PROTECTION OF PROPRIETARY INFORMATION- A Bell operating company and any entity it owns or otherwise controls shall protect the proprietary information submitted for procurement decisions from release not specifically authorized by the owner of such information.

`(f) ADMINISTRATION AND ENFORCEMENT AUTHORITY- For the purposes of administering and enforcing the provisions of this section and the regulations prescribed

thereunder, the Commission shall have the same authority, power, and functions with respect to any Bell operating company or any affiliate thereof as the Commission has in administering and enforcing the provisions of this title with respect to any common carrier subject to this Act.
`(g) ADDITIONAL RULES AND REGULATIONS- The Commission may prescribe such additional rules and regulations as the Commission determines are necessary to carry out the provisions of this section, and otherwise to prevent discrimination and cross-subsidization in a Bell operating company's dealings with its affiliate and with third parties.
`(h) DEFINITION- As used in this section, the term `manufacturing' has the same meaning as such term has under the AT&T Consent Decree.

`SEC. 274. ELECTRONIC PUBLISHING BY BELL OPERATING COMPANIES.
`(a) LIMITATIONS- No Bell operating company or any affiliate may engage in the provision of electronic publishing that is disseminated by means of such Bell operating company's or any of its affiliates' basic telephone service, except that nothing in this section shall prohibit a separated affiliate or electronic publishing joint venture operated in accordance with this section from engaging in the provision of electronic publishing.
`(b) SEPARATED AFFILIATE OR ELECTRONIC PUBLISHING JOINT VENTURE REQUIREMENTS- A separated affiliate or electronic publishing joint venture shall be operated independently from the Bell operating company. Such separated affiliate or joint venture and the Bell operating company with which it is affiliated shall--
`(1) maintain separate books, records, and accounts and prepare separate financial statements;
`(2) not incur debt in a manner that would permit a creditor of the separated affiliate or joint venture upon default to have recourse to the assets of the Bell operating company;
`(3) carry out transactions (A) in a manner consistent with such independence, (B) pursuant to written contracts or tariffs that are filed with the Commission and made publicly available, and (C) in a manner that is auditable in accordance with generally accepted auditing standards;
`(4) value any assets that are transferred directly or indirectly from the Bell operating company to a separated affiliate or joint venture, and record any transactions by which such assets are transferred, in accordance with such regulations as may be prescribed by the Commission or a State commission to prevent improper cross subsidies;
`(5) between a separated affiliate and a Bell operating company--
`(A) have no officers, directors, and employees in common after the effective date of this section; and
`(B) own no property in common;
`(6) not use for the marketing of any product or service of the separated affiliate or joint venture, the name, trademarks, or service marks of an existing Bell operating company except for names, trademarks, or service marks that are owned by the entity that owns or controls the Bell operating company;
`(7) not permit the Bell operating company--
`(A) to perform hiring or training of personnel on behalf of a separated affiliate;
`(B) to perform the purchasing, installation, or maintenance of equipment on behalf of a separated affiliate, except for telephone service that it provides under tariff or contract subject to the provisions of this section; or
`(C) to perform research and development on behalf of a separated affiliate;
`(8) each have performed annually a compliance review--

`(A) that is conducted by an independent entity for the purpose of determining compliance during the preceding calendar year with any provision of this section; and
`(B) the results of which are maintained by the separated affiliate or joint venture and the Bell operating company for a period of 5 years subject to review by any lawful authority; and
`(9) within 90 days of receiving a review described in paragraph (8), file a report of any exceptions and corrective action with the Commission and allow any person to inspect and copy such report subject to reasonable safeguards to protect any proprietary information contained in such report from being used for purposes other than to enforce or pursue remedies under this section.
`(c) JOINT MARKETING-
`(1) IN GENERAL- Except as provided in paragraph (2)--
`(A) a Bell operating company shall not carry out any promotion, marketing, sales, or advertising for or in conjunction with a separated affiliate; and
`(B) a Bell operating company shall not carry out any promotion, marketing, sales, or advertising for or in conjunction with an affiliate that is related to the provision of electronic publishing.
`(2) PERMISSIBLE JOINT ACTIVITIES-
`(A) JOINT TELEMARKETING- A Bell operating company may provide inbound telemarketing or referral services related to the provision of electronic publishing for a separated affiliate, electronic publishing joint venture, affiliate, or unaffiliated electronic publisher: [Italic->] Provided, [<-Italic] That if such services are provided to a separated affiliate, electronic publishing joint venture, or affiliate, such services shall be made available to all electronic publishers on request, on nondiscriminatory terms.
`(B) TEAMING ARRANGEMENTS- A Bell operating company may engage in nondiscriminatory teaming or business arrangements to engage in electronic publishing with any separated affiliate or with any other electronic publisher if (i) the Bell operating company only provides facilities, services, and basic telephone service information as authorized by this section, and (ii) the Bell operating company does not own such teaming or business arrangement.
`(C) ELECTRONIC PUBLISHING JOINT VENTURES- A Bell operating company or affiliate may participate on a nonexclusive basis in electronic publishing joint ventures with entities that are not a Bell operating company, affiliate, or separated affiliate to provide electronic publishing services, if the Bell operating company or affiliate has not more than a 50 percent direct or indirect equity interest (or the equivalent thereof) or the right to more than 50 percent of the gross revenues under a revenue sharing or royalty agreement in any electronic publishing joint venture. Officers and employees of a Bell operating company or affiliate participating in an electronic publishing joint venture may not have more than 50 percent of the voting control over the electronic publishing joint venture. In the case of joint ventures with small, local electronic publishers, the Commission for good cause shown may authorize the Bell operating company or affiliate to have a larger equity interest, revenue share, or voting control but not to exceed 80 percent. A Bell operating company participating in an electronic publishing joint venture may provide promotion, marketing, sales, or advertising personnel and services to such joint venture.
`(d) BELL OPERATING COMPANY REQUIREMENT- A Bell operating company under common ownership or control with a separated affiliate or electronic publishing joint venture shall provide network access and interconnections for basic telephone service to electronic publishers at just and reasonable rates that are tariffed (so long as rates for such services are

subject to regulation) and that are not higher on a per-unit basis than those charged for such services to any other electronic publisher or any separated affiliate engaged in electronic publishing.
`(e) PRIVATE RIGHT OF ACTION-
`(1) DAMAGES- Any person claiming that any act or practice of any Bell operating company, affiliate, or separated affiliate constitutes a violation of this section may file a complaint with the Commission or bring suit as provided in section 207 of this Act, and such Bell operating company, affiliate, or separated affiliate shall be liable as provided in section 206 of this Act; except that damages may not be awarded for a violation that is discovered by a compliance review as required by subsection (b)(7) of this section and corrected within 90 days.
`(2) CEASE AND DESIST ORDERS- In addition to the provisions of paragraph (1), any person claiming that any act or practice of any Bell operating company, affiliate, or separated affiliate constitutes a violation of this section may make application to the Commission for an order to cease and desist such violation or may make application in any district court of the United States of competent jurisdiction for an order enjoining such acts or practices or for an order compelling compliance with such requirement.
`(f) SEPARATED AFFILIATE REPORTING REQUIREMENT- Any separated affiliate under this section shall file with the Commission annual reports in a form substantially equivalent to the Form 10-K required by regulations of the Securities and Exchange Commission.
`(g) EFFECTIVE DATES-
`(1) TRANSITION- Any electronic publishing service being offered to the public by a Bell operating company or affiliate on the date of enactment of the Telecommunications Act of 1996 shall have one year from such date of enactment to comply with the requirements of this section.
`(2) SUNSET- The provisions of this section shall not apply to conduct occurring after 4 years after the date of enactment of the Telecommunications Act of 1996.
`(h) DEFINITION OF ELECTRONIC PUBLISHING-
`(1) IN GENERAL- The term `electronic publishing' means the dissemination, provision, publication, or sale to an unaffiliated entity or person, of any one or more of the following: news (including sports); entertainment (other than interactive games); business, financial, legal, consumer, or credit materials; editorials, columns, or features; advertising; photos or images; archival or research material; legal notices or public records; scientific, educational, instructional, technical, professional, trade, or other literary materials; or other like or similar information.
`(2) EXCEPTIONS- The term `electronic publishing' shall not include the following services:
`(A) Information access, as that term is defined by the AT&T Consent Decree.
`(B) The transmission of information as a common carrier.
`(C) The transmission of information as part of a gateway to an information service that does not involve the generation or alteration of the content of information, including data transmission, address translation, protocol conversion, billing management, introductory information content, and navigational systems that enable users to access electronic publishing services, which do not affect the presentation of such electronic publishing services to users.
`(D) Voice storage and retrieval services, including voice messaging and electronic mail services.

`(E) Data processing or transaction processing services that do not involve the generation or alteration of the content of information.
`(F) Electronic billing or advertising of a Bell operating company's regulated telecommunications services.
`(G) Language translation or data format conversion.
`(H) The provision of information necessary for the management, control, or operation of a telephone company telecommunications system.
`(I) The provision of directory assistance that provides names, addresses, and telephone numbers and does not include advertising.
`(J) Caller identification services.
`(K) Repair and provisioning databases and credit card and billing validation for telephone company operations.
`(L) 911-E and other emergency assistance databases.
`(M) Any other network service of a type that is like or similar to these network services and that does not involve the generation or alteration of the content of information.
`(N) Any upgrades to these network services that do not involve the generation or alteration of the content of information.
`(O) Video programming or full motion video entertainment on demand.
`(i) ADDITIONAL DEFINITIONS- As used in this section--
`(1) The term `affiliate' means any entity that, directly or indirectly, owns or controls, is owned or controlled by, or is under common ownership or control with, a Bell operating company. Such term shall not include a separated affiliate.
`(2) The term `basic telephone service' means any wireline telephone exchange service, or wireline telephone exchange service facility, provided by a Bell operating company in a telephone exchange area, except that such term does not include--
`(A) a competitive wireline telephone exchange service provided in a telephone exchange area where another entity provides a wireline telephone exchange service that was provided on January 1, 1984, or
`(B) a commercial mobile service.
`(3) The term `basic telephone service information' means network and customer information of a Bell operating company and other information acquired by a Bell operating company as a result of its engaging in the provision of basic telephone service.
`(4) The term `control' has the meaning that it has in 17 C.F.R. 240.12b-2, the regulations promulgated by the Securities and Exchange Commission pursuant to the Securities Exchange Act of 1934 (15 U.S.C. 78a et seq.) or any successor provision to such section.
`(5) The term `electronic publishing joint venture' means a joint venture owned by a Bell operating company or affiliate that engages in the provision of electronic publishing which is disseminated by means of such Bell operating company's or any of its affiliates' basic telephone service.
`(6) The term `entity' means any organization, and includes corporations, partnerships, sole proprietorships, associations, and joint ventures.
`(7) The term `inbound telemarketing' means the marketing of property, goods, or services by telephone to a customer or potential customer who initiated the call.
`(8) The term `own' with respect to an entity means to have a direct or indirect equity interest (or the equivalent thereof) of more than 10 percent of an entity, or the right to more than 10 percent of the gross revenues of an entity under a revenue sharing or royalty agreement.
`(9) The term `separated affiliate' means a corporation under common ownership or control

with a Bell operating company that does not own or control a Bell operating company and is not owned or controlled by a Bell operating company and that engages in the provision of electronic publishing which is disseminated by means of such Bell operating company's or any of its affiliates' basic telephone service.
`(10) The term `Bell operating company' has the meaning provided in section 3, except that such term includes any entity or corporation that is owned or controlled by such a company (as so defined) but does not include an electronic publishing joint venture owned by such an entity or corporation.

TITLE II--BROADCAST SERVICES
SEC. 201. BROADCAST SPECTRUM FLEXIBILITY.
Title III is amended by inserting after section 335 (47 U.S.C. 335) the following new section:

`SEC. 336. BROADCAST SPECTRUM FLEXIBILITY.
`(a) COMMISSION ACTION- If the Commission determines to issue additional licenses for advanced television services, the Commission--
`(1) should limit the initial eligibility for such licenses to persons that, as of the date of such issuance, are licensed to operate a television broadcast station or hold a permit to construct such a station (or both); and
`(2) shall adopt regulations that allow the holders of such licenses to offer such ancillary or supplementary services on designated frequencies as may be consistent with the public interest, convenience, and necessity.
`(b) CONTENTS OF REGULATIONS- In prescribing the regulations required by subsection (a), the Commission shall--
`(1) only permit such licensee or permittee to offer ancillary or supplementary services if the use of a designated frequency for such services is consistent with the technology or method designated by the Commission for the provision of advanced television services;
`(2) limit the broadcasting of ancillary or supplementary services on designated frequencies so as to avoid derogation of any advanced television services, including high definition television broadcasts, that the Commission may require using such frequencies;
`(3) apply to any other ancillary or supplementary service such of the Commission's regulations as are applicable to the offering of analogous services by any other person, except that no ancillary or supplementary service shall have any rights to carriage under section 614 or 615 or be deemed a multichannel video programming distributor for purposes of section 628;
`(4) adopt such technical and other requirements as may be necessary or appropriate to assure the quality of the signal used to provide advanced television services, and may adopt regulations that stipulate the minimum number of hours per day that such signal must be transmitted; and
`(5) prescribe such other regulations as may be necessary for the protection of the public interest, convenience, and necessity.
`(c) RECOVERY OF LICENSE- If the Commission grants a license for advanced television

services to a person that, as of the date of such issuance, is licensed to operate a television broadcast station or holds a permit to construct such a station (or both), the Commission shall, as a condition of such license, require that either the additional license or the original license held by the licensee be surrendered to the Commission for reallocation or reassignment (or both) pursuant to Commission regulation.
`(d) PUBLIC INTEREST REQUIREMENT- Nothing in this section shall be construed as relieving a television broadcasting station from its obligation to serve the public interest, convenience, and necessity. In the Commission's review of any application for renewal of a broadcast license for a television station that provides ancillary or supplementary services, the television licensee shall establish that all of its program services on the existing or advanced television spectrum are in the public interest. Any violation of the Commission rules applicable to ancillary or supplementary services shall reflect upon the licensee's qualifications for renewal of its license.
`(e) FEES-
`(1) SERVICES TO WHICH FEES APPLY- If the regulations prescribed pursuant to subsection (a) permit a licensee to offer ancillary or supplementary services on a designated frequency--
`(A) for which the payment of a subscription fee is required in order to receive such services, or
`(B) for which the licensee directly or indirectly receives compensation from a third party in return for transmitting material furnished by such third party (other than commercial advertisements used to support broadcasting for which a subscription fee is not required), the Commission shall establish a program to assess and collectfrom the licensee for such designated frequency an annual fee or other schedule or method of payment that promotes the objectives described in subparagraphs (A) and (B) of paragraph (2).
`(2) COLLECTION OF FEES- The program required by paragraph (1) shall--
`(A) be designed (i) to recover for the public a portion of the value of the public spectrum resource made available for such commercial use, and (ii) to avoid unjust enrichment through the method employed to permit such uses of that resource;
`(B) recover for the public an amount that, to the extent feasible, equals but does not exceed (over the term of the license) the amount that would have been recovered had such services been licensed pursuant to the provisions of section 309(j) of this Act and the Commission's regulations thereunder; and
`(C) be adjusted by the Commission from time to time in order to continue to comply with the requirements of this paragraph.
`(3) TREATMENT OF REVENUES-
`(A) GENERAL RULE- Except as provided in subparagraph (B), all proceeds obtained pursuant to the regulations required by this subsection shall be deposited in the Treasury in accordance with chapter 33 of title 31, United States Code.
`(B) RETENTION OF REVENUES- Notwithstanding subparagraph (A), the salaries and expenses account of the Commission shall retain as an offsetting collection such sums as may be necessary from such proceeds for the costs of developing and implementing the program required by this section and regulating and supervising advanced television services. Such offsetting collections shall be available for obligation subject to the terms and conditions of the receiving appropriations account, and shall be deposited in such accounts on a quarterly basis.

`(4) REPORT- Within 5 years after the date of enactment of the Telecommunications Act of 1996, the Commission shall report to the Congress on the implementation of the program required by this subsection, and shall annually thereafter advise the Congress on the amounts collected pursuant to such program.
`(f) EVALUATION- Within 10 years after the date the Commission first issues additional licenses for advanced television services, the Commission shall conduct an evaluation of the advanced television services program. Such evaluation shall include--
`(1) an assessment of the willingness of consumers to purchase the television receivers necessary to receive broadcasts of advanced television services;
`(2) an assessment of alternative uses, including public safety use, of the frequencies used for such broadcasts; and
`(3) the extent to which the Commission has been or will be able to reduce the amount of spectrum assigned to licensees.
`(g) DEFINITIONS- As used in this section:
`(1) ADVANCED TELEVISION SERVICES- The term `advanced television services' means television services provided using digital or other advanced technology as further defined in the opinion, report, and order of the Commission entitled `Advanced Television Systems and Their Impact Upon the Existing Television Broadcast Service', MM Docket 87-268, adopted September 17, 1992, and successor proceedings.
`(2) DESIGNATED FREQUENCIES- The term `designated frequency' means each of the frequencies designated by the Commission for licenses for advanced television services.
`(3) HIGH DEFINITION TELEVISION- The term `high definition television' refers to systems that offer approximately twice the vertical and horizontal resolution of receivers generally available on the date of enactment of the Telecommunications Act of 1996, as further defined in the proceedings described in paragraph (1) of this subsection.'.

SEC. 202. BROADCAST OWNERSHIP.
(a) NATIONAL RADIO STATION OWNERSHIP RULE CHANGES REQUIRED- The Commission shall modify section 73.3555 of its regulations (47 C.F.R. 73.3555) by eliminating any provisions limiting the number of AM or FM broadcast stations which may be owned or controlled by one entity nationally.
(b) LOCAL RADIO DIVERSITY-
(1) APPLICABLE CAPS- The Commission shall revise section 73.3555(a) of its regulations (47 C.F.R. 73.3555) to provide that--
(A) in a radio market with 45 or more commercial radio stations, a party may own, operate, or control up to 8 commercial radio stations, not more than 5 of which are in the same service (AM or FM);
(B) in a radio market with between 30 and 44 (inclusive) commercial radio stations, a party may own, operate, or control up to 7 commercial radio stations, not more than 4 of which are in the same service (AM or FM);
(C) in a radio market with between 15 and 29 (inclusive) commercial radio stations, a party may own, operate, or control up to 6 commercial radio stations, not more than 4 of which are in the same service (AM or FM); and
(D) in a radio market with 14 or fewer commercial radio stations, a party may own, operate, or control up to 5 commercial radio stations, not more than 3 of which are in the same service (AM or FM), except that a party may not own, operate, or control more than 50 percent of the stations in such market.

(2) EXCEPTION- Notwithstanding any limitation authorized by this subsection, the Commission may permit a person or entity to own, operate, or control, or have a cognizable interest in, radio broadcast stations if the Commission determines that such ownership, operation, control, or interest will result in an increase in the number of radio broadcast stations in operation.
(c) TELEVISION OWNERSHIP LIMITATIONS-
(1) NATIONAL OWNERSHIP LIMITATIONS- The Commission shall modify its rules for multiple ownership set forth in section 73.3555 of its regulations (47 C.F.R. 73.3555)--
(A) by eliminating the restrictions on the number of television stations that a person or entity may directly or indirectly own, operate, or control, or have a cognizable interest in, nationwide; and
(B) by increasing the national audience reach limitation for television stations to 35 percent.
(2) LOCAL OWNERSHIP LIMITATIONS- The Commission shall conduct a rulemaking proceeding to determine whether to retain, modify, or eliminate its limitations on the number of television stations that a person or entity may own, operate, or control, or have a cognizable interest in, within the same television market.
(d) RELAXATION OF ONE-TO-A-MARKET- With respect to its enforcement of its one-to-a-market ownership rules under section 73.3555 of its regulations, the Commission shall extend its waiver policy to any of the top 50 markets, consistent with the public interest, convenience, and necessity.
(e) DUAL NETWORK CHANGES- The Commission shall revise section 73.658(g) of its regulations (47 C.F.R. 658(g)) to permit a television broadcast station to affiliate with a person or entity that maintains 2 or more networks of television broadcast stations unless such dual or multiple networks are composed of--
(1) two or more persons or entities that, on the date of enactment of the Telecommunications Act of 1996, are `networks' as defined in section 73.3613(a)(1) of the Commission's regulations (47 C.F.R. 73.3613(a)(1)); or
(2) any network described in paragraph (1) and an English-language program distribution service that, on such date, provides 4 or more hours of programming per week on a national basis pursuant to network affiliation arrangements with local television broadcast stations in markets reaching more than 75 percent of television homes (as measured by a national ratings service).
(f) CABLE CROSS OWNERSHIP-
(1) ELIMINATION OF RESTRICTIONS- The Commission shall revise section 76.501 of its regulations (47 C.F.R. 76.501) to permit a person or entity to own or control a network of broadcast stations and a cable system.
(2) SAFEGUARDS AGAINST DISCRIMINATION- The Commission shall revise such regulations if necessary to ensure carriage, channel positioning, and nondiscriminatory treatment of nonaffiliated broadcast stations by a cable system described in paragraph (1).
(g) LOCAL MARKETING AGREEMENTS- Nothing in this section shall be construed to prohibit the origination, continuation, or renewal of any television local marketing agreement that is in compliance with the regulations of the Commission.
(h) FURTHER COMMISSION REVIEW- The Commission shall review its rules adopted pursuant to this section and all of its ownership rules biennially as part of its regulatory reform review under section 11 of the Communications Act of 1934 and shall determine whether any of such rules are necessary in the public interest as the result of competition. The Commission shall repeal or modify any regulation it determines to be no longer in the

public interest.
(i) ELIMINATION OF STATUTORY RESTRICTION- Section 613(a) (47 U.S.C. 533(a)) is amended--
(1) by striking paragraph (1);
(2) by redesignating paragraph (2) as subsection (a);
(3) by redesignating subparagraphs (A) and (B) as paragraphs (1) and (2), respectively;
(4) by striking `and' at the end of paragraph (1) (as so redesignated);
(5) by striking the period at the end of paragraph (2) (as so redesignated) and inserting `; and'; and
(6) by adding at the end the following new paragraph:
`(3) shall not apply the requirements of this subsection to any cable operator in any franchise area in which a cable operator is subject to effective competition as determined under section 623(l).'.

SEC. 203. TERM OF LICENSES.
Section 307(c) (47 U.S.C. 307(c)) is amended to read as follows:
`(c) TERMS OF LICENSES-
`(1) INITIAL AND RENEWAL LICENSES- Each license granted for the operation of a broadcasting station shall be for a term of not to exceed 8 years. Upon application therefor, a renewal of such license may be granted from time to time for a term of not to exceed 8 years from the date of expiration of the preceding license, if the Commission finds that public interest, convenience, and necessity would be served thereby. Consistent with the foregoing provisions of this subsection, the Commission may by rule prescribe the period or periods for which licenses shall be granted and renewed for particular classes of stations, but the Commission may not adopt or follow any rule which would preclude it, in any case involving a station of a particular class, from granting or renewing a license for a shorter period than that prescribed for stations of such class if, in its judgment, the public interest, convenience, or necessity would be served by such action.
`(2) MATERIALS IN APPLICATION- In order to expedite action on applications for renewal of broadcasting station licenses and in order to avoid needless expense to applicants for such renewals, the Commission shall not require any such applicant to file any information which previously has been furnished to the Commission or which is not directly material to the considerations that affect the granting or denial of such application, but the Commission may require any new or additional facts it deems necessary to make its findings.
`(3) CONTINUATION PENDING DECISION- Pending any hearing and final decision on such an application and the disposition of any petition for rehearing pursuant to section 405, the Commission shall continue such license in effect.'.

SEC. 204. BROADCAST LICENSE RENEWAL PROCEDURES.
(a) RENEWAL PROCEDURES-
(1) AMENDMENT- Section 309 (47 U.S.C. 309) is amended by adding at the end thereof the following new subsection:
`(k) BROADCAST STATION RENEWAL PROCEDURES-
`(1) STANDARDS FOR RENEWAL- If the licensee of a broadcast station submits an application to the Commission for renewal of such license, the Commission shall grant the application if it finds, with respect to that station, during the preceding term of its license--
`(A) the station has served the public interest, convenience, and necessity;

`(B) there have been no serious violations by the licensee of this Act or the rules and regulations of the Commission; and
`(C) there have been no other violations by the licensee of this Act or the rules and regulations of the Commission which, taken together, would constitute a pattern of abuse.
`(2) CONSEQUENCE OF FAILURE TO MEET STANDARD- If any licensee of a broadcast station fails to meet the requirements of this subsection, the Commission may deny the application for renewal in accordance with paragraph (3), or grant such application on terms and conditions as are appropriate, including renewal for a term less than the maximum otherwise permitted.
`(3) STANDARDS FOR DENIAL- If the Commission determines, after notice and opportunity for a hearing as provided in subsection (e), that a licensee has failed to meet the requirements specified in paragraph (1) and that no mitigating factors justify the imposition of lesser sanctions, the Commission shall--
`(A) issue an order denying the renewal application filed by such licensee under section 308; and
`(B) only thereafter accept and consider such applications for a construction permit as may be filed under section 308 specifying the channel or broadcasting facilities of the former licensee.
`(4) COMPETITOR CONSIDERATION PROHIBITED- In making the determinations specified in paragraph (1) or (2), the Commission shall not consider whether the public interest, convenience, and necessity might be served by the grant of a license to a person other than the renewal applicant.'.

SEC. 205. DIRECT BROADCAST SATELLITE SERVICE.
(a) DBS SIGNAL SECURITY- Section 705(e)(4) (47 U.S.C. 605(e)(4)) is amended by inserting `or direct-to-home satellite services,' after `programming,'.
(b) FCC JURISDICTION OVER DIRECT-TO-HOME SATELLITE SERVICES- Section 303 (47 U.S.C. 303) is amended by adding at the end thereof the following new subsection:
`(v) Have exclusive jurisdiction to regulate the provision of direct-to-home satellite services. As used in this subsection, the term `direct-to-home satellite services' means the distribution or broadcasting of programming or services by satellite directly to the subscriber's premises without the use of ground receiving or distribution equipment, except at the subscriber's premises or in the uplink process to the satellite.'

SEC. 207. RESTRICTIONS ON OVER-THE-AIR RECEPTION DEVICES.
Within 180 days after the date of enactment of this Act, the Commission shall, pursuant to section 303 of the Communications Act of 1934, promulgate regulations to prohibit restrictions that impair a viewer's ability to receive video programming services through devices designed for over-the-air reception of television broadcast signals, multichannel multipoint distribution service, or direct broadcast satellite services.

TITLE III--CABLE SERVICES
SEC. 301. CABLE ACT REFORM.

(a) DEFINITIONS-
(1) DEFINITION OF CABLE SERVICE- Section 602(6)(B) (47 U.S.C. 522(6)(B)) is amended by inserting `or use' after `the selection'.
(2) CHANGE IN DEFINITION OF CABLE SYSTEM- Section 602(7) (47 U.S.C. 522(7)) is amended by striking `(B) a facility that serves only subscribers in 1 or more multiple unit dwellings under common ownership, control, or management, unless such facility or facilities uses any public right-of-way;' and inserting `(B) a facility that serves subscribers without using any public right-of-way;'.
(b) RATE DEREGULATION-
(1) UPPER TIER REGULATION- Section 623(c) (47 U.S.C. 543(c)) is amended--
(A) in paragraph (1)(B), by striking `subscriber, franchising authority, or other relevant State or local government entity' and inserting `franchising authority (in accordance with paragraph (3))';
(B) in paragraph (1)(C), by striking `such complaint' and inserting `the first complaint filed with the franchising authority under paragraph (3)'; and
(C) by striking paragraph (3) and inserting the following:
`(3) REVIEW OF RATE CHANGES- The Commission shall review any complaint submitted by a franchising authority after the date of enactment of the Telecommunications Act of 1996 concerning an increase in rates for cable programming services and issue a final order within 90 days after it receives such a complaint, unless the parties agree to extend the period for such review. A franchising authority may not file a complaint under this paragraph unless, within 90 days after such increase becomes effective it receives subscriber complaints.
`(4) SUNSET OF UPPER TIER RATE REGULATION- This subsection shall not apply to cable programming services provided after March 31, 1999.'.
(2) SUNSET OF UNIFORM RATE STRUCTURE IN MARKETS WITH EFFECTIVE COMPETITION- Section 623(d) (47 U.S.C. 543(d)) is amended by adding at the end thereof the following: `This subsection does not apply to (1) a cable operator with respect to the provision of cable service over its cable system in any geographic area in which the video programming services offered by the operator in that area are subject to effective competition, or (2) any video programming offered on a per channel or per program basis. Bulk discounts to multiple dwelling units shall not be subject to this subsection, except that a cable operator of a cable system that is not subject to effective competition may not charge predatory prices to a multiple dwelling unit. Upon a prima facie showing by a complainant that there are reasonable grounds to believe that the discounted price is predatory, the cable system shall have the burden of showing that its discounted price is not predatory.'.
(3) EFFECTIVE COMPETITION- Section 623(l)(1) (47 U.S.C. 543(l)(1)) is amended--
(A) by striking `or' at the end of subparagraph (B);
(B) by striking the period at the end of subparagraph (C) and inserting `; or'; and
(C) by adding at the end the following:
`(D) a local exchange carrier or its affiliate (or any multichannel video programming distributor using the facilities of such carrier or its affiliate) offers video programming services directly to subscribers by any means (other than direct-to-home satellite services) in the franchise area of an unaffiliated cable operator which is providing cable service in that franchise area, but only if the video programming services so offered in that area are comparable to the video programming services provided by the unaffiliated cable operator in that area.'.
(c) GREATER DEREGULATION FOR SMALLER CABLE COMPANIES- Section 623

(47 U.S.C 543) is amended by adding at the end thereof the following:
`(m) SPECIAL RULES FOR SMALL COMPANIES-
`(1) IN GENERAL- Subsections (a), (b), and (c) do not apply to a small cable operator with respect to--
`(A) cable programming services, or
`(B) a basic service tier that was the only service tier subject to regulation as of December 31, 1994, in any franchise area in which that operator services 50,000 or fewer subscribers.
`(2) DEFINITION OF SMALL CABLE OPERATOR- For purposes of this subsection, the term `small cable operator' means a cable operator that, directly or through an affiliate, serves in the aggregate fewer than 1 percent of all subscribers in the United States and is not affiliated with any entity or entities whose gross annual revenues in the aggregate exceed $250,000,000.'.
(d) MARKET DETERMINATIONS-
(1) MARKET DETERMINATIONS; EXPEDITED DECISIONMAKING- Section 614(h)(1)(C) (47 U.S.C. 534(h)(1)(C)) is amended--
(A) by striking `in the manner provided in section 73.3555(d)(3)(i) of title 47, Code of Federal Regulations, as in effect on May 1, 1991,' in clause (i) and inserting `by the Commission by regulation or order using, where available, commercial publications which delineate television markets based on viewing patterns,'; and
(B) by striking clause (iv) and inserting the following:
`(iv) Within 120 days after the date on which a request is filed under this subparagraph (or 120 days after the date of enactment of the Telecommunications Act of 1996, if later), the Commission shall grant or deny the request.'.
(2) APPLICATION TO PENDING REQUESTS- The amendment made by paragraph (1) shall apply to--
(A) any request pending under section 614(h)(1)(C) of the Communications Act of 1934 (47 U.S.C. 534(h)(1)(C)) on the date of enactment of this Act; and
(B) any request filed under that section after that date.
(e) TECHNICAL STANDARDS- Section 624(e) (47 U.S.C. 544(e)) is amended by striking the last two sentences and inserting the following: `No State or franchising authority may prohibit, condition, or restrict a cable system's use of any type of subscriber equipment or any transmission technology.'.
(f) CABLE EQUIPMENT COMPATIBILITY- Section 624A (47 U.S.C. 544A) is amended--
(1) in subsection (a) by striking `and' at the end of paragraph (2), by striking the period at the end of paragraph (3) and inserting `; and'; and by adding at the end the following new paragraph:
`(4) compatibility among televisions, video cassette recorders, and cable systems can be assured with narrow technical standards that mandate a minimum degree of common design and operation, leaving all features, functions, protocols, and other product and service options for selection through open competition in the market.';
(2) in subsection (c)(1)--
(A) by redesignating subparagraphs (A) and (B) as subparagraphs (B) and (C), respectively; and
(B) by inserting before such redesignated subparagraph
(B) the following new subparagraph:
`(A) the need to maximize open competition in the market for all features, functions, protocols, and other product and service options of converter boxes and other cable converters

unrelated to the descrambling or decryption of cable television signals;'; and
(3) in subsection (c)(2)--
(A) by redesignating subparagraphs (D) and (E) as subparagraphs (E) and (F), respectively; and
(B) by inserting after subparagraph (C) the following new subparagraph:
`(D) to ensure that any standards or regulations developed under the authority of this section to ensure compatibility between televisions, video cassette recorders, and cable systems do not affect features, functions, protocols, and other product and service options other than those specified in paragraph (1)(B), including telecommunications interface equipment, home automation communications, and computer network services;'.
(g) SUBSCRIBER NOTICE- Section 632 (47 U.S.C. 552) is amended--
(1) by redesignating subsection (c) as subsection (d); and
(2) by inserting after subsection (b) the following new subsection:
`(c) SUBSCRIBER NOTICE- A cable operator may provide notice of service and rate changes to subscribers using any reasonable written means at its sole discretion. Notwithstanding section 623(b)(6) or any other provision of this Act, a cable operator shall not be required to provide prior notice of any rate change that is the result of a regulatory fee, franchise fee, or any other fee, tax, assessment, or charge of any kind imposed by any Federal agency, State, or franchising authority on the transaction between the operator and the subscriber.'.

SEC. 302. CABLE SERVICE PROVIDED BY TELEPHONE COMPANIES.
(a) PROVISIONS FOR REGULATION OF CABLE SERVICE PROVIDED BY TELEPHONE COMPANIES- Title VI (47 U.S.C. 521 et seq.) is amended by adding at the end the following new part:

`PART V--VIDEO PROGRAMMING SERVICES PROVIDED BY TELEPHONE COMPANIES
`SEC. 651. REGULATORY TREATMENT OF VIDEO PROGRAMMING SERVICES.
`(a) LIMITATIONS ON CABLE REGULATION-
`(1) RADIO-BASED SYSTEMS- To the extent that a common carrier (or any other person) is providing video programming to subscribers using radio communication, such carrier (or other person) shall be subject to the requirements of title III and section 652, but shall not otherwise be subject to the requirements of this title.
`(2) COMMON CARRIAGE OF VIDEO TRAFFIC- To the extent that a common carrier is providing transmission of video programming on a common carrier basis, such carrier shall be subject to the requirements of title II and section 652, but shall not otherwise be subject to the requirements of this title. This paragraph shall not affect the treatment under section 602(7)(C) of a facility of a common carrier as a cable system.
`(3) CABLE SYSTEMS AND OPEN VIDEO SYSTEMS- To the extent that a common carrier is providing video programming to its subscribers in any manner other than that described in paragraphs (1) and (2)--
`(A) such carrier shall be subject to the requirements of this title, unless such programming is provided by means of an open video system for which the Commission has approved a certification under section 653; or
`(B) if such programming is provided by means of an open video system for which the

Commission has approved a certification under section 653, such carrier shall be subject to the requirements of this part, but shall be subject to parts I through IV of this title only as provided in 653(c).
`(4) ELECTION TO OPERATE AS OPEN VIDEO SYSTEM- A common carrier that is providing video programming in a manner described in paragraph (1) or (2), or a combination thereof, may elect to provide such programming by means of an open video system that complies with section 653. If the Commission approves such carrier's certification under section 653, such carrier shall be subject to the requirements of this part, but shall be subject to parts I through IV of this title only as provided in 653(c).
`(b) LIMITATIONS ON INTERCONNECTION OBLIGATIONS- A local exchange carrier that provides cable service through an open video system or a cable system shall not be required, pursuant to title II of this Act, to make capacity available on a nondiscriminatory basis to any other person for the provision of cable service directly to subscribers.
`(c) ADDITIONAL REGULATORY RELIEF- A common carrier shall not be required to obtain a certificate under section 214 with respect to the establishment or operation of a system for the delivery of video programming.

`SEC. 652. PROHIBITION ON BUY OUTS.
`(a) ACQUISITIONS BY CARRIERS- No local exchange carrier or any affiliate of such carrier owned by, operated by, controlled by, or under common control with such carrier may purchase or otherwise acquire directly or indirectly more than a 10 percent financial interest, or any management interest, in any cable operator providing cable service within the local exchange carrier's telephone service area.
`(b) ACQUISITIONS BY CABLE OPERATORS- No cable operator or affiliate of a cable operator that is owned by, operated by, controlled by, or under common ownership with such cable operator may purchase or otherwise acquire, directly or indirectly, more than a 10 percent financial interest, or any management interest, in any local exchange carrier providing telephone exchange service within such cable operator's franchise area.
`(c) JOINT VENTURES- A local exchange carrier and a cable operator whose telephone service area and cable franchise area, respectively, are in the same market may not enter into any joint venture or partnership to provide video programming directly to subscribers or to provide telecommunications services within such market.
`(d) EXCEPTIONS-
`(1) RURAL SYSTEMS- Notwithstanding subsections (a), (b), and (c) of this section, a local exchange carrier (with respect to a cable system located in its telephone service area) and a cable operator (with respect to the facilities of a local exchange carrier used to provide telephone exchange service in its cable franchise area) may obtain a controlling interest in, management interest in, or enter into a joint venture or partnership with the operator of such system or facilities for the use of such system or facilities to the extent that--
`(A) such system or facilities only serve incorporated or unincorporated--
`(i) places or territories that have fewer than 35,000 inhabitants; and
`(ii) are outside an urbanized area, as defined by the Bureau of the Census; and
`(B) in the case of a local exchange carrier, such system, in the aggregate with any other system in which such carrier has an interest, serves less than 10 percent of the households in the telephone service area of such carrier.
`(2) JOINT USE- Notwithstanding subsection (c), a local exchange carrier may obtain, with the concurrence of the cable operator on the rates, terms, and conditions, the use of that part

of the transmission facilities of a cable system extending from the last multi-user terminal to the premises of the end user, if such use is reasonably limited in scope and duration, as determined by the Commission.

`(3) ACQUISITIONS IN COMPETITIVE MARKETS- Notwithstanding subsections (a) and (c), a local exchange carrier may obtain a controlling interest in, or form a joint venture or other partnership with, or provide financing to, a cable system (hereinafter in this paragraph referred to as `the subject cable system'), if--

`(A) the subject cable system operates in a television market that is not in the top 25 markets, and such market has more than 1 cable system operator, and the subject cable system is not the cable system with the most subscribers in such television market;

`(B) the subject cable system and the cable system with the most subscribers in such television market held on May 1, 1995, cable television franchises from the largest municipality in the television market and the boundaries of such franchises were identical on such date;

`(C) the subject cable system is not owned by or under common ownership or control of any one of the 50 cable system operators with the most subscribers as such operators existed on May 1, 1995; and

`(D) the system with the most subscribers in the television market is owned by or under common ownership or control of any one of the 10 largest cable system operators as such operators existed on May 1, 1995.

`(4) EXEMPT CABLE SYSTEMS- Subsection (a) does not apply to any cable system if--

`(A) the cable system serves no more than 17,000 cable subscribers, of which no less than 8,000 live within an urban area, and no less than 6,000 live within a nonurbanized area as of June 1, 1995;

`(B) the cable system is not owned by, or under common ownership or control with, any of the 50 largest cable system operators in existence on June 1, 1995; and

`(C) the cable system operates in a television market that was not in the top 100 television markets as of June 1, 1995.

`(5) SMALL CABLE SYSTEMS IN NONURBAN AREAS- Notwithstanding subsections (a) and (c), a local exchange carrier with less than $100,000,000 in annual operating revenues (or any affiliate of such carrier owned by, operated by, controlled by, or under common control with such carrier) may purchase or otherwise acquire more than a 10 percent financial interest in, or any management interest in, or enter into a joint venture or partnership with, any cable system within the local exchange carrier's telephone service area that serves no more than 20,000 cable subscribers, if no more than 12,000 of those subscribers live within an urbanized area, as defined by the Bureau of the Census.

`(6) WAIVERS- The Commission may waive the restrictions of subsections (a), (b), or (c) only if--

`(A) the Commission determines that, because of the nature of the market served by the affected cable system or facilities used to provide telephone exchange service--

`(i) the affected cable operator or local exchange carrier would be subjected to undue economic distress by the enforcement of such provisions;

`(ii) the system or facilities would not be economically viable if such provisions were enforced; or

`(iii) the anticompetitive effects of the proposed transaction are clearly outweighed in the public interest by the probable effect of the transaction in meeting the convenience and needs of the community to be served; and

`(B) the local franchising authority approves of such waiver.

`(e) DEFINITION OF TELEPHONE SERVICE AREA- For purposes of this section, the term `telephone service area' when used in connection with a common carrier subject in whole or in part to title II of this Act means the area within which such carrier provided telephone exchange service as of January 1, 1993, but if any common carrier after such date transfers its telephone exchange service facilities to another common carrier, the area to which such facilities provide telephone exchange service shall be treated as part of the telephone service area of the acquiring common carrier and not of the selling common carrier.

`SEC. 653. ESTABLISHMENT OF OPEN VIDEO SYSTEMS.

`(a) OPEN VIDEO SYSTEMS-

`(1) CERTIFICATES OF COMPLIANCE- A local exchange carrier may provide cable service to its cable service subscribers in its telephone service area through an open video system that complies with this section. To the extent permitted by such regulations as the Commission may prescribe consistent with the public interest, convenience, and necessity, an operator of a cable system or any other person may provide video programming through an open video system that complies with this section. An operator of an open video system shall qualify for reduced regulatory burdens under subsection (c) of this section if the operator of such system certifies to the Commission that such carrier complies with the Commission's regulations under subsection (b) and the Commission approves such certification. The Commission shall publish notice of the receipt of any such certification and shall act to approve or disapprove any such certification within 10 days after receipt of such certification.

`(2) DISPUTE RESOLUTION- The Commission shall have the authority to resolve disputes under this section and the regulations prescribed thereunder. Any such dispute shall be resolved within 180 days after notice of such dispute is submitted to the Commission. At that time or subsequently in a separate damages proceeding, the Commission may, in the case of any violation of this section, require carriage, award damages to any person denied carriage, or any combination of such sanctions. Any aggrieved party may seek any other remedy available under this Act.

`(b) COMMISSION ACTIONS-

`(1) REGULATIONS REQUIRED- Within 6 months after the date of enactment of the Telecommunications Act of 1996, the Commission shall complete all actions necessary (including any reconsideration) to prescribe regulations that--

`(A) except as required pursuant to section 611, 614, or 615, prohibit an operator of an open video system from discriminating among video programming providers with regard to carriage on its open video system, and ensure that the rates, terms, and conditions for such carriage are just and reasonable, and are not unjustly or unreasonably discriminatory;

`(B) if demand exceeds the channel capacity of the open video system, prohibit an operator of an open video system and its affiliates from selecting the video programming services for carriage on more than one-third of the activated channel capacity on such system, but nothing in this subparagraph shall be construed to limit the number of channels that the carrier and its affiliates may offer to provide directly to subscribers;

`(C) permit an operator of an open video system to carry on only one channel any video programming service that is offered by more than one video programming provider (including the local exchange carrier's video programming affiliate): [Italic->] Provided, [<-Italic] That subscribers have ready and immediate access to any such video programming service;

`(D) extend to the distribution of video programming over open video systems the Commission's regulations concerning sports exclusivity (47 C.F.R. 76.67), network

nonduplication (47 C.F.R. 76.92 et seq.), and syndicated exclusivity (47 C.F.R. 76.151 et seq.); and
`(E)(i) prohibit an operator of an open video system from unreasonably discriminating in favor of the operator or its affiliates with regard to material or information (including advertising) provided by the operator to subscribers for the purposes of selecting programming on the open video system, or in the way such material or information is presented to subscribers;
`(ii) require an operator of an open video system to ensure that video programming providers or copyright holders (or both) are able suitably and uniquely to identify their programming services to subscribers;
`(iii) if such identification is transmitted as part of the programming signal, require the carrier to transmit such identification without change or alteration; and
`(iv) prohibit an operator of an open video system from omitting television broadcast stations or other unaffiliated video programming services carried on such system from any navigational device, guide, or menu.
`(2) CONSUMER ACCESS- Subject to the requirements of paragraph (1) and the regulations thereunder, nothing in this section prohibits a common carrier or its affiliate from negotiating mutually agreeable terms and conditions with over-the-air broadcast stations and other unaffiliated video programming providers to allow consumer access to their signals on any level or screen of any gateway, menu, or other program guide, whether provided by the carrier or its affiliate.
`(c) REDUCED REGULATORY BURDENS FOR OPEN VIDEO SYSTEMS-
`(1) IN GENERAL- Any provision that applies to a cable operator under--
`(A) sections 613 (other than subsection (a) thereof), 616, 623(f), 628, 631, and 634 of this title, shall apply,
`(B) sections 611, 614, and 615 of this title, and section 325 of title III, shall apply in accordance with the regulations prescribed under paragraph (2), and
`(C) sections 612 and 617, and parts III and IV (other than sections 623(f), 628, 631, and 634), of this title shall not apply, to any operator of an open video system for which the Commission has approved a certification under this section.
`(2) IMPLEMENTATION-
`(A) COMMISSION ACTION- In the rulemaking proceeding to prescribe the regulations required by subsection (b)(1), the Commission shall, to the extent possible, impose obligations that are no greater or lesser than the obligations contained in the provisions described in paragraph (1)(B) of this subsection. The Commission shall complete all action (including any reconsideration) to prescribe such regulations no later than 6 months after the date of enactment of the Telecommunications Act of 1996.
has approved before the date of enactment of this Act.

SEC. 303. PREEMPTION OF FRANCHISING AUTHORITY REGULATION OF TELECOMMUNICATIONS SERVICES.
(a) PROVISION OF TELECOMMUNICATIONS SERVICES BY A CABLE OPERATOR- Section 621(b) (47 U.S.C. 541(b)) is amended by adding at the end thereof the following new paragraph:
`(3)(A) If a cable operator or affiliate thereof is engaged in the provision of telecommunications services--
`(i) such cable operator or affiliate shall not be required to obtain a franchise under this title

for the provision of telecommunications services; and
`(ii) the provisions of this title shall not apply to such cable operator or affiliate for the provision of telecommunications services.
`(B) A franchising authority may not impose any requirement under this title that has the purpose or effect of prohibiting, limiting, restricting, or conditioning the provision of a telecommunications service by a cable operator or an affiliate thereof.
`(C) A franchising authority may not order a cable operator or affiliate thereof--
`(i) to discontinue the provision of a telecommunications service, or
`(ii) to discontinue the operation of a cable system, to the extent such cable system is used for the provision of a telecommunications service, by reason of the failure of such cable operator or affiliate thereof to obtain a franchise or franchise renewal under this title with respect to the provision of such telecommunications service.
`(D) Except as otherwise permitted by sections 611 and 612, a franchising authority may not require a cable operator to provide any telecommunications service or facilities, other than institutional networks, as a condition of the initial grant of a franchise, a franchise renewal, or a transfer of a franchise.'.
(b) FRANCHISE FEES- Section 622(b) (47 U.S.C. 542(b)) is amended by inserting `to provide cable services' immediately before the period at the end of the first sentence thereof.

`SEC. 629. COMPETITIVE AVAILABILITY OF NAVIGATION DEVICES.
`(a) COMMERCIAL CONSUMER AVAILABILITY OF EQUIPMENT USED TO ACCESS SERVICES PROVIDED BY MULTICHANNEL VIDEO PROGRAMMING DISTRIBUTORS- The Commission shall, in consultation with appropriate industry standard-setting organizations, adopt regulations to assure the commercial availability, to consumers of multichannel video programming and other services offered over multichannel video programming systems, of converter boxes, interactive communications equipment, and other equipment used by consumers to access multichannel video programming and other services offered over multichannel video programming systems, from manufacturers, retailers, and other vendors not affiliated with any multichannel video programming distributor. Such regulations shall not prohibit any multichannel video programming distributor from also offering converter boxes, interactive communications equipment, and other equipment used by consumers to access multichannel video programming and other services offered over multichannel video programming systems, to consumers, if the system operator's charges to consumers for such devices and equipment are separately stated and not subsidized by charges for any such service.
`(b) PROTECTION OF SYSTEM SECURITY- The Commission shall not prescribe regulations under subsection (a) which would jeopardize security of multichannel video programming and other services offered over multichannel video programming systems, or impede the legal rights of a provider of such services to prevent theft of service.
`(c) WAIVER- The Commission shall waive a regulation adopted under subsection (a) for a limited time upon an appropriate showing by a provider of multichannel video programming and other services offered over multichannel video programming systems, or an equipment provider, that such waiver is necessary to assist the development or introduction of a new or improved multichannel video programming or other service offered over multichannel video programming systems, technology, or products. Upon an appropriate showing, the

Commission shall grant any such waiver request within 90 days of any application filed under this subsection, and such waiver shall be effective for all service providers and products in that category and for all providers of services and products.
`(d) AVOIDANCE OF REDUNDANT REGULATIONS-
`(1) COMMERCIAL AVAILABILITY DETERMINATIONS- Determinations made or regulations prescribed by the Commission with respect to commercial availability to consumers of converter boxes, interactive communications equipment, and other equipment used by consumers to access multichannel video programming and other services offered over multichannel video programming systems, before the date of enactment of the Telecommunications Act of 1996 shall fulfill the requirements of this section.
`(2) REGULATIONS- Nothing in this section affects section 64.702(e) of the Commission's regulations (47 C.F.R. 64.702(e)) or other Commission regulations governing interconnection and competitive provision of customer premises equipment used in connection with basic common carrier communications services.
`(e) SUNSET- The regulations adopted under this section shall cease to apply when the Commission determines that--
`(1) the market for the multichannel video programming distributors is fully competitive;
`(2) the market for converter boxes, and interactive communications equipment, used in conjunction with that service is fully competitive; and
`(3) elimination of the regulations would promote competition and the public interest.
`(f) COMMISSION'S AUTHORITY- Nothing in this section shall be construed as expanding or limiting any authority that the Commission may have under law in effect before the date of enactment of the Telecommunications Act of 1996.'.

`SEC. 713. VIDEO PROGRAMMING ACCESSIBILITY.
`(a) COMMISSION INQUIRY- Within 180 days after the date of enactment of the Telecommunications Act of 1996, the Federal Communications Commission shall complete an inquiry to ascertain the level at which video programming is closed captioned. Such inquiry shall examine the extent to which existing or previously published programming is closed captioned, the size of the video programming provider or programming owner providing closed captioning, the size of the market served, the relative audience shares achieved, or any other related factors. The Commission shall submit to the Congress a report on the results of such inquiry.
`(b) ACCOUNTABILITY CRITERIA- Within 18 months after such date of enactment, the Commission shall prescribe such regulations as are necessary to implement this section. Such regulations shall ensure that--
`(1) video programming first published or exhibited after the effective date of such regulations is fully accessible through the provision of closed captions, except as provided in subsection (d); and
`(2) video programming providers or owners maximize the accessibility of video programming first published or exhibited prior to the effective date of such regulations through the provision of closed captions, except as provided in subsection (d).
`(c) DEADLINES FOR CAPTIONING- Such regulations shall include an appropriate schedule of deadlines for the provision of closed captioning of video programming.
`(d) EXEMPTIONS- Notwithstanding subsection (b)--

`(1) the Commission may exempt by regulation programs, classes of programs, or services for which the Commission has determined that the provision of closed captioning would be economically burdensome to the provider or owner of such programming;
`(2) a provider of video programming or the owner of any program carried by the provider shall not be obligated to supply closed captions if such action would be inconsistent with contracts in effect on the date of enactment of the Telecommunications Act of 1996, except that nothing in this section shall be construed to relieve a video programming provider of its obligations to provide services required by Federal law; and
`(3) a provider of video programming or program owner may petition the Commission for an exemption from the requirements of this section, and the Commission may grant such petition upon a showing that the requirements contained in this section would result in an undue burden.
`(e) UNDUE BURDEN- The term `undue burden' means significant difficulty or expense. In determining whether the closed captions necessary to comply with the requirements of this paragraph would result in an undue economic burden, the factors to be considered include--
`(1) the nature and cost of the closed captions for the programming;
`(2) the impact on the operation of the provider or program owner;
`(3) the financial resources of the provider or program owner; and
`(4) the type of operations of the provider or program owner.
`(f) VIDEO DESCRIPTIONS INQUIRY- Within 6 months after the date of enactment of the Telecommunications Act of 1996, the Commission shall commence an inquiry to examine the use of video descriptions on video programming in order to ensure the accessibility of video programming to persons with visual impairments, and report to Congress on its findings. The Commission's report shall assess appropriate methods and schedules for phasing video descriptions into the marketplace, technical and quality standards for video descriptions, a definition of programming for which video descriptions would apply, and other technical and legal issues that the Commission deems appropriate.
`(g) VIDEO DESCRIPTION- For purposes of this section, `video description' means the insertion of audio narrated descriptions of a television program's key visual elements into natural pauses between the program's dialogue.
`(h) PRIVATE RIGHTS OF ACTIONS PROHIBITED- Nothing in this section shall be construed to authorize any private right of action to enforce any requirement of this section or any regulation thereunder. The Commission shall have exclusive jurisdiction with respect to any complaint under this section.'.

TITLE IV--REGULATORY REFORM
SEC. 401. REGULATORY FORBEARANCE.
Title I is amended by inserting after section 9 (47 U.S.C. 159) the following new section:

`SEC. 10. COMPETITION IN PROVISION OF TELECOMMUNICATIONS SERVICE.

SEC. 402. BIENNIAL REVIEW OF REGULATIONS; REGULATORY RELIEF.
(a) BIENNIAL REVIEW- Title I is amended by inserting after section 10 (as added by section 401) the following new section:

`SEC. 11. REGULATORY REFORM.

`(a) BIENNIAL REVIEW OF REGULATIONS- In every even-numbered year (beginning with 1998), the Commission--
`(1) shall review all regulations issued under this Act in effect at the time of the review that apply to the operations or activities of any provider of telecommunications service; and
`(2) shall determine whether any such regulation is no longer necessary in the public interest as the result of meaningful economic competition between providers of such service.
`(b) EFFECT OF DETERMINATION- The Commission shall repeal or modify any regulation it determines to be no longer necessary in the public interest.'.
(b) REGULATORY RELIEF-
(1) Streamlined procedures for changes in charges, classifications, regulations, or practices-
(A) Section 204(a) (47 U.S.C. 204(a)) is amended--
(i) by striking `12 months' the first place it appears in paragraph (2)(A) and inserting `5 months';
(ii) by striking `effective,' and all that follows in paragraph (2)(A) and inserting `effective.'; and
(iii) by adding at the end thereof the following:
`(3) A local exchange carrier may file with the Commission a new or revised charge, classification, regulation, or practice on a streamlined basis. Any such charge, classification, regulation, or practice shall be deemed lawful and shall be effective 7 days (in the case of a reduction in rates) or 15 days (in the case of an increase in rates) after the date on which it is filed with the Commission unless the Commission takes action under paragraph (1) before the end of that 7-day or 15-day period, as is appropriate.'.
(B) Section 208(b) (47 U.S.C. 208(b)) is amended--
(i) by striking `12 months' the first place it appears in paragraph (1) and inserting `5 months'; and
(ii) by striking `filed,' and all that follows in paragraph (1) and inserting `filed.'.
(2) EXTENSIONS OF LINES UNDER SECTION 214; ARMIS REPORTS- The Commission shall permit any common carrier--
(A) to be exempt from the requirements of section 214 of the Communications Act of 1934 for the extension of any line; and
(B) to file cost allocation manuals and ARMIS reports annually, to the extent such carrier is required to file such manuals or reports.
(3) FORBEARANCE AUTHORITY NOT LIMITED- Nothing in this subsection shall be construed to limit the authority of the Commission to waive, modify, or forbear from applying any of the requirements to which reference is made in paragraph (1) under any other provision of this Act or other law.
(4) EFFECTIVE DATE OF AMENDMENTS- The amendments made by paragraph (1) of this subsection shall apply with respect to any charge, classification, regulation, or practice filed on or after one year after the date of enactment of this Act.
(c) CLASSIFICATION OF CARRIERS- In classifying carriers according to section 32.11 of its regulations (47 C.F.R. 32.11) and in establishing reporting requirements pursuant to part 43 of its regulations (47 C.F.R. part 43) and section 64.903 of its regulations (47 C.F.R. 64.903), the Commission shall adjust the revenue requirements to account for inflation as of the release date of the Commission's Report and Order in CC Docket No. 91-141, and annually thereafter. This subsection shall take effect on the date of enactment of this Act.

SUBTITLE C--JUDICIAL REVIEW
SEC. 561. EXPEDITED REVIEW.
(a) THREE-JUDGE DISTRICT COURT HEARING- Notwithstanding any other provision of law, any civil action challenging the constitutionality, on its face, of this title or any amendment made by this title, or any provision thereof, shall be heard by a district court of 3 judges convened pursuant to the provisions of section 2284 of title 28, United States Code.
(b) APPELLATE REVIEW- Notwithstanding any other provision of law, an interlocutory or final judgment, decree, or order of the court of 3 judges in an action under subsection (a) holding this title or an amendment made by this title, or any provision thereof, unconstitutional shall be reviewable as a matter of right by direct appeal to the Supreme Court. Any such appeal shall be filed not more than 20 days after entry of such judgment, decree, or order.

TITLE VI--EFFECT ON OTHER LAWS
SEC. 601. APPLICABILITY OF CONSENT DECREES AND OTHER LAW.
(a) APPLICABILITY OF AMENDMENTS TO FUTURE CONDUCT-

(1) AT&T CONSENT DECREE- Any conduct or activity that was, before the date of enactment of this Act, subject to any restriction or obligation imposed by the AT&T Consent Decree shall, on and after such date, be subject to the restrictions and obligations imposed by the Communications Act of 1934 as amended by this Act and shall not be subject to the restrictions and the obligations imposed by such Consent Decree.
(2) GTE CONSENT DECREE- Any conduct or activity that was, before the date of enactment of this Act, subject to any restriction or obligation imposed by the GTE Consent Decree shall, on and after such date, be subject to the restrictions and obligations imposed by the Communications Act of 1934 as amended by this Act and shall not be subject to the restrictions and the obligations imposed by such Consent Decree.
(3) MCCAW CONSENT DECREE- Any conduct or activity that was, before the date of enactment of this Act, subject to any restriction or obligation imposed by the McCaw Consent Decree shall, on and after such date, be subject to the restrictions and obligations imposed by the Communications Act of 1934 as amended by this Act and subsection (d) of this section and shall not be subject to the restrictions and the obligations imposed by such Consent Decree.
(b) ANTITRUST LAWS-
(1) SAVINGS CLAUSE- Except as provided in paragraphs (2) and (3), nothing in this Act or the amendments made by this Act shall be construed to modify, impair, or supersede the applicability of any of the antitrust laws.
(2) REPEAL- Subsection (a) of section 221 (47 U.S.C. 221(a)) is repealed.
(3) CLAYTON ACT- Section 7 of the Clayton Act (15 U.S.C. 18) is amended in the last paragraph by striking `Federal Communications Commission,'.
(c) FEDERAL, STATE, AND LOCAL LAW-
(1) NO IMPLIED EFFECT- This Act and the amendments made by this Act shall not be construed to modify, impair, or supersede Federal, State, or local law unless expressly so provided in such Act or amendments.
(2) STATE TAX SAVINGS PROVISION- Notwithstanding paragraph (1), nothing in this Act or the amendments made by this Act shall be construed to modify, impair, or supersede, or authorize the modification, impairment, or supersession of, any State or local law pertaining to taxation, except as provided in sections 622 and 653(c) of the Communications Act of 1934 and section 602 of this Act.
(d) COMMERCIAL MOBILE SERVICE JOINT MARKETING- Notwithstanding section 22.903 of the Commission's regulations (47 C.F.R. 22.903) or any other Commission regulation, a Bell operating company or any other company may, except as provided in sections 271(e)(1) and 272 of the Communications Act of 1934 as amended by this Act as they relate to wireline service, jointly market and sell commercial mobile services in conjunction with telephone exchange service, exchange access, intraLATA telecommunications service, interLATA telecommunications service, and information services.
(e) DEFINITIONS- As used in this section:
(1) AT&T CONSENT DECREE- The term `AT&T Consent Decree' means the order entered August 24, 1982, in the antitrust action styled United States v. Western Electric, Civil Action No. 82-0192, in the United States District Court for the District of Columbia, and includes any judgment or order with respect to such action entered on or after August 24, 1982.
(2) GTE CONSENT DECREE- The term `GTE Consent Decree' means the order entered December 21, 1984, as restated January 11, 1985, in the action styled United States v. GTE

Corp., Civil Action No. 83-1298, in the United States District Court for the District of Columbia, and any judgment or order with respect to such action entered on or after December 21, 1984.
(3) MCCAW CONSENT DECREE- The term `McCaw Consent Decree' means the proposed consent decree filed on July 15, 1994, in the antitrust action styled United States v. AT&T Corp. and McCaw Cellular Communications, Inc., Civil Action No. 94-01555, in the United States District Court for the District of Columbia. Such term includes any stipulation that the parties will abide by the terms of such proposed consent decree until it is entered and any order entering such proposed consent decree.
(4) ANTITRUST LAWS- The term `antitrust laws' has the meaning given it in subsection (a) of the first section of the Clayton Act (15 U.S.C. 12(a)), except that such term includes the Act of June 19, 1936 (49 Stat. 1526; 15 U.S.C. 13 et seq.), commonly known as the Robinson-Patman Act, and section 5 of the Federal Trade Commission Act (15 U.S.C. 45) to the extent that such section 5 applies to unfair methods of competition.

SEC. 602. PREEMPTION OF LOCAL TAXATION WITH RESPECT TO DIRECT-TO-HOME SERVICES.
(a) PREEMPTION- A provider of direct-to-home satellite service shall be exempt from the collection or remittance, or both, of any tax or fee imposed by any local taxing jurisdiction on direct-to-home satellite service.
(b) DEFINITIONS- For the purposes of this section--
(1) DIRECT-TO-HOME SATELLITE SERVICE- The term `direct-to-home satellite service' means only programming transmitted or broadcast by satellite directly to the subscribers' premises without the use of ground receiving or distribution equipment, except at the subscribers' premises or in the uplink process to the satellite.
(2) PROVIDER OF DIRECT-TO-HOME SATELLITE SERVICE- For purposes of this section, a `provider of direct-to-home satellite service' means a person who transmits, broadcasts, sells, or distributes direct-to-home satellite service.
(3) LOCAL TAXING JURISDICTION- The term `local taxing jurisdiction' means any municipality, city, county, township, parish, transportation district, or assessment jurisdiction, or any other local jurisdiction in the territorial jurisdiction of the United States with the authority to impose a tax or fee, but does not include a State.
(4) STATE- The term `State' means any of the several States, the District of Columbia, or any territory or possession of the United States.
(5) TAX OR FEE- The terms `tax' and `fee' mean any local sales tax, local use tax, local intangible tax, local income tax, business license tax, utility tax, privilege tax, gross receipts tax, excise tax, franchise fees, local telecommunications tax, or any other tax, license, or fee that is imposed for the privilege of doing business, regulating, or raising revenue for a local taxing jurisdiction.
(c) PRESERVATION OF STATE AUTHORITY- This section shall not be construed to prevent taxation of a provider of direct-to-home satellite service by a State or to prevent a local taxing jurisdiction from receiving revenue derived from a tax or fee imposed and collected by a State.

TITLE VII--MISCELLANEOUS PROVISIONS
SEC. 701. PREVENTION OF UNFAIR BILLING PRACTICES FOR INFORMATION OR SERVICES PROVIDED OVER TOLL-FREE TELEPHONE CALLS.

(a) PREVENTION OF UNFAIR BILLING PRACTICES-
(1) IN GENERAL- Section 228(c) (47 U.S.C. 228(c)) is amended--
(A) by striking out subparagraph (C) of paragraph (7) and inserting in lieu thereof the following:
`(C) the calling party being charged for information conveyed during the call unless--
`(i) the calling party has a written agreement (including an agreement transmitted through electronic medium) that meets the requirements of paragraph (8); or
`(ii) the calling party is charged for the information in accordance with paragraph (9); or';
(B)(i) by striking `or' at the end of subparagraph (C) of such paragraph;
(ii) by striking the period at the end of subparagraph (D) of such paragraph and inserting a semicolon and `or'; and
(iii) by adding at the end thereof the following:
`(E) the calling party being assessed, by virtue of being asked to connect or otherwise transfer to a pay-per-call service, a charge for the call.'; and
(C) by adding at the end the following new paragraphs:
`(8) SUBSCRIPTION AGREEMENTS FOR BILLING FOR INFORMATION PROVIDED VIA TOLL-FREE CALLS-
`(A) IN GENERAL- For purposes of paragraph (7)(C)(i), a written subscription does not meet the requirements of this paragraph unless the agreement specifies the material terms and conditions under which the information is offered and includes--
`(i) the rate at which charges are assessed for the information;
`(ii) the information provider's name;
`(iii) the information provider's business address;
`(iv) the information provider's regular business telephone number;
`(v) the information provider's agreement to notify the subscriber at least one billing cycle in advance of all future changes in the rates charged for the information; and
`(vi) the subscriber's choice of payment method, which may be by direct remit, debit, prepaid account, phone bill, or credit or calling card.
`(B) BILLING ARRANGEMENTS- If a subscriber elects, pursuant to subparagraph (A)(vi), to pay by means of a phone bill--
`(i) the agreement shall clearly explain that the subscriber will be assessed for calls made to the information service from the subscriber's phone line;
`(ii) the phone bill shall include, in prominent type, the following disclaimer:
`Common carriers may not disconnect local or long distance telephone service for failure to pay disputed charges for information services.'; and
`(iii) the phone bill shall clearly list the 800 number dialed.
`(C) USE OF PINS TO PREVENT UNAUTHORIZED USE- A written agreement does not meet the requirements of this paragraph unless it--
`(i) includes a unique personal identification number or other subscriber-specific identifier and requires a subscriber to use this number or identifier to obtain access to the information provided and includes instructions on its use; and
`(ii) assures that any charges for services accessed by use of the subscriber's personal identification number or subscriber-specific identifier be assessed to subscriber's source of payment elected pursuant to subparagraph (A)(vi).
`(D) EXCEPTIONS- Notwithstanding paragraph (7)(C), a written agreement that meets the requirements of this paragraph is not required--
`(i) for calls utilizing telecommunications devices for the deaf;

`(ii) for directory services provided by a common carrier or its affiliate or by a local exchange carrier or its affiliate; or
`(iii) for any purchase of goods or of services that are not information services.
`(E) TERMINATION OF SERVICE- On receipt by a common carrier of a complaint by any person that an information provider is in violation of the provisions of this section, a carrier shall--
`(i) promptly investigate the complaint; and
`(ii) if the carrier reasonably determines that the complaint is valid, it may terminate the provision of service to an information provider unless the provider supplies evidence of a written agreement that meets the requirements of this section.
`(F) TREATMENT OF REMEDIES- The remedies provided in this paragraph are in addition to any other remedies that are available under title V of this Act.
`(9) CHARGES BY CREDIT, PREPAID, DEBIT, CHARGE, OR CALLING CARD IN ABSENCE OF AGREEMENT- For purposes of paragraph (7)(C)(ii), a calling party is not charged in accordance with this paragraph unless the calling party is charged by means of a credit, prepaid, debit, charge, or calling card and the information service provider includes in response to each call an introductory disclosure message that--
`(A) clearly states that there is a charge for the call;
`(B) clearly states the service's total cost per minute and any other fees for the service or for any service to which the caller may be transferred;
`(C) explains that the charges must be billed on either a credit, prepaid, debit, charge, or calling card;
`(D) asks the caller for the card number;
`(E) clearly states that charges for the call begin at the end of the introductory message; and
`(F) clearly states that the caller can hang up at or before the end of the introductory message without incurring any charge whatsoever.
`(10) BYPASS OF INTRODUCTORY DISCLOSURE MESSAGE- The requirements of paragraph (9) shall not apply to calls from repeat callers using a bypass mechanism to avoid listening to the introductory message: [Italic->] Provided, [<-Italic] That information providers shall disable such a bypass mechanism after the institution of any price increase and for a period of time determined to be sufficient by the Federal Trade Commission to give callers adequate and sufficient notice of a price increase.
`(11) DEFINITION OF CALLING CARD- As used in this subsection, the term `calling card' means an identifying number or code unique to the individual, that is issued to the individual by a common carrier and enables the individual to be charged by means of a phone bill for charges incurred independent of where the call originates.'.
(2) REGULATIONS- The Federal Communications Commission shall revise its regulations to comply with the amendment made by paragraph (1) not later than 180 days after the date of enactment of this Act.
(3) EFFECTIVE DATE- The amendments made by paragraph (1) shall take effect on the date of enactment of this Act.
(b) CLARIFICATION OF `PAY-PER-CALL SERVICES'-
(1) TELEPHONE DISCLOSURE AND DISPUTE RESOLUTION ACT- Section 204(1) of the Telephone Disclosure and Dispute Resolution Act (15 U.S.C. 5714(1)) is amended to read as follows:
`(1) The term `pay-per-call services' has the meaning provided in section 228(i) of the Communications Act of 1934, except that the Commission by rule may, notwithstanding

subparagraphs (B) and (C) of section 228(i)(1) of such Act, extend such definition to other similar services providing audio information or audio entertainment if the Commission determines that such services are susceptible to the unfair and deceptive practices that are prohibited by the rules prescribed pursuant to section 201(a).'.
(2) COMMUNICATIONS ACT- Section 228(i)(2) (47 U.S.C. 228(i)(2)) is amended by striking `or any service the charge for which is tariffed,'.

SEC. 702. PRIVACY OF CUSTOMER INFORMATION.
Title II is amended by inserting after section 221 (47 U.S.C. 221) the following new section:

`SEC. 222. PRIVACY OF CUSTOMER INFORMATION.
`(a) IN GENERAL- Every telecommunications carrier has a duty to protect the confidentiality of proprietary information of, and relating to, other telecommunication carriers, equipment manufacturers, and customers, including telecommunication carriers reselling telecommunications services provided by a telecommunications carrier.
`(b) CONFIDENTIALITY OF CARRIER INFORMATION- A telecommunications carrier that receives or obtains proprietary information from another carrier for purposes of providing any telecommunications service shall use such information only for such purpose, and shall not use such information for its own marketing efforts.
`(c) CONFIDENTIALITY OF CUSTOMER PROPRIETARY NETWORK INFORMATION-
`(1) PRIVACY REQUIREMENTS FOR TELECOMMUNICATIONS CARRIERS- Except as required by law or with the approval of the customer, a telecommunications carrier that receives or obtains customer proprietary network information by virtue of its provision of a telecommunications service shall only use, disclose, or permit access to individually identifiable customer proprietary network information in its provision of (A) the telecommunications service from which such information is derived, or (B) services necessary to, or used in, the provision of such telecommunications service, including the publishing of directories.
`(2) DISCLOSURE ON REQUEST BY CUSTOMERS- A telecommunications carrier shall disclose customer proprietary network information, upon affirmative written request by the customer, to any person designated by the customer.
`(3) AGGREGATE CUSTOMER INFORMATION- A telecommunications carrier that receives or obtains customer proprietary network information by virtue of its provision of a telecommunications service may use, disclose, or permit access to aggregate customer information other than for the purposes described in paragraph (1). A local exchange carrier may use, disclose, or permit access to aggregate customer information other than for purposes described in paragraph (1) only if it provides such aggregate information to other carriers or persons on reasonable and nondiscriminatory terms and conditions upon reasonable request therefor.
`(d) EXCEPTIONS- Nothing in this section prohibits a telecommunications carrier from using, disclosing, or permitting access to customer proprietary network information obtained from its customers, either directly or indirectly through its agents--
`(1) to initiate, render, bill, and collect for telecommunications services;
`(2) to protect the rights or property of the carrier, or to protect users of those services and other carriers from fraudulent, abusive, or unlawful use of, or subscription to, such services; or
`(3) to provide any inbound telemarketing, referral, or administrative services to the customer

for the duration of the call, if such call was initiated by the customer and the customer approves of the use of such information to provide such service.
`(e) SUBSCRIBER LIST INFORMATION- Notwithstanding subsections (b), (c), and (d), a telecommunications carrier that provides telephone exchange service shall provide subscriber list information gathered in its capacity as a provider of such service on a timely and unbundled basis, under nondiscriminatory and reasonable rates, terms, and conditions, to any person upon request for the purpose of publishing directories in any format.
`(f) DEFINITIONS- As used in this section:
`(1) CUSTOMER PROPRIETARY NETWORK INFORMATION- The term `customer proprietary network information' means--
`(A) information that relates to the quantity, technical configuration, type, destination, and amount of use of a telecommunications service subscribed to by any customer of a telecommunications carrier, and that is made available to the carrier by the customer solely by virtue of the carrier-customer relationship; and
`(B) information contained in the bills pertaining to telephone exchange service or telephone toll service received by a customer of a carrier; except that such term does not include subscriber list information.
`(2) AGGREGATE INFORMATION- The term `aggregate customer information' means collective data that relates to a group or category of services or customers, from which individual customer identities and characteristics have been removed.
`(3) SUBSCRIBER LIST INFORMATION- The term `subscriber list information' means any information--
`(A) identifying the listed names of subscribers of a carrier and such subscribers' telephone numbers, addresses, or primary advertising classifications (as such classifications are assigned at the time of the establishment of such service), or any combination of such listed names, numbers, addresses, or classifications; and`(B) that the carrier or an affiliate has published, caused to be published, or accepted for publication in any directory format.'.

SEC. 705. MOBILE SERVICES DIRECT ACCESS TO LONG DISTANCE CARRIERS.
Section 332(c) (47 U.S.C. 332(c)) is amended by adding at the end the following new paragraph:
`(8) MOBILE SERVICES ACCESS- A person engaged in the provision of commercial mobile services, insofar as such person is so engaged, shall not be required to provide equal access to common carriers for the provision of telephone toll services. If the Commission determines that subscribers to such services are denied access to the provider of telephone toll services of the subscribers' choice, and that such denial is contrary to the public interest, convenience, and necessity, then the Commission shall prescribe regulations to afford subscribers unblocked access to the provider of telephone toll services of the subscribers' choice through the use of a carrier identification code assigned to such provider or other mechanism. The requirements for unblocking shall not apply to mobile satellite services unless the Commission finds it to be in the public interest to apply such requirements to such services.'

SEC. 706. ADVANCED TELECOMMUNICATIONS INCENTIVES.
(a) IN GENERAL- The Commission and each State commission with regulatory jurisdiction over telecommunications services shall encourage the deployment on a reasonable and timely basis of advanced telecommunications capability to all Americans (including, in particular, elementary and secondary schools and classrooms) by utilizing, in a manner consistent with the public interest, convenience, and necessity, price cap regulation, regulatory forbearance, measures that promote competition in the local telecommunications market, or other regulating methods that remove barriers to infrastructure investment.
(b) INQUIRY- The Commission shall, within 30 months after the date of enactment of this Act, and regularly thereafter, initiate a notice of inquiry concerning the availability of advanced telecommunications capability to all Americans (including, in particular, elementary and secondary schools and classrooms) and shall complete the inquiry within 180 days after its initiation. In the inquiry, the Commission shall determine whether advanced telecommunications capability is being deployed to all Americans in a reasonable and timely fashion. If the Commission's determination is negative, it shall take immediate action to accelerate deployment of such capability by removing barriers to infrastructure investment and by promoting competition in the telecommunications market.
(c) DEFINITIONS- For purposes of this subsection:
(1) ADVANCED TELECOMMUNICATIONS CAPABILITY- The term `advanced telecommunications capability' is defined, without regard to any transmission media or technology, as high-speed, switched, broadband telecommunications capability that enables users to originate and receive high-quality voice, data, graphics, and video telecommunications using any technology.
(2) ELEMENTARY AND SECONDARY SCHOOLS- The term `elementary and secondary schools' means elementary and secondary schools, as defined in paragraphs (14) and (25), respectively, of section 14101 of the Elementary and Secondary Education Act of 1965 (20 U.S.C. 8801).

`SEC. 714. TELECOMMUNICATIONS DEVELOPMENT FUND.
`(a) PURPOSE OF SECTION- It is the purpose of this section--
`(1) to promote access to capital for small businesses in order to enhance competition in the telecommunications industry;
`(2) to stimulate new technology development, and promote employment and training; and
`(3) to support universal service and promote delivery of telecommunications services to underserved rural and urban areas.
`(b) ESTABLISHMENT OF FUND- There is hereby established a body corporate to be known as the Telecommunications Development Fund, which shall have succession until dissolved. The Fund shall maintain its principal office in the District of Columbia and shall be deemed, for purposes of venue and jurisdiction in civil actions, to be a resident and citizen thereof.
`(c) BOARD OF DIRECTORS-
`(1) COMPOSITION OF BOARD; CHAIRMAN- The Fund shall have a Board of Directors which shall consist of 7 persons appointed by the Chairman of the Commission. Four of such directors shall be representative of the private sector and three of such directors shall be representative of the Commission, the Small Business Administration, and the Department of

the Treasury, respectively. The Chairman of the Commission shall appoint one of the representatives of the private sector to serve as chairman of the Fund within 30 days after the date of enactment of this section, in order to facilitate rapid creation and implementation of the Fund. The directors shall include members with experience in a number of the following areas: finance, investment banking, government banking, communications law and administrative practice, and public policy.

`(2) TERMS OF APPOINTED AND ELECTED MEMBERS- The directors shall be eligible to serve for terms of 5 years, except of the initial members, as designated at the time of their appointment--

`(A) 1 shall be eligible to service for a term of 1 year;

`(B) 1 shall be eligible to service for a term of 2 years;

`(C) 1 shall be eligible to service for a term of 3 years;

`(D) 2 shall be eligible to service for a term of 4 years; and

`(E) 2 shall be eligible to service for a term of 5 years (1 of whom shall be the Chairman). Directors may continue to serve until their successors have been appointed and have qualified.

`(3) MEETINGS AND FUNCTIONS OF THE BOARD- The Board of Directors shall meet at the call of its Chairman, but at least quarterly. The Board shall determine the general policies which shall govern the operations of the Fund. The Chairman of the Board shall, with the approval of the Board, select, appoint, and compensate qualified persons to fill the offices as may be provided for in the bylaws, with such functions, powers, and duties as may be prescribed by the bylaws or by the Board of Directors, and such persons shall be the officers of the Fund and shall discharge all such functions, powers, and duties.

`(d) ACCOUNTS OF THE FUND- The Fund shall maintain its accounts at a financial institution designated for purposes of this section by the Chairman of the Board (after consultation with the Commission and the Secretary of the Treasury). The accounts of the Fund shall consist of--

`(1) interest transferred pursuant to section 309(j)(8)(C) of this Act;

`(2) such sums as may be appropriated to the Commission for advances to the Fund;

`(3) any contributions or donations to the Fund that are accepted by the Fund; and

`(4) any repayment of, or other payment made with respect to, loans, equity, or other extensions of credit made from the Fund.

`(e) USE OF THE FUND- All moneys deposited into the accounts of the Fund shall be used solely for--

`(1) the making of loans, investments, or other extensions of credits to eligible small businesses in accordance with subsection (f);

`(2) the provision of financial advice to eligible small businesses;

`(3) expenses for the administration and management of the Fund (including salaries, expenses, and the rental or purchase of office space for the fund);

`(4) preparation of research, studies, or financial analyses;and

`(5) other services consistent with the purposes of this section.

`(f) LENDING AND CREDIT OPERATIONS- Loans or other extensions of credit from the Fund shall be made available in accordance with the requirements of the Federal Credit Reform Act of 1990 (2 U.S.C. 661 et seq.) and any other applicable law to an eligible small business on the basis of--

`(1) the analysis of the business plan of the eligible small business;

`(2) the reasonable availability of collateral to secure the loan or credit extension;

`(3) the extent to which the loan or credit extension promotes the purposes of this section; and

`(4) other lending policies as defined by the Board.
`(g) RETURN OF ADVANCES- Any advances appropriated pursuant to subsection (d)(2) shall be disbursed upon such terms and conditions (including conditions relating to the time or times of repayment) as are specified in any appropriations Act providing such advances.
`(h) GENERAL CORPORATE POWERS- The Fund shall have power--
`(1) to sue and be sued, complain and defend, in its corporate name and through its own counsel;
`(2) to adopt, alter, and use the corporate seal, which shall be judicially noticed;
`(3) to adopt, amend, and repeal by its Board of Directors, bylaws, rules, and regulations as may be necessary for the conduct of its business;
`(4) to conduct its business, carry on its operations, and have officers and exercise the power granted by this section in any State without regard to any qualification or similar statute in any State;
`(5) to lease, purchase, or otherwise acquire, own, hold, improve, use, or otherwise deal in and with any property, real, personal, or mixed, or any interest therein, wherever situated, for the purposes of the Fund;
`(6) to accept gifts or donations of services, or of property, real, personal, or mixed, tangible or intangible, in aid of any of the purposes of the Fund;
`(7) to sell, convey, mortgage, pledge, lease, exchange, and otherwise dispose of its property and assets;
`(8) to appoint such officers, attorneys, employees, and agents as may be required, to determine their qualifications, to define their duties, to fix their salaries, require bonds for them, and fix the penalty thereof; and
`(9) to enter into contracts, to execute instruments, to incur liabilities, to make loans and equity investment, and to do all things as are necessary or incidental to the proper management of its affairs and the proper conduct of its business.
`(i) ACCOUNTING, AUDITING, AND REPORTING- The accounts of the Fund shall be audited annually. Such audits shall be conducted in accordance with generally accepted auditing standards by independent certified public accountants. A report of each such audit shall be furnished to the Secretary of the Treasury and the Commission. The representatives of the Secretary and the Commission shall have access to all books, accounts, financial records, reports, files, and all other papers, things, or property belonging to or in use by the Fund and necessary to facilitate the audit.
`(j) REPORT ON AUDITS BY TREASURY- A report of each such audit for a fiscal year shall be made by the Secretary of the Treasury to the President and to the Congress not later than 6 months following the close of such fiscal year. The report shall set forth the scope of the audit and shall include a statement of assets and liabilities, capital and surplus or deficit; a statement of surplus or deficit analysis; a statement of income and expense; a statement of sources and application of funds; and such comments and information as may be deemed necessary to keep the President and the Congress informed of the operations and financial condition of the Fund, together with such recommendations with respect thereto as the Secretary may deem advisable.
`(k) DEFINITIONS- As used in this section:
`(1) ELIGIBLE SMALL BUSINESS- The term `eligible small business' means business enterprises engaged in the telecommunications industry that have $50,000,000 or less in annual revenues, on average over the past 3 years prior to submitting the application under

this section.
`(2) FUND- The term `Fund' means the Telecommunications Development Fund established pursuant to this section.
`(3) TELECOMMUNICATIONS INDUSTRY- The term `telecommunications industry' means communications businesses using regulated or unregulated facilities or services and includes broadcasting, telecommunications, cable, computer, data transmission, software, programming, advanced messaging, and electronics businesses.'.

Section B: Current Investments in Innovation in the Information Technology Sector: Statistical Background (Abridged Version for Strategists)

CBO MEMORANDUM

CURRENT INVESTMENTS IN INNOVATION
IN THE
INFORMATION TECHNOLOGY SECTOR:
STATISTICAL BACKGROUND

(Abridged Version for Strategists)

April 1999

The President's budget for fiscal year 2000 includes a proposal to increase federal spending on research and development (R&D) of computer and data communications by $366 million. That proposal, called Information Technology for the Twenty-First Century, or IT^2, coincides with a significant upswing in private investment in such R&D. The outpouring of private funds raises the question of how much the IT^2 initiative can contribute to the nation's overall technology development effort.

The House Committee on Science has asked the Congressional Budget Office (CBO) to conduct two analyses of the IT^2 initiative: a statistical analysis of baseline spending in the technology areas that would be funded by the initiative and a policy analysis of the overlap between federal and private efforts in the information technology sector. This CBO memorandum responds to the first request. It provides background statistics about current private and public spending on R&D in the technology areas funded by the initiative. A later CBO analysis will address the substantive policy issues. In keeping with CBO's mandate to provide objective, impartial analysis, this memorandum makes no recommendations.

Author's note: These excerpts are culled from the Congressional Budget Office's website at cbo.gov. The full report can be obtained under the title indicated above within this website. Data outlined below are referred to in Chapters 1, 2, 4 and 8.

CONGRESSIONAL BUDGET OFFICE
SECOND AND D STREETS, S.W.
WASHINGTON, D.C. 20515

CONTENTS/

The President's Information Technology Advisory Committee has proposed a five-year federal initiative that would double the current federal funding of roughly $1.5 billion for developing computer and data communications technology. Following the committee's general recommendations, the President's budget submission requests $366 million in new funds for fiscal year 2000 to stimulate innovation in that area. The Administration's proposal, called Information Technology for the Twenty-First Century, or IT^2, would distribute those funds among several agencies, with the National Science Foundation receiving the largest share.

This Congressional Budget Office (CBO) memorandum provides background statistics on the current level of spending by federal, industrial, and academic organizations on research and development in the computer and data communications—or information technology (IT)—sector. It also provides statistics on the human resources available to that sector. A later CBO analysis will explore in depth the substantive policy issues raised by the IT^2 initiative.

The statistics and data in this memorandum provide a snapshot of the nation's innovation capabilities in the IT area. Four actors dominate that sector: federal agencies, private companies, and venture capitalists, which are the main sources of funding for IT research and development; and academic researchers, who are net recipients of that funding. This analysis focuses on the funding of research, development, and innovation from the point of view of those who provide it, ignoring the crucial role that many of those participants also play in creating and disseminating information technology by buying and defining that technology as consumers. This memorandum also focuses only on research and development spending by U.S. entities; it excludes investments by foreign companies, governments, and universities.

TOTAL SPENDING ON RESEARCH AND DEVELOPMENT IN THE INFORMATION TECHNOLOGY SECTOR

In 1997, the most recent year for which complete information exists, private industry provided the lion's share of funding for research and development (R&D) in the information technology sector (see Table 1). Funds from venture capital firms ran a distant second, and the federal government contributed the smallest amount. Those sources tend to spend their money on different things, however. Private companies and venture capitalists focus more of their spending on short-term development of products. Venture capital funding also covers such non-R&D expenses as capital investment and marketing. Federal funding, by contrast, is more concentrated in long-term basic research. Consequently, despite its small size, federal funding may have a disproportionately large effect on the direction that information technology takes in the long run. (CBO could find no information

TABLE 1. SPENDING ON INFORMATION TECHNOLOGY RESEARCH AND DEVELOPMENT, BY SOURCE OF FUNDS, 1997

Source of Funds	Funding Level (Billions of dollars)	Typical Activity
Industry	44.2	Developing new computers or data communications devices or software

Venture Capital	6.9	Funding new companies to bring new products to market
Federal[a]	1.2	Using high-end computers to solve problems in basic science

SOURCE: Congressional Budget Office based on data from National Science Foundation, *Research and Development in Industry, 1997* (1999); National Coordination Office for Computing, Information, and Communications, *Computing, Information, and Communications. Networked Computing for the 21st Century* (August 1998), and *High Performance Computing and Communications: FY 1998 Implementation Plan* (September 1998); and Venture Economics Information Services.

NOTE: Industry and venture capital spending is for calendar year 1997. Federal spending is for fiscal year 1997.

a. Includes only technology development programs from the High Performance Computing and Communications initiative and the Department of Energy.

about the amount of IT funding that universities, colleges, and independent research institutes contribute from their endowments or from state appropriations.)

In absolute terms, the information technology sector now receives one-third of all corporate spending on R&D in the U.S. economy. Similarly, the venture capital funds going to the IT sector represent more than half of all venture capital funding in the economy.

WHAT CONSTITUTES THE INFORMATION TECHNOLOGY SECTOR?

As this memorandum defines it, the information technology sector comprises five main industries:

- Computer, accounting, and business machines;
- Communications equipment;
- Electronic components;
- Communications services, including telephone services; and
- Computer and data-processing services.[1]

Many analysts do not include the industry that produces electronic components, such as semiconductors and printed circuit boards, in their definition of the IT sector. However, that exclusion fails to recognize the changing relationship between the computer industry and the suppliers of its components. Many manufacturers of personal computers do little research and development of their own, relying instead on the R&D performed by component manufacturers for their technological impetus. Only part of that R&D benefits the IT sector, however; the automotive industry and consumer electronics other than computers and

1. Those are the standard industrial classification designations that the Bureau of the Census uses in reporting on industrial activity.

communications equipment are also large direct consumers of electronic components.

According to data from the Bureau of the Census, the output of the U.S. information technology sector totaled $845 billion in 1997, the last year for which complete data are available (see Table 2).[2] Excluding electronic components, that output was $717 billion. Between 1993 and 1997, the IT sector grew at an average rate of 12.3 percent a year (or 11.7 percent if electronic components are excluded). Because prices in the sector have largely been declining, the growth rate of output adjusted for inflation is probably even higher than that. Moreover, although prices have been falling, product quality has been improving markedly, further increasing the sector's total output.

By way of comparison, the nation's inflation-adjusted gross domestic product (GDP) grew at an average rate of just 3 percent per year between 1993 and 1997. Of course, since several IT industries produce inputs for each other, their net contribution to GDP would be different from their total output.

TABLE 2. OUTPUT OF INFORMATION TECHNOLOGY INDUSTRIES, CALENDAR YEARS 1993-1997 (In billions of dollars)

Industry	1993	1994	1995	1996	1997
Office, Computing, and Accounting Machines	55	65	76	83	82
Electronic Components	71	93	119	122	128
Communications Equipment	40	47	53	62	71
Communications Services	247	267	291	322	348
Computer and Data-Processing Services	118	136	156	184	216
Total	531	607	696	775	845
Total Excluding Electronic Components	460	514	577	653	717

SOURCE: Congressional Budget Office based on data from Bureau of the Census, *Current Industrial Reports, 1997 Service Annual Survey*, and *1997 Annual Survey of Communication Services*.

2. The Census Bureau uses the value of shipments to measure output from manufacturing industries and total revenue to measure output from service industries, which do not "ship" anything. The IT sector includes both manufacturing and services; consequently, its output estimate includes both measures.

FEDERAL FUNDING FOR INFORMATION TECHNOLOGY R&D

As noted above, the President's Information Technology Advisory Committee has proposed doubling the current level of federal funding for IT development over five years. That funding doubled between 1991 and 1995, but since then its growth has been slower, only 35 percent. The federal funding for information technology R&D discussed in this memorandum is only funding that forms part of the High Performance Computing and Communications program and the Department of Energy's strategic stockpile management programs. Although those programs are the main federal initiatives for technology development, other federal programs may be investing in IT development as part of their mission.

Different Measures of Federal Spending

Funding for the major federal IT initiatives totals $1.4 billion to $1.5 billion a year. Within that total, spending on the High Performance Computing and Communications program has been stable for five years, whereas spending on the Department of Energy's Accelerated Strategic Computing Initiative and stockpile computing program has been growing since those programs began in 1996 and 1998, respectively (see Table 3).

Those figures miss some federal spending on information technology, however. An alternative measure of federal efforts to promote IT research comes from the annual surveys of federal research managers conducted by the National Science Foundation (NSF)—particularly those of research managers in two IT-related fields, electrical engineering and computer science.[3] The data from those surveys, which are available through 1996, reflect funding for both applied and basic research for all entities that perform such research: federal, industry, academic, and other facilities.

TABLE 3. FUNDING FOR CURRENT FEDERAL PROGRAMS THAT FOCUS ON INFORMATION TECHNOLOGY (By fiscal year, in millions of dollars of budget authority)

	1991	1992	1993	1994	1995	1996	1997	1998	1999
High Performance Computing and Communications Initiative	489	655	795	938	1,129	1,043	1,009	1,070	1,041
Department of Energy Accelerated Strategic Computing Initiative	a	a	a	a	a	86	152	224	301
Stockpile Computing	a	a	a	a	a	a	a	151	183

3. Other fields of science obviously play a role in the development of information technology, but their relationship to that area is more difficult to separate out.

SOURCE: Congressional Budget Office based on National Coordination Office for Computing, Information, and Communications, *Computing, Information, and Communications: Networked Computing for the 21st Century* (August 1998), and *High Performance Computing and Communications: FY 1998 Implementation Plan* (September 1998).

NOTE: Excludes some mission-related research and development performed by other federal programs.

a. This program did not exist in this year.

Federal spending on computer science research, whether applied or basic, has risen substantially over the past 20 years (see Figure 1). During the 1990s, it more than tripled for applied research. By contrast, federal spending on electrical engineering has fluctuated—rising and falling several times since the mid-1980s in the case of applied research, and rising more steadily but at a much lower rate in the case of basic research. Adjusted for inflation, federal spending on both IT-related fields presents a simple story: funding for computer science research is increasing, whereas funding for electrical engineering research is stagnant or declining.

FIGURE 1. TOTAL FEDERAL FUNDING FOR RESEARCH RELATED TO INFORMATION TECHNOLOGY, FISCAL YEARS 1976-1996

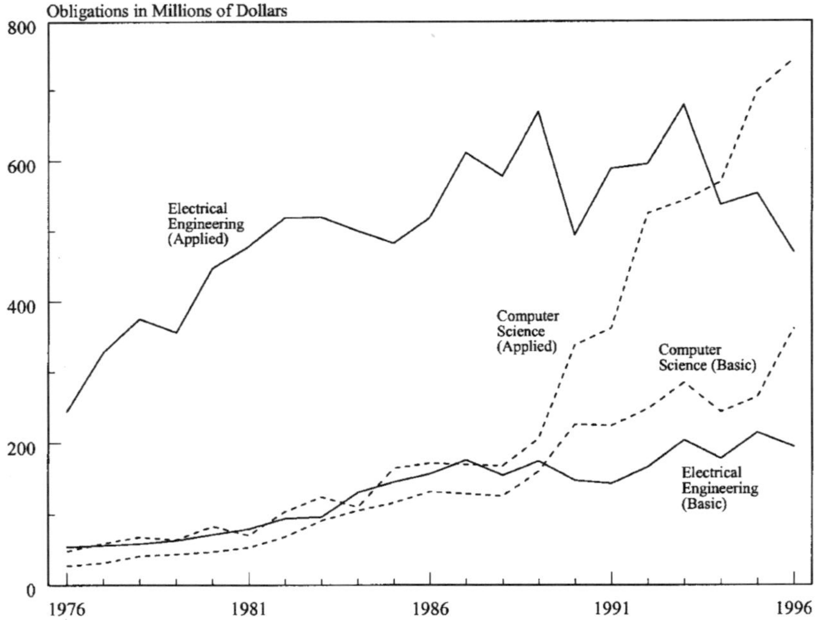

SOURCE: Congressional Budget Office based on data from National Science Foundation, *Federal Funds for Research and Development: Fiscal Years 1996, 1997, and 1998*, vol. 46 (1998).

A comparison of the data in Table 3 and Figure 1 suggests that the program totals in Table 3 probably undercount the total federal IT effort. According to the

NSF surveys, federal agencies spent $1.1 billion on computer science in 1996—roughly the same amount that the main federal IT programs spent that year. The major difference is in the area of electrical engineering, especially hardware development involving electronic components and other things below the computer system level. The bulk of the computer science spending shown in Figure 1 is probably included in the numbers in Table 3, but only a portion of the federal spending on electrical engineering is.[4]

Federal spending on electrical engineering research is only partly related to IT development, however. A great deal of federal research in that area is focused on developing sensors and other specialized military equipment that may not have any role in information technology. But other federal research on electronic components—such as the long-standing effort to develop high-performance substitutes for silicon in semiconductors—may end up playing an important role in civilian wireless communications.

Federal Spending by Agency

Federal funding for applied research in computer science has been driven by the needs of the Department of Defense (DoD) until quite recently (see Table 4). In 1990, for example, DoD research accounted for 80 percent of all such federal funding. By 1996, however, that share had fallen to 55 percent. Department of Energy programs account for most of the growth in nondefense spending.[5]

TABLE 4. FEDERAL FUNDING FOR COMPUTER SCIENCE RESEARCH, BY AGENCY, FISCAL YEARS 1986-1996 (Obligations in millions of dollars)

	Department of Defense	Department of Energy	National Aeronautics and Space Administration	National Science Foundation	Other Agencies	Total
			Applied Research			
1986	124	5	20	1	21	171
1987	111	5	27	2	24	169
1988	113	5	23	3	23	167
1989	146	5	26	4	25	205
1990	270	5	23	12	27	337
1991	219	64	27	8	44	361
1992	335	92	25	20	53	524
1993	361	83	7	9	93	552
1994	381	73	8	15	92	568

4. Of course, not all federal spending on computer science is contained within the programs listed in Table 3, but the overlap is large.

5. The data in Table 4 may understate the National Aeronautics and Space Administration's (NASA's) role in information technology spending. A disproportionate amount of NASA's spending is in a catchall category called Mathematics and Computer Science Not Elsewhere Classified. The precise allocation of those funds is not clear, but including them would not change the picture materially.

1995	443	80	9	9	171	711
1996	407	187	8	15	140	757
			Basic Research			
1986	40	25	20	44	3	131
1987	38	10	19	58	3	129
1988	42	11	17	52	3	125
1989	55	12	22	67	3	160
1990	85	18	22	97	3	225
1991	70	23	24	99	8	224
1992	92	25	19	104	8	248
1993	112	21	17	118	16	284
1994	74	17	18	131	3	243
1995	100	17	18	116	13	264
1996	67	14	18	231	31	361

SOURCE: Congressional Budget Office based on data from National Science Foundation, *Federal Funds Survey, Fields of Science and Engineering Research Historical Tables, Fiscal Years 1970-98* (1998).

For basic research in computer science, federal funding was divided almost in half between DoD and the National Science Foundation at the beginning of the decade, with other agencies playing lesser roles (see Table 4). By 1996, however, NSF spending for computer science accounted for almost two-thirds of federal spending in that area—a consequence of the High Performance Computing and Communications initiative.

CORPORATE SPENDING ON INFORMATION TECHNOLOGY R&D

Data on IT research and development by private industry can be helpful to policymakers trying to decide whether to double the current level of federal funding over the next five years. Is the IT sector investing enough in R&D, both in absolute terms and relative to its size? Furthermore, are those funds being invested in the kinds of research that will produce benefits for the long run, or are they focused on the details of product development? Available data suggest that the IT sector is investing sufficiently overall, but it is disproportionately focusing on short-run investments in product development.

Current Data on Industry R&D

The National Science Foundation collects R&D data by company rather than by establishment (such as an individual factory) as the Census Bureau does for its industrial output statistics. In the NSF data series, all of a company's R&D is allocated to the single largest activity (measured by payroll) in which the company is engaged. Thus, a company that conducts a variety of activities in information technology would be classified as conducting only one. Small changes in a large company's payroll can therefore result in apparently large changes and discontinuities in industry data series that stem mainly from the method of classification, not from changes in allocation of R&D resources. Because large companies typically shift among industries within the IT sector—say, from communications equipment to communications services—the data for the total sector should not be affected by those shifts in classification, but changes in individual industries may be misleading.

Overall R&D Levels

Companies in the IT sector spent $44.2 billion on research and development in 1997, the most recent year for which data exist (see Table 5). That spending represents about one-third of all corporate spending on R&D in the U.S. economy. Between 1993 and 1997, the IT sector increased its investment in R&D by an

TABLE 5. NONFEDERAL SPENDING ON INDUSTRIAL R&D, BY INDUSTRY, CALENDAR YEARS 1993-1997 (In billions of dollars)

Industry	1993	1994	1995	1996	1997
Office, Computing, and Accounting Machines	4.9	4.1	4.7	8.1	12.8
Electronic Components	5.1	5.9	9.6	12.5	10.8
Communications Equipment	4.0	4.9	3.8	4.4	7.4
Communications Services	4.3	4.2	4.8	4.0	1.9
Computer and Data-Processing Services	6.9	5.7	8.5	10.0	11.3
Total	25.2	24.8	31.5	39.0	44.2
Memorandum: All Industries	94.6	97.1	108.7	121.0	133.6

SOURCE: Congressional Budget Office based on data from National Science Foundation, *Research and Development in Industry, 1997* (1999).

average of 15 percent a year, or a total of $19 billion. If the electronic components industry is excluded from the IT sector, growth in R&D averaged 14 percent a year, and total corporate R&D funding in 1997 was $33.4 billion. Because of the swings in company classification, however, that subtotal is less reliable than the total for the entire IT sector; the data suggest that some companies were reclassified between 1996 and 1997.

R&D Relative to Industry Sales

According to the National Science Foundation, the information technology sector spends much more on research and development, relatively speaking, than industry as a whole does—6.1 percent of net sales in 1997 compared with less than 3 percent (see Table 6).[6] That comparison is possible only for years since 1995, however. Before that, the NSF did not report IT industries separately from nonmanufacturing industries in general.

6. Net sales should not be confused with the value of shipments. One is on a company basis; the other uses establishments. Furthermore, the NSF data are for net sales (that is, sales minus returns and other adjustments) only of companies performing R&D; the shipment data are for all establishments in the industry.

Although the general level of funding might indicate which industries are more R&D intensive, a fluctuation or even a long-term decline in the ratio of R&D to sales does not necessarily mean that the sector is decreasing its commitment to research and development. Information goods are typically characterized by high fixed costs and low variable costs. A certain level of R&D is required to develop a computer, a software package, or some other IT product. Once that product has been developed, the company can sell 1,000 or 1 million copies without performing additional research and development. The ratio of R&D to sales will decline as the number of copies sold increases, but that does not mean that the company is any less technologically adept, just that its market has grown. Since the IT sector is

TABLE 6. CORPORATE R&D SPENDING AS A PERCENTAGE OF NET SALES OF ALL COMPANIES PERFORMING R&D, BY INDUSTRY, CALENDAR YEARS 1995-1997

Industry	1995	1996	1997
Office, Computing, and Accounting Machines	8.1	9.9	9.2
Electronic Components	8.0	8.5	8.1
Communications Equipment	8.0	8.5	8.0
Communications Services	2.2	1.9	0.7
Computer and Data-Processing Services	11.1	12.4	13.3
Sales-Weighted Average	6.0	6.8	6.1
Memorandum: All Industries	2.8	3.0	2.9

SOURCE: Congressional Budget Office based on data from National Science Foundation, *Research and Development in Industry, 1997* (1999).

growing very rapidly, its ratio would be expected to vary and even decline over time even as IT firms increased their investment in R&D.

To provide some background, Figure 2 presents a 10-year time series of the ratio of R&D spending to net sales for the manufacturing segments of the IT sector: computers and office equipment, communications equipment, and electronic components. Clearly, those three industries have moved in very different directions.

FIGURE 2. CORPORATE R&D SPENDING AS A PERCENTAGE OF NET SALES OF ALL COMPANIES PERFORMING R&D, BY INDUSTRY, CALENDAR YEARS 1987-1997

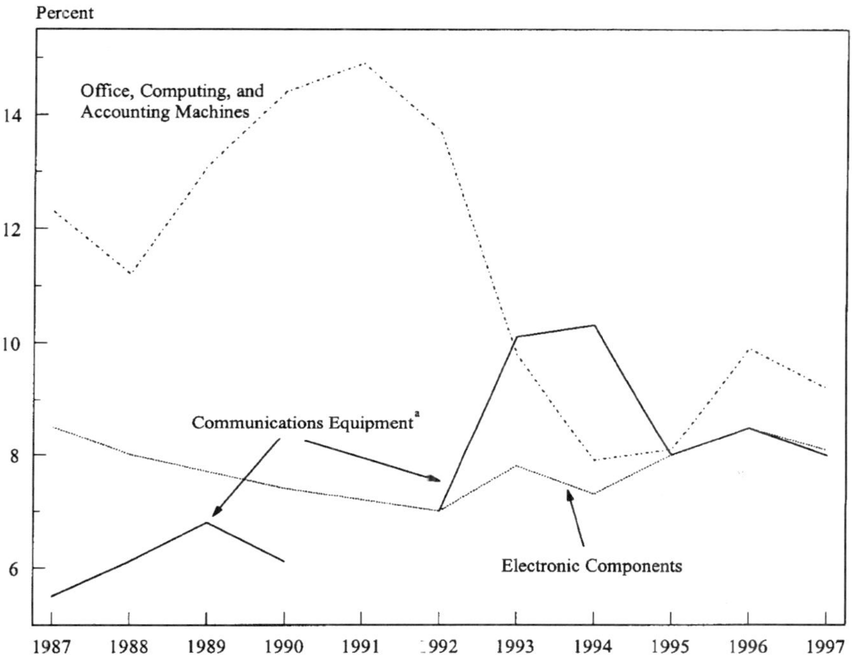

SOURCE: Congressional Budget Office based on data from National Science Foundation, *Research and Development in Industry, 1997* (1999).

a. Data for communications equipment were not available for 1991.

The most dramatic change is the steep decline in the ratio of R&D to sales in the computer industry (labeled "Office, Computing, and Accounting Machines" in Figure 2). The decline in the market share of the International Business Machine (IBM) Corporation and other integrated computer manufacturers explains a large part of that drop.[7] When those companies enjoyed dominant market shares, they were able to capture much of the economic benefits (most notably the profits) generated by the innovative products that their R&D created. When other firms with less R&D overhead moved in and secured a larger part of the market for computers, IBM and the other integrated computer companies were less able to capture the benefits of their R&D investments. They therefore reduced the share of sales they spent on research and development.

The increase in the R&D-to-sales ratio in the communications equipment industry is probably correlated with a decline in market concentration associated with the breakup of the Bell network. That breakup moved the net sales revenue of regional Bell operating companies into communications services and out of communications equipment. Census Bureau rules preclude disclosing which companies

7. Integrated computer manufacturers such as IBM and Digital Equipment Corporation made many of their own components and sold a broad range of computer types.

are classified in which industries, so the above explanations involve some speculation.

R&D by Category

Some analysts argue that IT companies are underinvesting in basic research, even when their overall investment in R&D appears strong. Because the market for information technology is growing so rapidly, those analysts contend, resources that might have gone into long-term research are being diverted into projects with higher prospects for short-term payoffs.

The division between basic and applied research is always somewhat arbitrary but especially so in the IT sector. Because the intent of IT research is to improve information technology, such research is almost always, by definition, applied. Some policy analysts therefore distinguish research by time frame rather than by type: research that is likely to be brought to the marketplace in less than five to seven years versus research that has a longer time frame.[8] The NSF data, however, do not permit such an analysis.

As noted above, those data indicate that the IT companies invested $44.2 billion in R&D in 1997. Of that amount, $31.7 billion went for development and $12.4 billion for research (see Table 7). Within the category of research, the IT sector (excluding communications services) spent $1.4 billion on basic research and $10.3 billion on applied research. (In the interests of confidentiality, the NSF did

TABLE 7. CATEGORIES OF NONFEDERAL SPENDING ON R&D FOR THE INFORMATION TECHNOLOGY SECTOR, CALENDAR YEAR 1997
(In billions of dollars)

Industry	Research: Basic	Research: Applied	Research: Total	Development	Total R&D
Office, Computing, and Accounting Machines	0.2	3.8	4.0	8.8	12.8
Electronic Components	0.4	4.2	4.6	6.2	10.8
Communications Equipment	0.1	1.1	1.3	6.1	7.4
Communications Services	a	a	0.7	1.1	1.9
Computer and Data-Processing Services	0.6	1.2	1.9	9.4	11.3
Total	1.4[b]	10.3[b]	12.4	31.7	44.2

8. Many analysts argue that with the Internet, IT firms now have an even shorter time horizon, measured in months rather than years.

Memorandum:					
All Industries	8.8	29.8	38.5	95.1	133.6

SOURCE: Congressional Budget Office based on data from National Science Foundation, *Research and Development in Industry, 1997* (1999).

a. Not published separately.

b. Excludes communications services.

not break down data for the communications services industry further because doing so might reveal too much about the activities of a single company.)

Although overall, the IT sector's effort in basic and applied research in 1997 represented almost a third of all research performed by industry, in percentage terms, the sector performed less research and more development than industry as a whole. Similarly, within the research category, the IT sector accounted for 16 percent of all basic research by industry but performed less basic research than its overall level of R&D would have suggested (see Table 7).

VENTURE CAPITAL INVESTMENT IN THE INFORMATION TECHNOLOGY SECTOR

Venture capital firms play an important role in the IT sector. They typically raise funds from private investors—usually organizations and wealthy individuals—and invest those funds in often-risky new ventures that offer the prospect of a high payoff. An individual or group trying to start a company to sell a new product or service typically offers part ownership to a venture capital firm in exchange for funding. Besides funding, venture capital firms often provide management and other nontechnical assistance. Those firms hope to profit when the new company is sold to a larger one or starts selling stock on the major exchanges. The U.S. market for venture capital is the envy of the industrialized world, and its ability to move resources rapidly into areas perceived as "hot" is legendary.

According to Venture Economics Information Services, a company that tracks venture capital firms, such firms invested $46.6 billion in start-up companies in all industries between January 1, 1995, and December 31, 1998. The IT sector accounted for $26 billion—or 56 percent—of that total (see Table 8). Within the sector, roughly half of the venture capital investment went to computer software

TABLE 8. INVESTMENT BY VENTURE CAPITALISTS IN INFORMATION TECHNOLOGY AND OTHER INDUSTRIES, JANUARY 1, 1995, THROUGH DECEMBER 31, 1998

Industry	Amount of Investment (Millions of dollars)
Information Technology	
Computer software and services	12,722
Communications	8,054
Semiconductors and other electronic components	2,659
Computer hardware	2,529

Subtotal	25,964
Other Industries	
Medical and health related	6,624
Other products	4,786
Consumer related	4,000
Biotechnology	3,670
Industrial and energy	1,593
Total[a]	46,636

SOURCE: Congressional Budget Office based on data from Venture Economics Information Services.

a. Includes disbursements in a miscellaneous category that total less than $1 million.

and services companies, with electronic components and computer hardware companies accounting for much smaller fractions.

Between 1994 and 1998, venture capital firms quadrupled their investment in the IT sector—from a little less than $2 billion to an estimated $8 billion (see Figure 3). The jump may have been even greater than that: estimates for the latest years are likely to be revised upward as more funds are reported.

FIGURE 3. INVESTMENT BY VENTURE CAPITALISTS IN THE INFORMATION TECHNOLOGY SECTOR, CALENDAR YEARS 1994-1998

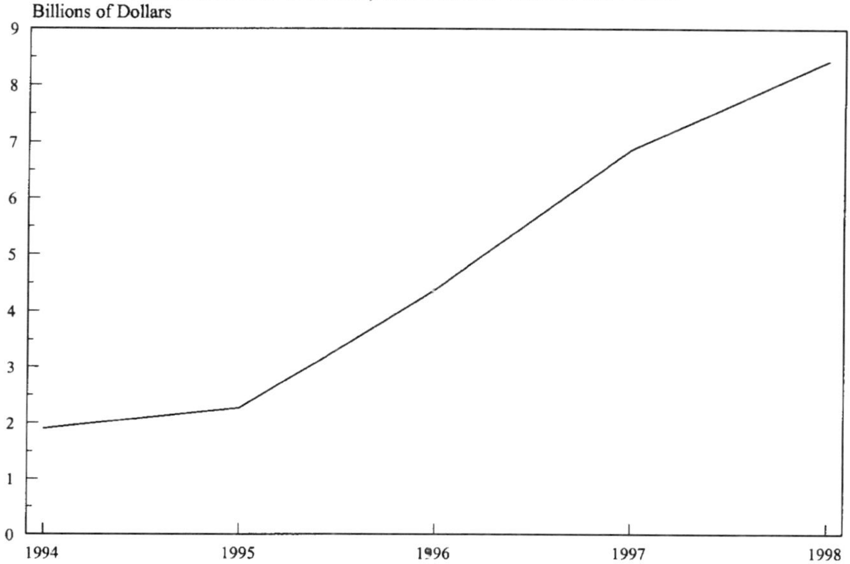

SOURCE: Congressional Budget Office based on data from Venture Economics Information Services.

Not all of the IT funding provided by venture capitalists is devoted to technology development efforts; some is used to produce and sell start-up firms' new products. However, CBO has no way of estimating what share of total investment goes toward technology development.

Some analysts argue that venture capital does not represent funding for new R&D for industry. They contend that the technology pursued by most venture capital firms has been conceived and sometimes proved at another corporation or an academic laboratory—the new firm is not creating new ideas.

But even if that is the case, venture capital complements the investment capital of existing companies. Existing companies commercialize the technological breakthroughs they deem most profitable, using either technology they have developed or technology they obtain from academia or government laboratories. By funding ideas that existing companies will not back, venture capitalists are second-guessing the judgment of existing companies about the economic potential of new technologies. Consequently, venture capitalists increase the number of new ideas introduced into the economy from the stock of ideas generated in the laboratory. In other words, they raise the efficiency of existing R&D by raising the rate at which ideas developed in the laboratory are brought to market. Thus, venture capital funds may be more important as a technological lubricant—making the process of getting to market easier—than as a source of funding for new R&D.

Section C: World Trade Organization Basic Telecommunications Service Agreements/International Trade in Telecommunications Products and Services

WTO Basic Telecommunications Services Agreement
Summary of Country Commitments
(73 Countries)

Source: Department of Commerce (March, 1999)

All commitments are effective as of February 5, 1998, unless otherwise noted. Unlimited foreign investment permitted unless otherwise noted. The "reference paper" is a set of pro-competitive regulatory principles.

Antigua & Barbuda -- Market access and national treatment for enhanced service suppliers, mobile, pcs, trunked radio, and closed user groups. Mobile and fixed satellite transport services may be provided through arrangements with exclusive operator. Market access and national treatment for all other services as of 2012. No commitments on regulatory principles. MFN exemption for treatment accorded to CARICOM members with regard to terrestrial-based mobile services.

Argentina -- Market access and national treatment for domestic data and telex, domestic and international fax, paging, trunk radio and leased circuits (with a preference given to existing supplier until 11/8/2000). Duopoly for cellular. Access for PCS to be decided in "The light of present and future needs." Market access and national treatment for all other services as of 11/8/2000, including those provided via non-geostationary, non-fixed satellite services. Adopted the reference paper. MFN exemption for access to geostationary fixed satellite systems.

Australia -- Market access and national treatment for all services. Majority Australian ownership required in Vodafone. Foreign ownership in Telstra of up to 35% of initial sale (11.7% of total equity) allowed, with a limit of 5% per individual or associated group of foreign investors. Also, limits on the amount of individual foreign ownership in Optus. Adopted the reference paper.

Austria -- Market access and national treatment for all services. Adopted the reference paper.

Bangladesh -- Two private operators provide local and long distance services in competition with government operator. Four licenses issued for cellular. All service suppliers must use facilities of network operator, including VSAT and gateway earth station services. No regulatory commitments. MFN exemption on accounting rates. Market access and national treatment for telecommunications terminal equipment rental, maintenance, connection, repair and consulting services.

Barbados - Market access and national treatment for enhanced services (including Internet), VSAT services, and mobile terrestrial and satellite services. All other services reserved to exclusive suppliers until 1/1/2012. Adopted the reference paper.

Belgium -- Market access and national treatment for all services. Adopted the reference paper.

Belize -- Market access (but not national treatment) for trunked radio services and teleconferencing as of 1/1/2003. Trunked radio services must be provided through joint venture with Belizean national. Market access (but not national treatment) for enhanced services and paging as of 1/1/2008 over the facilities provided by the incumbent operator. Paging must be provided through joint venture with Belizean national. Adopted the reference paper.

Bolivia -- Market access and national treatment for all services as of 11/28/2001. Market access and national treatment as of 1/1/98 for local data, telex, telegraph, fax, closed user groups and mobile services (cellular, paging, PCS, mobile satellite.) Local services limited to existing companies. Adopted limited set of regulatory principles.

Brazil -- Market access and national treatment for enhanced services, paging and nonpublic domestic and international services for closed user groups. Duopoly for analog/digital cellular mobile service. 100% foreign ownership of nonpublic service providers; 49% limit on cellular and satellite until 7/20/99, thereafter none. Commitment to bind outcome of future reform legislation which is expected to cover public and nonpublic services within one year of enactment. Use of foreign-licensed GSO space segment facilities allowed whenever they offer better technical, operational and commercial conditions. MFN exemption for telecommunications services supplied for distribution of radio or television programming for direct reception by service consumers. **Has not yet ratified the Agreement.**

Brunei -- Local service reserved for up to 10 years after privatization of government operator. Two operators have exclusivity in international services until 2010. New licenses in cellular may be issued in 2010. Adopted the reference paper.

Bulgaria -- Market access and national treatment for closed user groups and VSAT (not connected to the public switched network), data, paging, mobile data. Market access and national treatment on 1/1/2003 for resale of public voice, telegraph/telex, and cellular voice. Market access and national treatment for all other services and facilities on 1/1/2005. Adopted the reference paper.

Canada -- Market access and national treatment (with a limit of 46.7% foreign ownership) for all services except fixed satellite services and submarine cables. Market access and national treatment for fixed satellite services in 2000, with 100% foreign ownership, and submarine cable landings as of 10/1/98, with

100% foreign ownership. 100% foreign ownership also permitted for resellers. No routing requirements on mobile satellite services between points in Canada and between Canada and points in the United States. Routing requirements on international mobile satellite services ends on 10/1/98. Routing requirements on international services ends on 1/1/2000, except for fixed satellite services between Canada and points in the United States. Routing requirements on all satellites ends as of 3/1/2000. No date fixed for end to routing restrictions on Canada to Canada points (other than by satellite). Licenses to operate earth stations for provision of Canada-United States fixed satellite service may be limited until 3/1/2000. Adopted the reference paper. (Note: Through domestic regulatory action, Canada has abolished routing requirements on international services, including those between Canada and the United States, and restrictions on Canada to Canada points (other than by satellite.)

Chile -- Market access and national treatment for long-distance and international wireline and wireless (including satellites). No commitment on local service, one-way transmission by satellite of direct-to-home, direct broadcast satellite and digital audio services. Adopted the reference paper.

Colombia -- Market access and national treatment for private networks for voice and data, paging, trunked radio, geostationary satellite and local public voice services. Market access and national treatment for long-distance and international services but subject to an economic needs test. 70% foreign ownership permitted for all services. Market access and national treatment for cellular services on 9/1/99 and PCS on 1/1/2000 but subject to an economic needs test. Adopted the reference paper.

Côte D'Ivoire -- Market access and national treatment for data, fax, telegraph, private circuits, mobile (voice and data), and PCS. Open for all satellite services except domestic and international public voice service. Market access and national treatment for local, long distance, and international voice and telex on 1/1/2005. Adopted the reference paper.

Cyprus - No commitments made (except for "consumption abroad") for domestic and international services pending review for regulatory study and decision of Council of Ministers on specific steps to gradually liberalize the market.

Czech Republic -- Market access and national treatment for domestic and international data, telex, telegraph, fax, leased circuits, closed user groups, domestic mobile services, and video and audio transport services. Market access and national treatment for domestic and international voice services on 1/1/2001. Adopted the reference paper.

Denmark -- Market access and national treatment for all services. Adopted the reference paper.

Dominica -- Market access and national treatment for non-public data, fixed and mobile satellite systems, enhanced services (including Internet access), teleconferencing and telecommunications equipment sales, rental, maintenance, connection, repair and consulting services. Adopted the reference paper. **Has not yet ratified the Agreement.**

Dominican Republic -- Market access (but no national treatment) for all services (but must establish a commercial presence). Adopted the reference paper.

Ecuador -- Market access and national treatment for domestic cellular services. No commitments on regulatory principles.

El Salvador -- Market access and national treatment for all services. Adopted the reference paper.

Finland -- Mmarket access and national treatment for all services. Adopted the reference paper.

France -- Market access and national treatment for all services. Indirect investment is unlimited, but there is 20% direct investment limit for radio-based networks and limit on investment in France Telecom. Adopted the reference paper.

Germany -- Market access and national treatment for all services. Adopted the reference paper.

Ghana -- Duopoly for domestic and international voice for five years, after which additional suppliers may be licensed. Market access and national treatment for data, telex/telegraph, fax, closed user groups (not connected to public switched network), internet access services (excluding voice), teleconferencing, trunked radio, domestic fixed satellite and global mobile satellite services but must provide through joint ventures with Ghanaian nationals. Market access and national treatment for telecommunications equipment rental, maintenance, connection, repair and consulting services through joint venture with Ghanian nationals. Adopted reference paper.

Greece -- Market access and national treatment for all services other than public voice telephony and facilities-based services which will be open on 1/1/2003. Adopted the reference paper.

Grenada -- Market access and national treatment for enhanced services and mobile and fixed satellite services through arrangements with the incumbent operator. Market access and national treatment for all other services as of 1/1/2006. Adopted reference paper.

Guatemala -- Market access and national treatment for all services. Adopted the reference paper. **Has not yet ratified the Agreement.**

Hong Kong -- Local wireline and wireless network services limited to current four providers. Market access and national treatment for switched resale, international data and fax by resale, call back, resale-based virtual private network service (not connected to the public switched network), mobile satellite services, and self-provision of external satellite circuits by a company or closed user group permitted. Hong Kong now allows external service based competition (such as international simple resale) and will allow international facilities based competition on 1/1/2000. Adopted the reference paper.

Hungary -- Market access and national treatment for data, telex/telegraph, fax, leased circuits, satellite services (other than public voice). Market access and national treatment for international and long distance public voice as of 1/1/2003 and local public voice as of 1/1/2004. Public land mobile services limited to 1 operator for NMT 450 MHz; 2 operators for GSM until 2003. Paging limited to 1 analog operator and 2 for pan-European service. 100% foreign ownership permitted except 75% limit for Matav and Antenna Hungaria Rt. Adopted the reference paper.

Iceland -- Market access and national treatment for all services. Adopted the reference paper.

India -- Duopoly for local and long distance wireline, duopoly for cellular, and 25% foreign investment limit. No resale allowed. Additional licenses may be issued based on economic needs test. Will review market access for national long distance in 1999 and international in 2004. GSM technology mandated for cellular services. MFN exemption for accounting rates. Adopted some of the reference paper.

Indonesia -- Local services (public switched voice, circuit switched data, and teleconferencing) limited to provision by PT Telkom and 5 regional joint operation scheme operators. Long distance exclusive to PT Telkom. Packet-switched public data provided by 3 operators on a non-exclusive basis. Domestic telex/telegraph services must be provided through joint venture or joint operation. Selection of new mobile cellular and PCS operators is subject to economic needs test and frequency availability and must be through joint venture (and, in case of PCS, with state-owned company). New paging operators subject to economic needs test, frequency availability, and public interest test and a joint venture or joint operation is required. Duopoly for international public switched voice, circuit switched data, teleconferencing services, and Internet access services (until 2005). Duopoly for domestic and international satellite services and international telex/telegraph. Call back prohibited. 35% foreign ownership limit. Adopted the reference paper. Government will review number of suppliers permitted to provide international services in 2005, long distance services in 2006 and local services in 2011.

Ireland -- Market access and national treatment for all services other than public voice telephony and facilities-based services which will be open as of 1/1/2000. Mobile operators may interconnect with networks other than the state-controlled company to supply international services from 1/1/99. Adopted the reference paper. (Note: The Irish Prime Minister has announced that Ireland's market would be fully liberalized on 12/1/1998.)

Israel -- Market access and national treatment for private networks, cellular, paging, satellite voice and data, and international fax, with limit of 80% foreign investment for cellular. Market access and national treatment for all other services as of 1/1/2002 with limitation of 74% foreign ownership. Permits 80% foreign investment for wireless service providers. Adopted the reference paper.

Italy -- Market access and national treatment for all services, with limitation on investment in Stet. Adopted the reference paper.

Jamaica -- Market access and national treatment for data transmission and enhanced services, including Internet access, mobile data, PCS and paging services, satellite-based video transmission services (excluding video telephony), and telecom equipment, sales, rental, maintenance, connection, repair and consulting services. Market access and national treatment for trunked radio, teleconferencing and international voice, data and video transmission services supplied to firms involved information processing located within freezones but restrictions on use of facilities and interconnection apply. Call back services and refile are not permitted. Voice telephony reserved to exclusive supply until September 2013. Adopted the reference paper.

Japan -- Market access and national treatment for all services, with a 20% limit on foreign investment in KDD and NTT. Adopted the reference paper.

Korea -- Market access and national treatment for all services. Foreign investment limitations -- 33% investment limit on facilities until 1/1/2001 then 49% (with individual shareholdings limited to 33% for wireless and 10% for wire-line); 20% for Korea Telecom until 1/1/2001 then 33% with individual shareholdings limited to 3%; 49% for resellers until 1/1/2001, 100% thereafter. Adopted the reference paper.

Luxembourg -- Market access and national treatment for all services. Adopted the reference paper.

Malaysia -- Market access and national treatment for all services only through acquisition of up to 30% of the shares of existing licensed public telecom operators. No resale permitted. Adopted some of the reference paper.

Mauritius -- Market access and national treatment for fax, paging, private mobile radio, GMPCS and telecommunications equipment rental, sales and maintenance services. Market access and national treatment for other services market by 2004 with no foreign investment restrictions other than company registration requirement. Committed to adopt reference paper in future.

Mexico -- Market access and national treatment for all services except for requirement to use Mexican satellites for the provision of domestic services until 2002. 100% foreign ownership for cellular services; 49% for all other services. Adopted the reference paper.

Morocco -- Market access (but no national treatment) for domestic packet switched data, frame relay, mobile, paging and PCS. Market access to "point-to-point voice telephone service" as of 1/1/2002. Adopted own set of regulatory commitments on interconnection and regulatory authority.

Netherlands -- Market access and national treatment for all services. Adopted the reference paper.

New Zealand -- Market access and national treatment for all services, except limit of 49.9% foreign investment in NZ Telecom. Adopted the reference paper.

Norway -- Market access and national treatment for all services. Adopted the reference paper.

Pakistan -- Market access and national treatment for domestic data, VSAT, telex, fax, video conferencing, tele-medicine, tele-education terminal end services. Market access (but no national treatment) for other services on 1/1/2004. Adopted own regulatory principles. MFN exception on accounting rates.

Papua New Guinea -- Will review exclusivity for all telecom services two years prior to expiration of license in 2002. Call back not allowed. Adopted the reference paper. **Has not yet ratified the Agreement.**

Peru -- Market access and national treatment for local data, telex/telegraph, fax, private leased circuits, cellular, PCS, paging, trunking services, and mobile data. Market access and national treatment for all other services as of June 1999. Adopted the reference paper. (Note: Peru ended its voice monopoly on 8/1/1998.)

Philippines -- Market access and national treatment for facilities-based services except satellites, subject to a limit of 40% foreign ownership. No commitment on resale or leased circuits/closed user groups. Partial adoption of the reference paper. **Has not yet ratified the Agreement.**

Poland -- Market access and national treatment for local public voice (wireline), domestic and international data, international fax, cellular, pan-European paging systems, and international private leased circuits. Market access and national treatment for telex and telegraph as of 1/1/2000. Market access and national treatment for all other services as of 1/1/2003, subject to a limit of 49% foreign ownership for international and domestic long-distance services including cellular. Public digital cellular must be provided thru TPSA's international network until 1/1/2003. Adopted the reference paper.

Portugal -- Market access and national treatment for resale of all services. Market access and national treatment for alternative facilities services as of 1/1/1999. Market access and national treatment for public voice telephony, telex and telegraph as of 1/1/2000. Adopted the reference paper.

Romania -- Market access and national treatment for non-public voice, data, telex/telegraph, fax, paging, and VSAT (not connected to PSN). Market access for analogue cellular as of 4/1/2002. Market access and national treatment for all other services as of 1/1/2003. Adopted the reference paper.

Senegal -- Market access and national treatment for all services (except cellular and mobile satellites) as of 2006, although may be earlier due to government commitment to consider additional access in 2003. Three cellular providers as of 1998. Adopted the reference paper.

Singapore -- Market access and national treatment for local and international resale, resale of cellular mobile and paging services, paging, mobile data and trunked radio. Market access and national treatment for all other services as of 4/1/2000. 74% (combined direct and indirect) foreign ownership permitted. Adopted the reference paper.

Slovak Republic -- Market access and national treatment for data, telex/telegraph, fax, private leased circuits, paging, trunked radio, domestic mobile and PCS (excluding analog cellular voice) and closed user groups (not connected to the PSN). Market access and national treatment for all other services as of 1/1/ 2003. Adopted the reference paper.

South Africa -- Market access and national treatment for paging, PCS and trunked radio services. Duopoly for local, long-distance, international, data, telex, fax, and private leased circuits services as of 1/1/2004. Duopoly for mobile but will license one more supplier within two years. No commitment on satellite-based services but will schedule commitments within one year of adoption of legislation in this area. Limitation of 30% foreign ownership. Liberalization of resale services to take place between 2000 and 2003. Government will consider feasibility of additional suppliers of public switched services and satellites services by 12/31/2003. Government will consider feasibility of additional cellular suppliers by 12/31/98. Adopted the reference paper.

Spain -- Market access and national treatment for all services from 12/1/98. Foreign investment limit for the government-owned firms (Telefonica and Retevision). Adopted the reference paper.

Sri Lanka -- Market access and national treatment for data, paging and mobile cellular but number of operators limited. Duopoly for international services allowed as of 1/1/2000. Foreign ownership in excess of 40% requires government approval. Adopted the reference paper. MFN exception for accounting rates.

Suriname -- Market access and national treatment, using leased lines, for data, Internet and Internet access services, closed user groups, mobile data, paging, and trunked radio services, telecom equipment sales and rental, maintenance, connection, repair and consulting services. Foreign investment limited to 40%. Market access and national treatment for other services as of 1/1/2003. Adopted the reference paper.

Sweden -- Market access and national treatment for all services. Adopted the reference paper.

Switzerland -- Market access and national treatment for all services. Adopted the reference paper.

Thailand -- Market access and national treatment for all services as of 1/1/2006. Until then, market access for local services provided only with case-by-case cabinet approval. Foreign investment limited to 20%. No regulatory commitments.

Trinidad & Tobago -- Market access and national treatment for enhanced services, trunked radio, cellular, mobile data, paging, PCS, telecommunications equipment sales, rental, maintenance, connection, repair and consulting services. Market access and national treatment for all other services as of 2010. Adopted the reference paper.

Tunisia -- Market access and national treatment for telex, packet-switched data and teleconference services as of 1/1/99; mobile telephone, paging, teleconferencing and frame relay from 1/1/2000 and local voice telephony as of 1/1/2003. Foreign ownership limited to 49% (10% in Tunisia Telecom). No regulatory commitments.

Turkey -- Market access and national treatment for mobile, paging, and private data networks, with a limitation of 49% foreign ownership. Market access and national treatment for all other services as of 1/1/2006. Adopted some additional commitments on regulatory principles. MFN exception for accounting rates and fees for transit land connections and use of satellite ground stations.

Uganda -- Market access and national treatment for closer user groups (resale of excess capacity is not allowed), Internet data, paging, private mobile and trunked radio, and global mobile personal satellite communications services. Voice duopoly subject to review after 2003 with no foreign investment restrictions except that neither of the exclusive providers can hold ownership in the other. Until 2003, two more cellular operators will be licensed in addition to the one existing operator. Adopted the reference paper.

United Kingdom -- Market access and national treatment for all services. Adopted the reference paper.

United States -- Market access and national treatment for all services, with a limit of 20% direct foreign investment. Adopted the reference paper. MFN exception for one-way satellite transmission of direct-to-home, direct broadcast satellite and digital audio transmission services.

Venezuela -- Market access and national treatment for all services as of 11/27/2000. Adopted some of the reference paper.

INTERNATIONAL TRADE IN TELECOMMUNICATIONS PRODUCTS AND SERVICES

TELECOMMUNICATIONS IMPORT SOURCES, 1997-98
(millions of U.S. dollars)

1997			1998		
Country	Value	Percent	Country	Value	Percent
Canada	$2,845	19%	Canada	$3,348	19%
Japan	2,355	16	Mexico	2,884	16
Mexico	2,080	14	Japan	2,719	15
China	1,598	11	China	1,917	11
Taiwan	1,061	7	Taiwan	1,370	8
Malaysia	799	5	Korea	964	5
Philippines	688	5	Malaysia	753	4
Israel	546	4	Israel	614	3
Korea	456	3	Philippines	424	2
UK	297	2	Thailand	401	2
Thailand	295	2	UK	369	2
Sweden	233	2	Sweden	261	1
Indonesia	212	1	Germany	210	1
Germany	200	1	Indonesia	154	1
Finland	179	1	France	156	1
Costa Rica	140	1	Costa Rica	148	1
EU Total	1,250	9	EU Total	1,469	8
Total	$14,678	100%	Total	$17,581	100%

U.S. TELECOMMUNICATIONS EQUIPMENT TRADE IN 1998

The U.S. continued to register a surplus in telecommunications equipment trade in 1998 for the sixth consecutive year. However the surplus in 1998 dropped 75 percent from that registered the year before (from $5.1 billion in 1997 to $1.3 billion in 1998). This change resulted primarily from an increase in imported telecommunications equipment, which grew 20 percent from $14.7 billion 1/ in 1997 to $17.6 billion in 1998. U.S. telecommunications equipment exports also fell 5 percent from $19.8 billion in 1997 to $18.9 billion in 1998.

Exports by Country

The export figures offer a mixed picture, with exports of U.S. telecommunications equipment to most Asian countries significantly lower in 1998, while exports to Canada and Mexico were up from 1997. Canada remains our largest telecommunications export market, with U.S. telecommunications equipment exports to that market rising 9 percent ($231 million) to slightly over $2.9 billion, a record high. Canada accounted for 15 percent of total U.S. telecommunications equipment exports, only slightly above last year. Leading exports to Canada in 1998 were radio transceivers/other, parts for telephonic apparatus, and modems.

Exports to Mexico grew 35 percent in 1998, placing Mexico as our second largest export market and pushing Japan to the third position. Exports to Mexico were $1.95 billion, 10 percent of the total in 1998, up from $1.4 billion and 7 percent of the total in 1997. Leading exports to Mexico were parts for telephonic apparatus, cellular phones, radio transceivers/other and miscellaneous radio parts.

Exports to Japan fell $137 million, or almost 7 percent, to $1,933 million in 1998. This places Japan as our third largest export market, although it retains 10 percent of total U.S. telecommunications equipment exports (the same as in 1997). Leading exports to Japan were miscellaneous radio parts, parts for telephonic apparatus, telegraphic switching equipment, and optical fiber bundles and cables.

Exports to the European Union increased again in 1998, by $509 million (or 16 percent) to$3,698 million. The EU share of total U.S. telecommunications exports, after declining slightly in 1997 and 1996, moved back up to 19 percent of the total in 1998. The United Kingdom, the Netherlands, and Germany led with $1.2 billion, $490 million, and $438 million in U.S. telecommunications equipment exports, respectively. Of note, exports to Italy fell by about a third from 1997 and exports to France fell 14 percent in 1998. Principal exports to the EU were parts for telephonic apparatus, parts for telephonic switching equipment, modems, radio transceivers (over 400 MHz), and communications satellites.

Telecommunications equipment exports to Asia declined in 1998 by 26 percent, yet, on a regional basis, Asia remained the largest market for export sales at over $5.1 billion. The top export markets in Asia for U.S. telecommunications equipment were Japan, China, Hong Kong, Taiwan and Korea. Within the $5.1 billion total, exports to Taiwan increased 35 percent to $483 million and exports to India increased 32 percent to $123 million. Exports to most other Asian markets declined, however, including a decline of 35 percent in U.S. telecommunications equipment exports to Hong Kong (to $708 million), a decline of 53 percent to Korea (to $417 million), a decline of 61 percent to the Philippines (to $115 million), a decline of 66 percent in exports to Thailand (to $77 million), and a decline of 86 percent in exports to Indonesia (to $56 million). Exports to China were flat compared to last year, and exports to Japan and Hong Kong both declined (7 percent and 35 percent, respectively).

Telecommunications equipment exports to Latin America were flat compared to 1997, growing less than 1 percent to $5 billion. The top three export markets within the region were Mexico, Brazil and Argentina. Exports to Mexico increased by 35 percent to $1.95 billion, although exports to Brazil fell by 22 percent to $957 million and exports to Argentina fell by 16 percent to $386 million. Despite these declines, these two countries still each imported from the U.S. significantly more telecommunications equipment in 1998 than two years ago in 1996. Exports to Chile were up considerably, by 59 percent to a total of $189 million.

Exports to the Middle East and North Africa increased 8 percent to $879 million. Exports to Israel, the largest market in the region for U.S. exports, declined by 5 percent (to $484 million). U.S. telecommunications equipment exports to sub-Saharan Africa increased 39 percent to $206 million, the third year of increased telecommunications exports to that region. Exports to the largest market in the region, South Africa, more than doubled from $64 million to $133 million.

Imports by Country

Canada was again the largest source of U.S. telecommunications equipment imports in 1998. Imports from Canada grew by $503 million, or 18 percent to $2,922 million. Telephonic carrier line equipment, radio and television transmission apparatus, and parts for telephonic apparatus were the leading telecommunications equipment import categories from Canada in 1998.

Imports from Mexico grew 39 percent (to $2.9 billion) in comparison to 16 percent growth (to $2.7 billion) in imports from Japan, which placed Mexico as the second largest source of imported telecommunications equipment and pushed Japan to third. Closed-circuit tv transmitters, TVROs, cellular telephones and radio paging receivers accounted for 43 percent of telecommunications imports from Mexico. Parts for telephonic apparatus, facsimile machines and cellular telephones accounted for 44 percent of the telecommunications imports from Japan.

Asia continued as a dominant supplier of U.S. telecommunications equipment imports, representing 50 percent of U.S. telecommunications imports, or $8.9 billion. The share of U.S. telecommunications imports that came from our traditional suppliers in East Asia -- Japan, Hong Kong, Korea, Singapore and Taiwan -- remained at 30 percent, consistent with figures from the last two years. China remained the fourth largest supplier of U.S. telecommunications equipment imports with 11 percent of the import share ($1.9 billion), the same percentage as in 1997. Leading imports from China were cordless telephone sets, single line telephone sets, cellular telephones, and closed circuit tv transmitters.

Of note, the share of imports from the Philippines and Indonesia fell dramatically. Imports from the Philippines fell from $688 million in 1997 to $424 million in 1998, a 39 percent decline. Imports from Indonesia fell 28 percent, from $212 million to $153 million. Imports from Korea rose 111 percent and imports from Sri Lanka rose 72,070 percent (from $3,440 to $2.5 million). $2.4 million of cable and closed-circuit television transmission apparatus accounted for 99 percent of U.S. telecommunications equipment exports to Sri Lanka in 1998.

Telecommunications equipment imports from the EU rose 18 percent in 1998, a smaller rate of growth than in 1997 (when imports from the EU rose 24 percent). Imports from the EU, $1.47 billion, represented 8 percent of total U.S. telecommunications equipment imports. Leading imports from the EU were parts for radio products, radio transceivers, communications satellites, and telephonic switching equipment. Leading country sources within the EU were the United Kingdom, Sweden, Germany and France.

Bilateral Trade Balances

The U.S. continues to have an overall telecom trade surplus, including more bilateral trade surpluses than deficits. The largest surplus in 1998, as was the case in 1997, was with Brazil, at $935 million. The surplus with Brazil, though, is 23 percent smaller than it was in 1997. The surplus with the United Kingdom in 1998 was $828 million; Hong Kong, $582 million; and the Netherlands was $453 million.

The largest bilateral telecom trade deficit in 1998 was with China, at $1.15 billion. The trade deficit with China grew 28 percent in 1998, after an increase of 36 percent in 1997. The trade deficit with Mexico was $937 million; with Taiwan, $887 million; and with Malaysia, $659 million. In 1998, our trade deficit with Japan -- $785 million -- almost tripled over that in 1997, when the deficit was at its lowest point since the 1980s.

Our bilateral telecommunications trade balances with many of the Asian nations worsened in 1998. The bilateral deficits with China, Taiwan, Malaysia and Thailand increased, while the bilateral surpluses with Hong Kong and Singapore decreased.

Composition of U.S. Telecom Trade

The United States again tallied a substantial deficit, $3.5 billion, in customer premises equipment. This is a sizeable increase from 1997 and larger, in fact, than the deficit in customer premises equipment registered in 1996. This was more than offset by the surplus in network/transmission equipment, which increased 4 percent in 1998 to $3.96 billion. The United States also tallied a surplus of $1.1 billion in "other" telecommunications equipment (largely wireless communications equipment, radio and television broadcasting, and basket categories of telephone equipment which cannot be allocated to either network equipment or customer premises equipment).

Parts continue to represent a substantial percentage of U.S. telecommunications equipment trade. In 1998, parts exports of $5.6 billion accounted for 30 percent of total telecommunications exports, about the same percentage as in 1997. Parts for radio equipment accounted for about half of U.S. telecommunications parts exports, with telephonic apparatus accounting for most of the rest. Parts imports of $4.2 billion were 25 percent of total telecommunications equipment imports, down slightly from 26 percent in 1997. Definitive comparisons to parts import and export activity in 1997 are complicated by new categories that were created in mid-1997.

Exports by Product

The largest export categories in 1998 were radio transceivers ($4.7 billion), parts for radio equipment ($2.2 billion); parts for telephonic apparatus ($1.9 billion); cellular phones ($1.7 billion); and line system equipment ($1.3 billion). The top ten categories of telecommunications product exports accounted for 89 percent of total telecommunications exports, in contrast to the years immediately preceding, in which these categories accounted for an average of 75 percent of total telecommunications exports. Within the top product export categories, seven categories increased by a combined total of $2.7 billion. The three other categories decreased by a combined total of almost $1 billion.

Imports by Product

The largest import categories in 1998 were parts for telephonic apparatus ($2.3 billion); cordless telephones ($2.1 billion); radio transceivers ($1.5 billion); cellular telephones ($1.4 billion); and miscellaneous radio parts ($1.3 billion). The fastest growing category of imports was set top boxes, which increased by 343 percent (to $628 million). Other categories with substantial increases were: cellular phones, up 37 percent ($375 million); radio transceivers, up 26 percent ($321 million); and parts for telephonic apparatus, up 16 percent ($316 million). Imports of cordless phones, line system equipment, TVROs, optical fiber/cable, and pagers each increased by significant percentages. Imports of TV transmitters declined by $119 million (9 percent) and imports of fax machines declined by $65 million (8 percent).

The attached tables provide detail about U.S. telecommunications equipment trade in 1998. For additional information or questions, please contact Linda Astor at (202) 482-4523 or Elizabeth Farrand at (202) 482-2953 in the Office of Telecommunications of the International Trade Administration, U.S. Department of Commerce.

Please note that definitive comparisons to import and export activity in 1997 are complicated by new categories that were created in mid-1997 due to implementation of the Information Technologies Agreement.

Source: Department of Commerce (April, 1999).

TELECOMMUNICATIONS EXPORT DESTINATIONS, 1997-98
(millions of U.S. dollars)

1997			1998		
Country	Value	Percent	Country	Value	Percent
Canada	$2,691	14%	Canada	$2,922	15%
Japan	2,070	10	Mexico	1,947	10
Mexico	1,439	7	Japan	1,933	10
Brazil	1,256	6	UK	1,197	6
UK	1,128	6	Brazil	957	5
Hong Kong	1,093	6	China	765	4
Korea	892	5	Hong Kong	708	4
China	795	4	Netherlands	490	3
Israel	512	3	Israel	484	3
Fr. Guiana	473	2	Taiwan	483	3
Argentina	461	2	Germany	438	2
France	454	2	Korea	418	2
Germany	434	2	Argentina	386	2
Indonesia	413	2	France	374	2
Singapore	398	2	Russia	344	2
Taiwan	357	2	Colombia	322	2
Netherlands	355	2	Venezuela	301	2
Australia	341	2	Singapore	270	1
Colombia	336	2	Fr. Guiana	229	1
Philippines	293	1	Saudi Arabia	227	1
EU Total	3,190	16	EU Total	3,698	19
Total	$19,760	100%	Total	$18,878	100%

U.S. TELECOMMUNICATIONS TRADE

U.S. IMPORT PERFORMANCE BY COUNTRY, 1997-98
(millions of U.S. dollars)

Gainers			Decliners		
Country	Increase	Percent	Country	Decrease	Percent
Mexico	$804.0	39%	Philippines	$266.8	39%
Korea	508.2	111			
Canada	503.1	18			
Japan	364.0	16			
China	319.0	20			
Taiwan	308.8	29			
Thailand	105.6	36			
EU	222.1	18			

U.S. TELECOMMUNICATIONS TRADE

U.S. EXPORT PERFORMANCE BY COUNTRY, 1997-98
(millions of U.S. dollars)

Gainers			Decliners		
Country	Increase	Percent	Country	Decrease	Percent
Mexico	$507.9	35%	Korea	$474.3	53%
Canada	231.4	9	Hong Kong	384.7	35
Netherlands	135.9	38	Indonesia	356.8	86
Russia	189.2	122	Brazil	268.3	22
Netherlands	136.0	38	French Guiana	243.5	52
Taiwan	126.1	35	Philippines	178.7	61
Belgium	108.6	94	Thailand	147.3	66
Luxembourg	104.3	2,522	Kazakhstan	131.3	88
			Singapore	127.9	32
EU	508.8	16			

U.S. TELECOMMUNICATIONS TRADE BALANCES 1994-98
(millions of U.S. dollars)

Deficits

Country	1994	1995	1996	1997	1998
China	$653.4	$525.0	$612.8	$832.5	$1,151.9
Mexico	n/a	706.3	1,217.3	641.0	937.1
Taiwan	335.9	407.8	617.6	704.5	887.3
Japan	1,752.2	941.6	395.6	275.3	785.4
Malaysia	754.0	751.3	771.4	647.2	658.8
Korea	n/a	n/a	n/a	n/a	546.8
Canada	67.3	87.0	236.3	154.1	425.8
Thailand	78.6	n/a	35.0	71.3	324.2
Philippines	166.5	152.8	327.5	394.9	310.0
Israel	99.3	n/a	n/a	33.8	130.5
Sweden	44.9	108.6	208.1	142.5	117.1
Indonesia	n/a	n/a	n/a	n/a	97.7
Singapore	21.1	21.0	n/a	n/a	n/a

U.S. TELECOMMUNICATIONS TRADE BALANCES 1994-98
(millions of U.S. dollars)

Surpluses

Country	1994	1995	1996	1997	1998
Brazil	477.7	615.1	731.0	1,220.1	934.5
UK	566.9	681.6	700.1	830.8	828.1
Hong Kong	212.1	688.3	602.8	991.9	582.2
Netherlands	210.3	299.8	252.6	313.5	453.0
Argentina	369.2	219.9	234.0	460.7	385.9
Colombia	107.3	191.3	178.8	335.5	321.6
Venezuela	119.9	84.7	140.9	292.6	300.7
Germany	185.9	267.7	199.0	233.6	228.1
Saudi Arabia	57.6	139.9	107.9	163.1	226.8
France	44.0	170.4	213.9	340.7	218.3
Australia	339.5	517.2	328.6	313.5	194.8
Singapore	n/a	121.3	215.1	287.3	147.4
Spain	154.1	250.3	255.4	92.9	142.3
Italy	123.2	217.0	159.7	143.8	55.4
Indonesia	n/a	69.1	34.6	200.4	n/a
Israel	n/a	105.2	72.2	n/a	n/a
Korea	309.5	250.9	580.2	435.7	n/a
EU	1,377.3	1,987.6	1,703.0	1,939.5	2,229.2

U.S. TRADE IN TELECOMMUNICATIONS EQUIPMENT, 1996-98
(millions of U.S. dollars)

	1996	1997	1998	% Change 1997-1998
		Brazil		
U.S. Exports	734.9	1,225.5	957.2	-21.9%
U.S. Imports	3.9	5.4	22.7	320.4
Balance	731.0	1,220.1	934.5	-23.4
		Canada		
U.S. Exports	2,264.0	2,690.5	2,921.9	8.6
U.S. Imports	2,500.3	2,844.6	3347.7	17.7
Balance	-236.3	-154.1	-425.8	176.3
		China		
U.S. Exports	761.2	756.1	764.7	1.1
U.S. Imports	1,374.0	1,597.6	1,916.6	20.0
Balance	-612.8	-832.5	1,151.9	36.9
		European Union 1/		
U.S. Exports	2,709.4	3,189.6	3,698.4	16.0
U.S. Imports	1,006.4	1,247.1	1,469.2	17.8
Balance	1,703.0	1,942.5	2,229.2	14.8
		France		
U.S. Exports	297.9	435.5	374.1	-14.1
U.S. Imports	84.0	112.8	155.8	38.1
Balance	213.9	340.7	218.3	-32.4
		Germany		
U.S. Exports	344.5	433.6	437.6	0.9
U.S. Imports	145.5	200.0	209.5	4.8
Balance	199.0	233.6	228.1	-2.4
		Hong Kong		
U.S. Exports	708.3	1,092.8	708.1	-35.2

U.S. Imports	105.5	100.9	125.9	24.8
Balance	602.8	991.9	582.2	-41.3
Italy				
U.S. Exports	183.3	229.1	153.0	-33.2
U.S. Imports	23.6	85.3	97.6	14.4
Balance	159.7	143.8	55.4	-61.5
		Japan		
U.S. Exports	1,907.8	2,070.4	1,933.3	-6.6
U.S. Imports	2,303.4	2,354.7	2,718.7	15.5
Balance	-395.6	-275.3	-785.4	176.3
		Korea		
U.S. Exports	928.3	891.9	417.6	-53.2
U.S. Imports	348.1	456.2	964.4	111.4
Balance	580.2	435.7	-546.8	-225.5
		Malaysia		
U.S. Exports	159.8	151.5	94.5	-37.6
U.S. Imports	931.2	798.7	753.3	-5.7
Balance	-771.4	-647.2	-658.8	1.8
		Mexico		
U.S. Exports	961.2	1,439.3	1,947.2	35.3
U.S. Imports	2,178.5	2,080.3	2,884.3	38.6
Balance	-1,217.3	-641.0	-937.1	46.2
		Netherlands		
U.S. Exports	290.7	354.5	490.4	38.3
U.S. Imports	38.1	41.0	37.4	-8.8
Balance	252.6	313.5	453.0	44.5
		Singapore		
U.S. Exports	334.0	398.3	270.4	-32.1
U.S. Imports	118.9	111.0	123.0	10.8
Balance	215.1	287.3	147.4	-48.7

Spain				
U.S. Exports	260.7	97.2	158.0	62.6
U.S. Imports	5.3	4.3	15.7	265.1
Balance	255.4	92.9	142.3	53.2
		Sweden		
U.S. Exports	96.1	90.7	143.8	58.5
U.S. Imports	304.2	233.2	260.9	11.9
Balance	-208.1	-142.5	-117.1	-17.8
		Taiwan		
U.S. Exports	222.0	356.9	482.9	35.3
U.S. Imports	839.6	1,061.4	1,370.2	29.1
Balance	-617.6	-704.5	-887.3	25.9
		Thailand		
U.S. Exports	227.2	224.1	76.8	-65.7
U.S. Imports	262.2	295.4	401.0	35.7
Balance	-35.0	-71.3	-324.2	354.7
		United Kingdom		
U.S. Exports	939.5	1,127.8	1,197.4	6.2
U.S. Imports	239.4	297.0	369.3	24.3
Balance	700.1	830.8	828.1	-0.3
		World		
U.S. Exports	15,738.6	19,760.3	18,878.2	-4.5
U.S. Imports	13,358.6	14,677.7	17,581.0	19.8
Balance	2,380.0	5,082.5	1,297.2	-74.5

1/ The data includes the enlarged European Union.
Exports: domestic exports, f.a.s. value basis; Imports: imports for consumption, c.i.f. value basis

KEY TELECOMMUNICATIONS PRODUCT IMPORTS, 1996-98
(millions of U.S. dollars)

Product	1996	1997	1998
Parts/Telephonic Apparatus	$1,780.4	$1,940.4	$2,256.2
Cordless Telephones	1,467.2	1,830.7	2,138.0
Radio Transceivers	1,084.6	1,220.0	1,529.3
Cellular Telephones	728.3	1,008.7	1,383.8
Misc. Radio Parts	1,018.2	1,247.7	1,339.3
TV Transmitters	1,560.4	1,317.4	1,198.9
Telephone Sets 2/	1,107.7	1,105.7	1,032.4
Line System Equipment	452.4	693.2	915.3
Facsimile Machines	740.7	827.4	762.4
Modems	517.9	521.8	649.6
Subtotal	10,457.8	11,713.0	11,821.4
Percent of total	78	80	68
Total Imports	13,358.6	14,677.7	17,328.1

1/ Listed in decreasing order for the latest year.
2/ Includes telephone handsets.

KEY TELECOMMUNICATIONS PRODUCT EXPORTS, 1996-98
(millions of U.S. dollars)

Product	1996	1997	1998
Radio Transceivers	$3,476.7	$3,029.4	$4,734.8
Misc. Radio Parts	2,197.9	2,887.9	2,207.0
Cellular Phones 2/	n/a	1,839.1	1,744.8
Parts/Telephonic Apparatus	1,309.5	1,450.5	1,889.7
Line System Equipment	639.8	1,062.5	1,312.9
Switching Equipment	827.5	1,153.6	1,261.1
Satellites	532.1	819.1	1,055.8
Parts/Switching Equipment	767.0	915.0	921.0
Modems	1,060.3	1,025.3	808.9
Fiber Optics	645.5	806.5	807.2
Subtotal	11,456.3	14,988.9	16,743.2
Percent of total	73	76	89
Total Exports	15,738.6	19,760.3	18,878.2

Data Source: Department of Commerce/International Trade Administration, 1999.

Section D: Telecommunications and the American Economy: Selected Charts and Graphs for Strategists

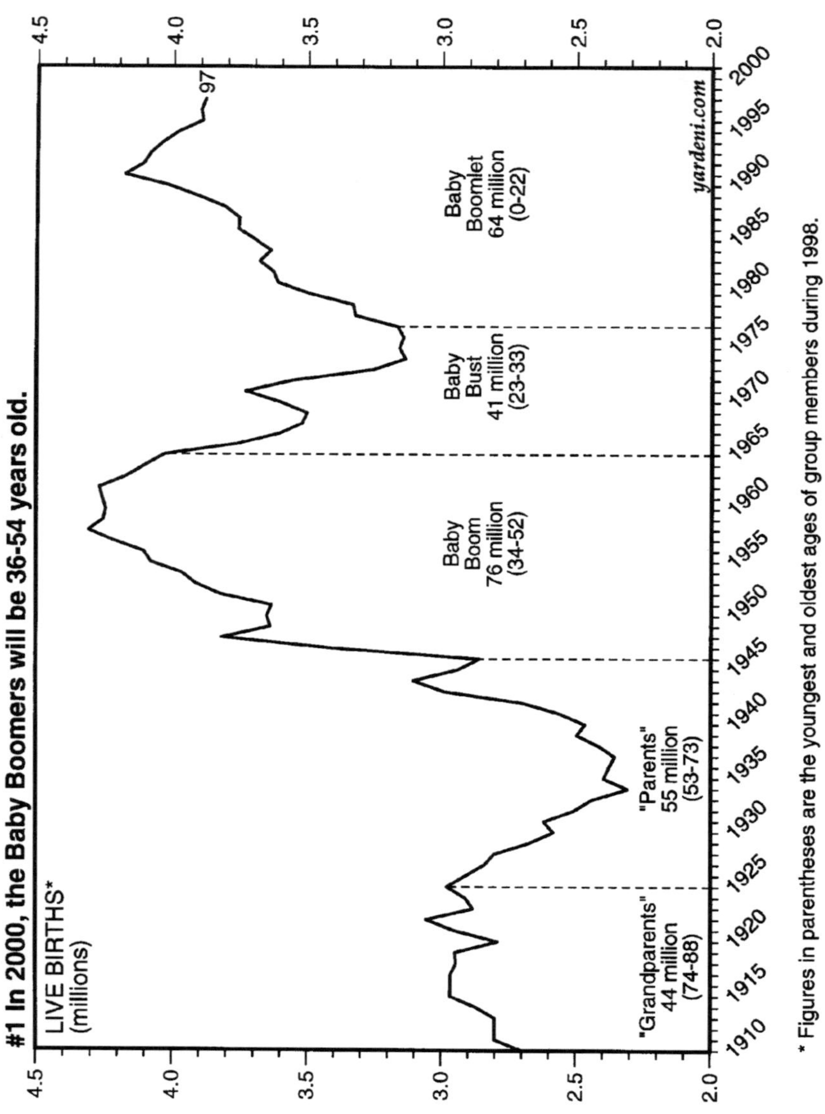
#1 In 2000, the Baby Boomers will be 36-54 years old.
LIVE BIRTHS*
(millions)
"Grandparents"
44 million
(74-88)
"Parents"
55 million
(53-73)
Baby
Boom
76 million
(34-52)
Baby
Bust
41 million
(23-33)
Baby
Boomlet
64 million
(0-22)
97
yardeni.com
4.5
4.0
3.5
3.0
2.5
2.0
1910
1915
1920
1925
1930
1935
1940
1945
1950
1955
1960
1965
1970
1975
1980
1985
1990
1995
2000
* Figures in parentheses are the youngest and oldest ages of group members during 1998.

Figure D.1

- The Age Wave -

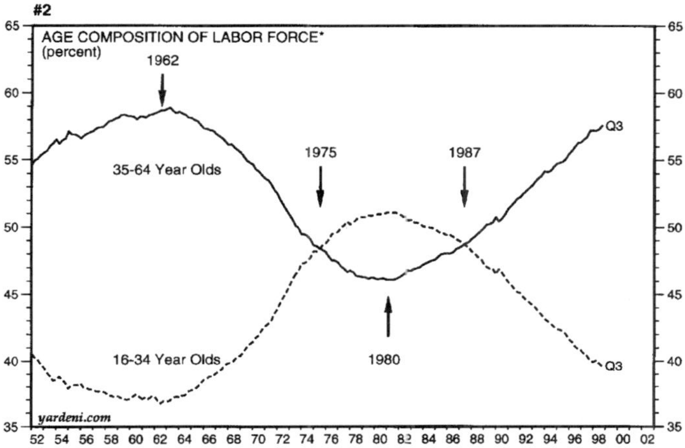

* Oldest and youngest Baby Boomers turned 16 in 1962 and 1980, respectively.

In 1980, the Baby Boomers all turned 16 years or older. Maturity will increase through 2010. The labor force is the most mature since the mid-1960s.

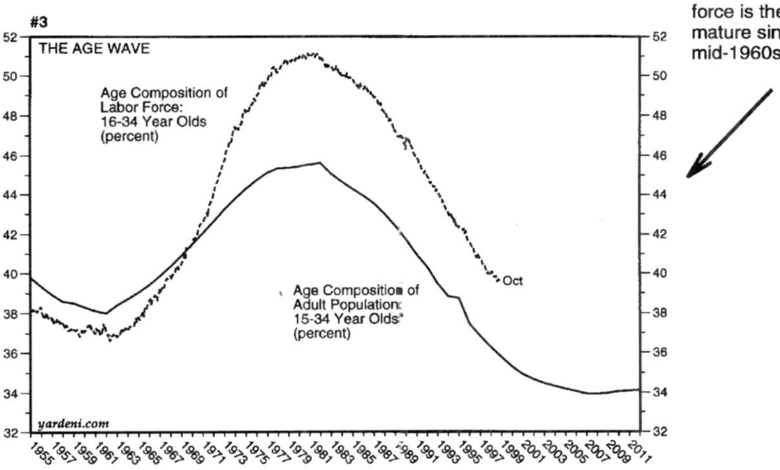

* Projected data start in 1994. Adult population is 15 years and older, and includes armed forces overseas.

Figure D.2

- The Age Wave -

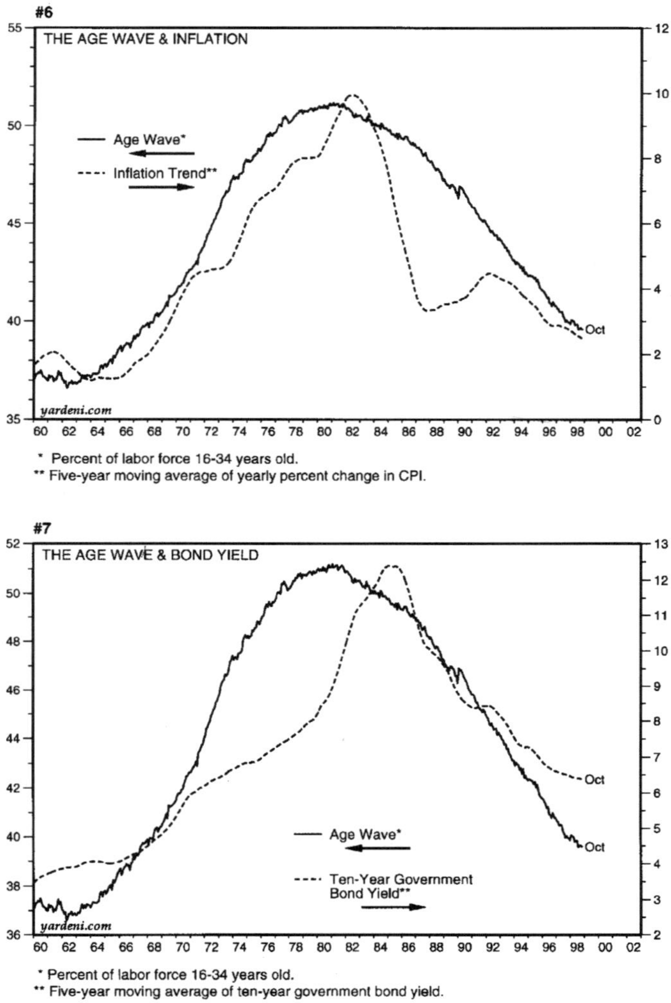

Age Wave is a big influence on inflation trend and suggests inflation will remain subdued at least through end of the century. The Baby Boomers will be 36-54 years old by the year 2000. That's bullish for bonds. We expect the government bond yield to fall to 3%-4% in 2000.

Figure D.3

- Population: Numbers -

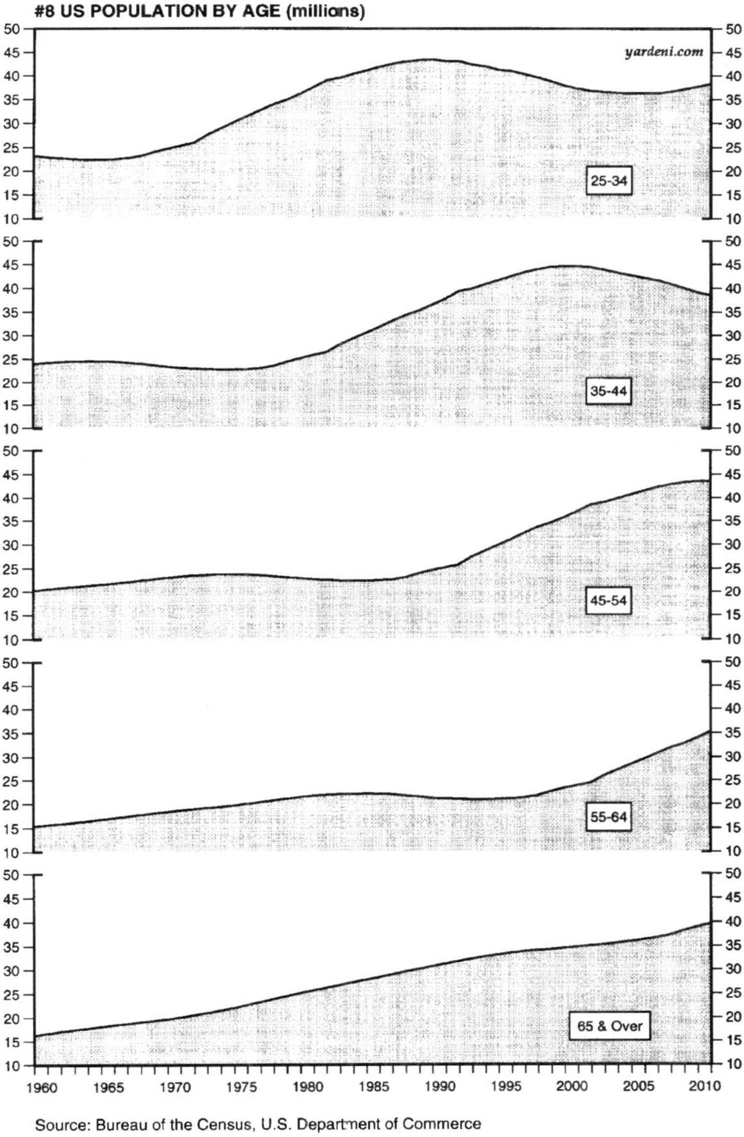

Figure D.4

- Population: Numbers -

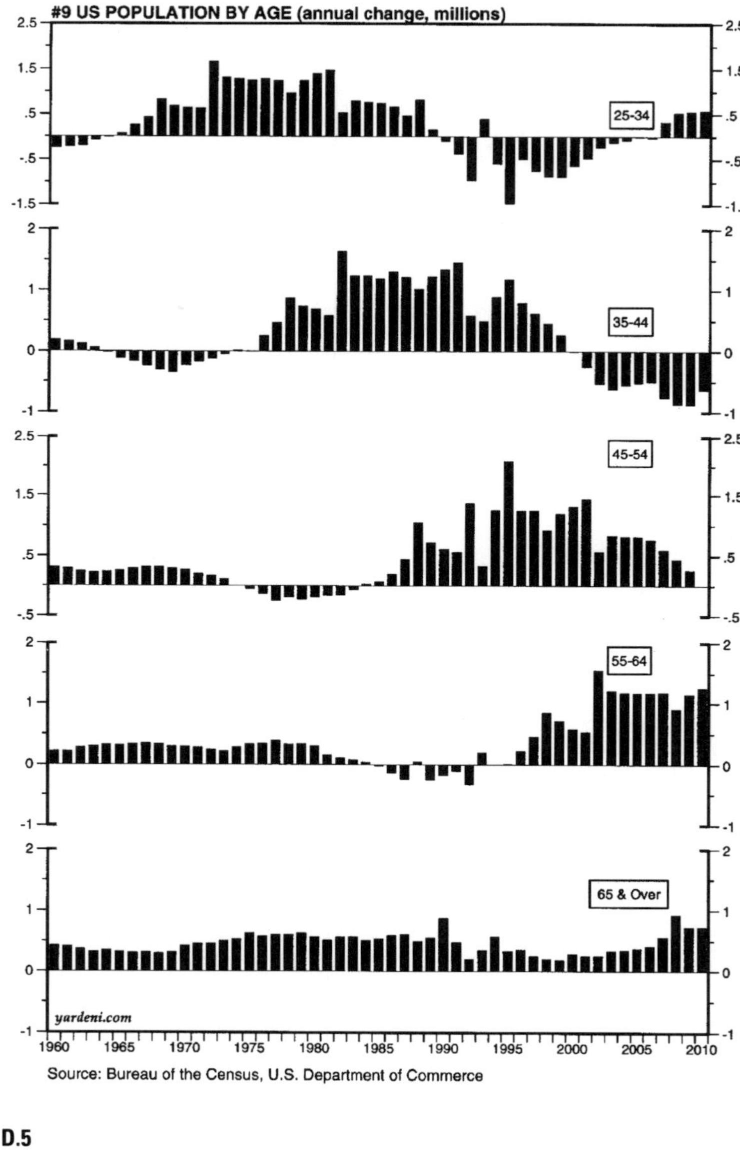

Figure D.5

- Population: Numbers -

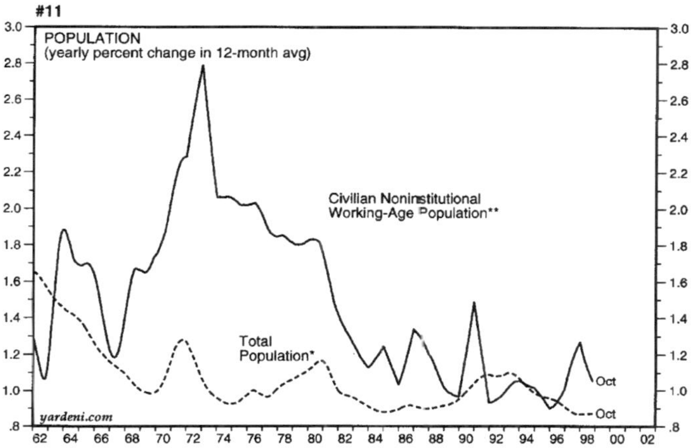

* Population used to calculate per capita income. Includes military in US.
Source: Census Bureau.
** Source: US Department of Labor, Bureau of Economic Analysis.

The working-age population grew much faster than the total population during the late 1960s and 1970s as the Baby Boomers flooded into the labor markets. The growth rate has been at record lows through most of the 1990s.

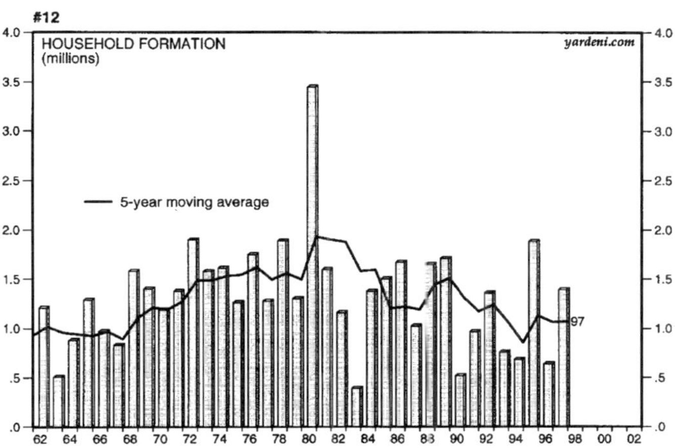

Source: US Department of Commerce, Bureau of the Census, Series P-20.

The Baby Boomers are no longer young adults. As a result, household formation has declined sharply from 1.6 million per year, on average, during the 1980s to 1.0 million per year from 1990 to 1997.

Figure D.6

- Population: Characteristics -

Since the mid-1960s there has been a significant downtrend in the average number of people in both family and household living units. However, both have flattened in recent years. Households are smaller than families, on average, because there are many single-person households: People are marrying later, getting divorced, and living longer.

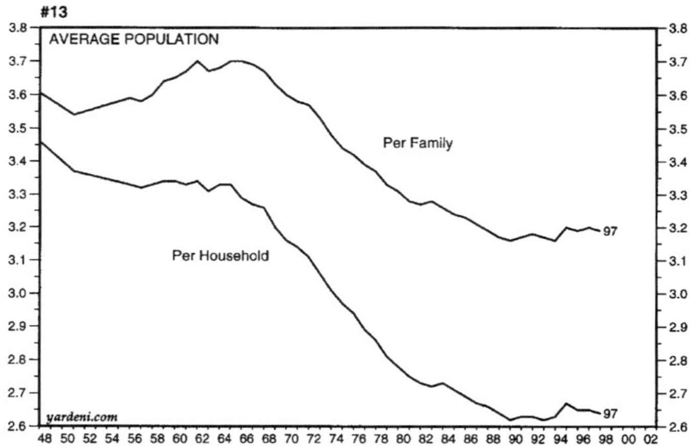

Source: Current Population Reports

In 1997, 31% of all households were not families, up from 18% in 1967. One-third of the nonfamily households are senior citizens.

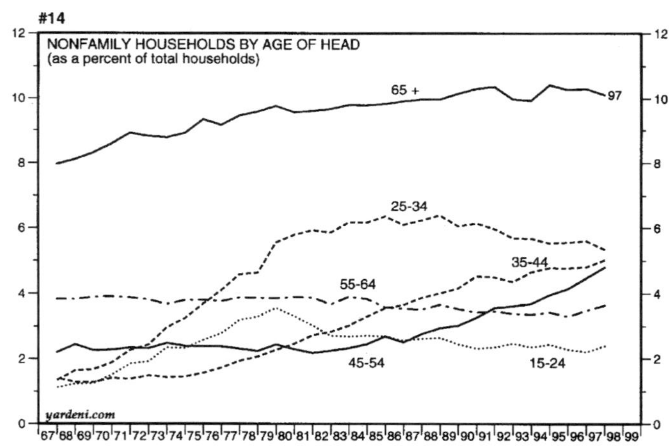

Source: US Department of Commerce, Bureau of the Census, Series P-20.

Figure D.7

- Population: Characteristics -

The median age of the population is 35, up from 29 twenty years ago, and it will continue to rise over the rest of the decade because the Baby Boomers are aging and senior citizens are living longer. Life expectancy is up over 76 years compared to about 70 years during the late 1960s.

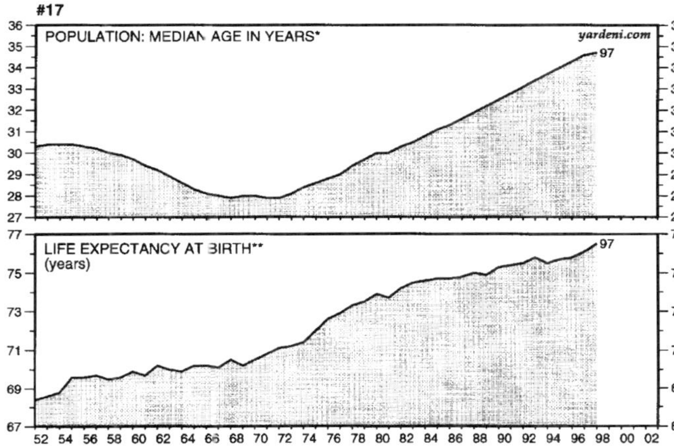

* Source: US Bureau of the Census, Population Division, release PPL-57, United States Population Estimates by Age, Sex, Race and Hispanic Origin.
** Source: US National Center for Health Statistics, Vital Statistics of the United States.

Older people tend to move less than younger ones.

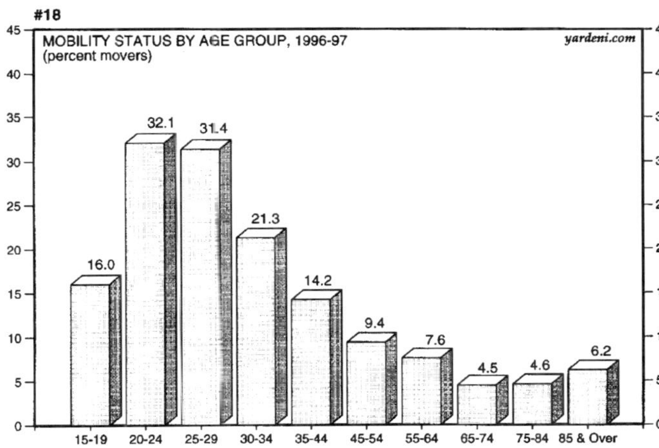

Source: Bureau of the Census, Geographical Mobility: March 1992 to March 1993, Current Population Reports, Series P-20.

Figure D.8

- Income & Spending -

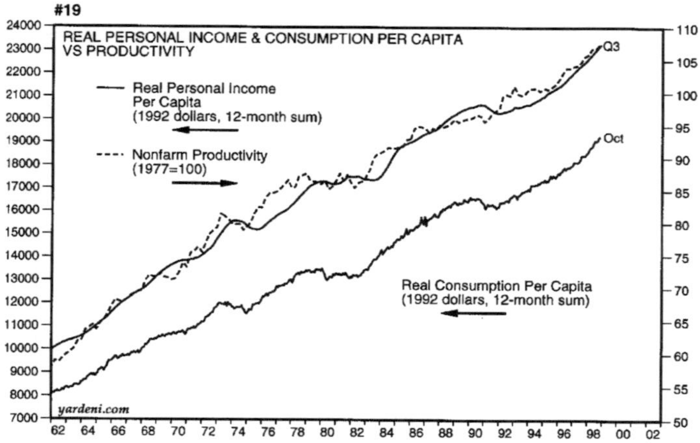

Per capita real personal income and consumption are at record highs. Productivity is the main determinant of these three measures of the standard of living and it is also at a record.

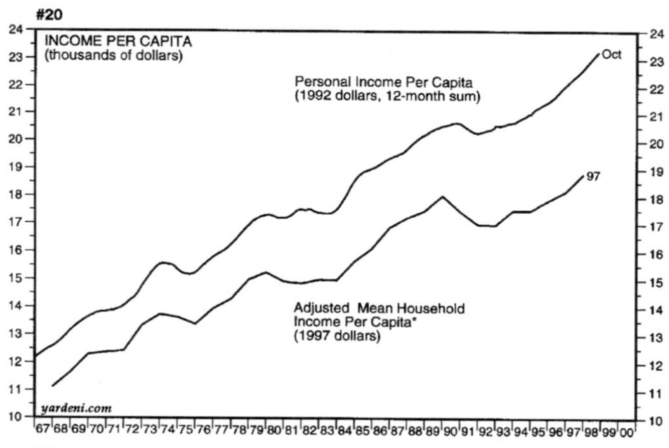

* Mean income of households divided by average population per household.

Both median and mean family household income data show virtually no growth over the past 25 years. Both measures need to be divided by the average size of the household unit, which has been falling. Adjusted this way, mean household income per capita has been growing in line with personal income per capita!

Figure D.9

- Income & Spending -

#21

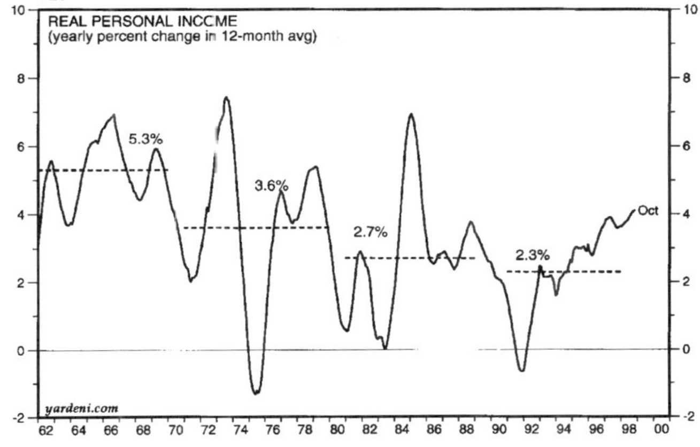

* Dashed lines show decades' annual averages.

Real personal income growth was on a downtrend over the past three decades as growth in productivity stagnated. This long-term trend is already getting reversed in the current decade.

#22

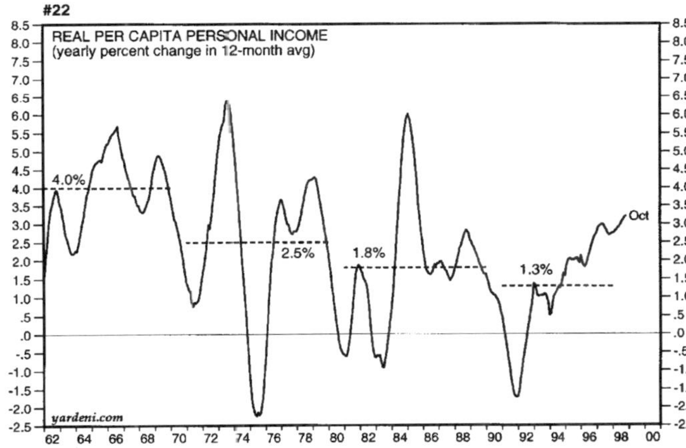

* Dashed lines show decades' annual averages.

Figure D.10

- Income & Spending -

Older workers tend to earn more than younger ones, presumably because they are more experienced, productive, and committed to their jobs. By the year 2000, the Baby Boomers will be 36-54 years old. So there will be more households earning more money than ever before.

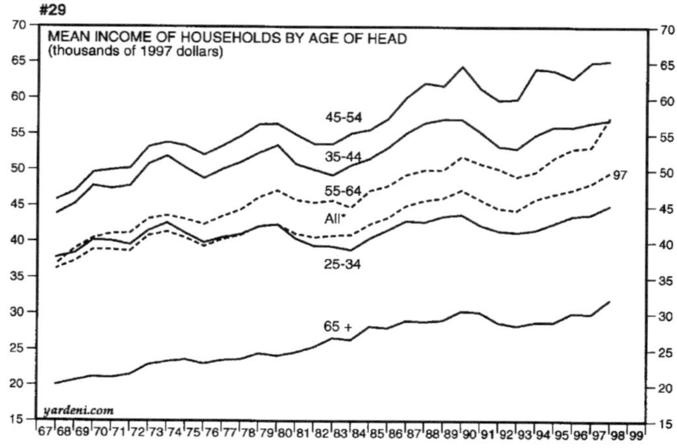

* Includes 15-24 year olds. Source: Bureau of the Census, Income Statistics Branch, unpublished data.

During the 1980s, all household age groups shown here increased in numbers except the 55-64 group. During the 1990s, only the 35-44 and 45-54 year olds are growing.

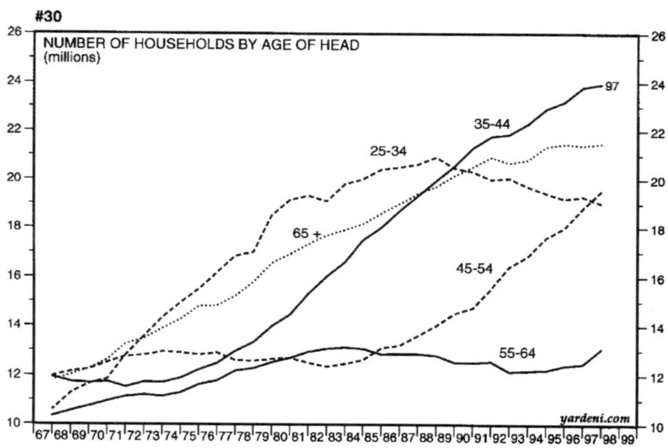

Source: Bureau of the Census, Income Statistics Branch, unpublished data.

Figure D.11

- Income & Spending -

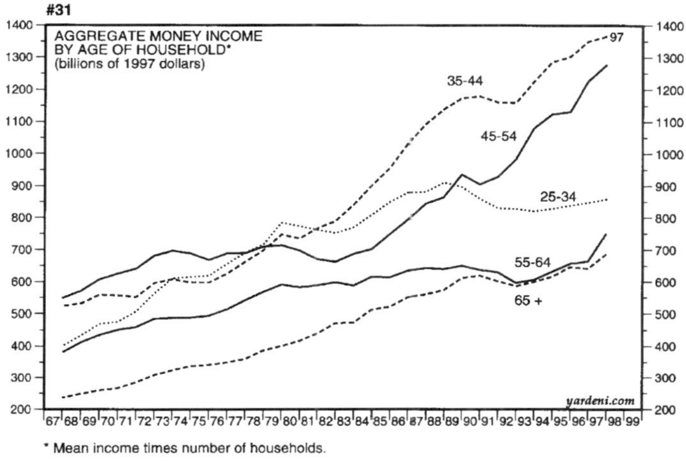

During the 1990s, the only age groups showing significant gains in total group money income are people 35-44 and 45-54.

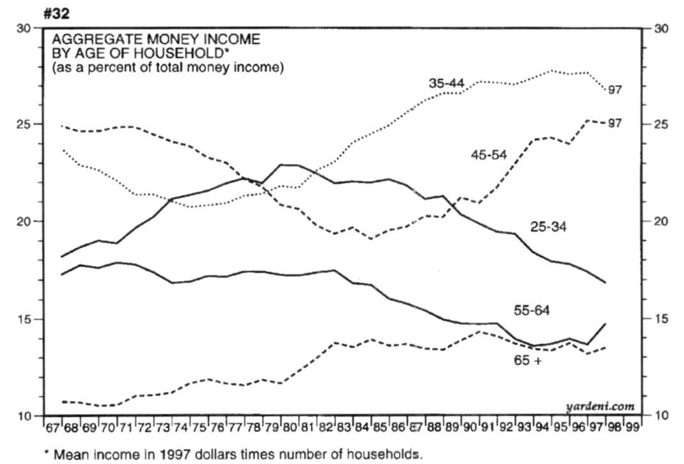

By the year 2000, 35-54 year olds will probably account for 65% of personal income, up from over 52% currently.

Figure D.12

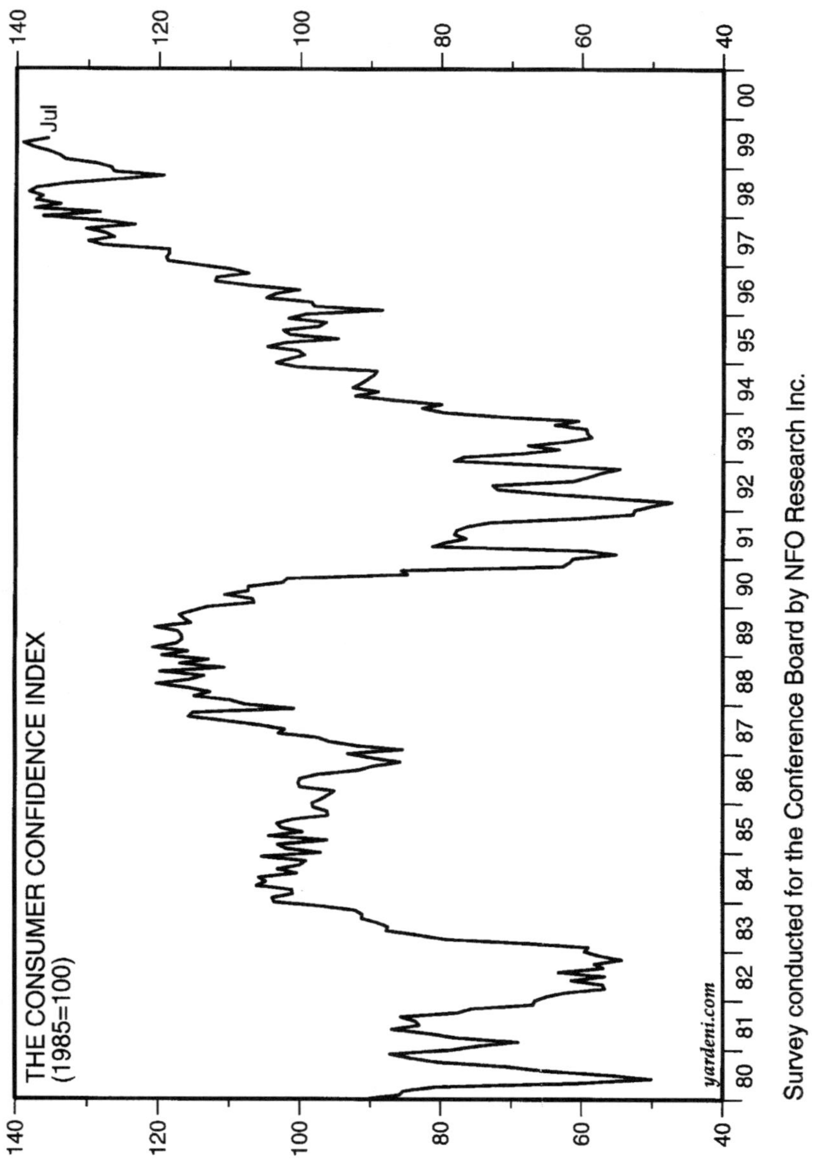
THE CONSUMER CONFIDENCE INDEX
(1985=100)
Jul
yardeni.com
140
120
100
80
60
40
80
81
82
83
84
85
86
87
88
89
90
91
92
93
94
95
96
97
98
99
00
Survey conducted for the Conference Board by NFO Research Inc.

Figure D.13

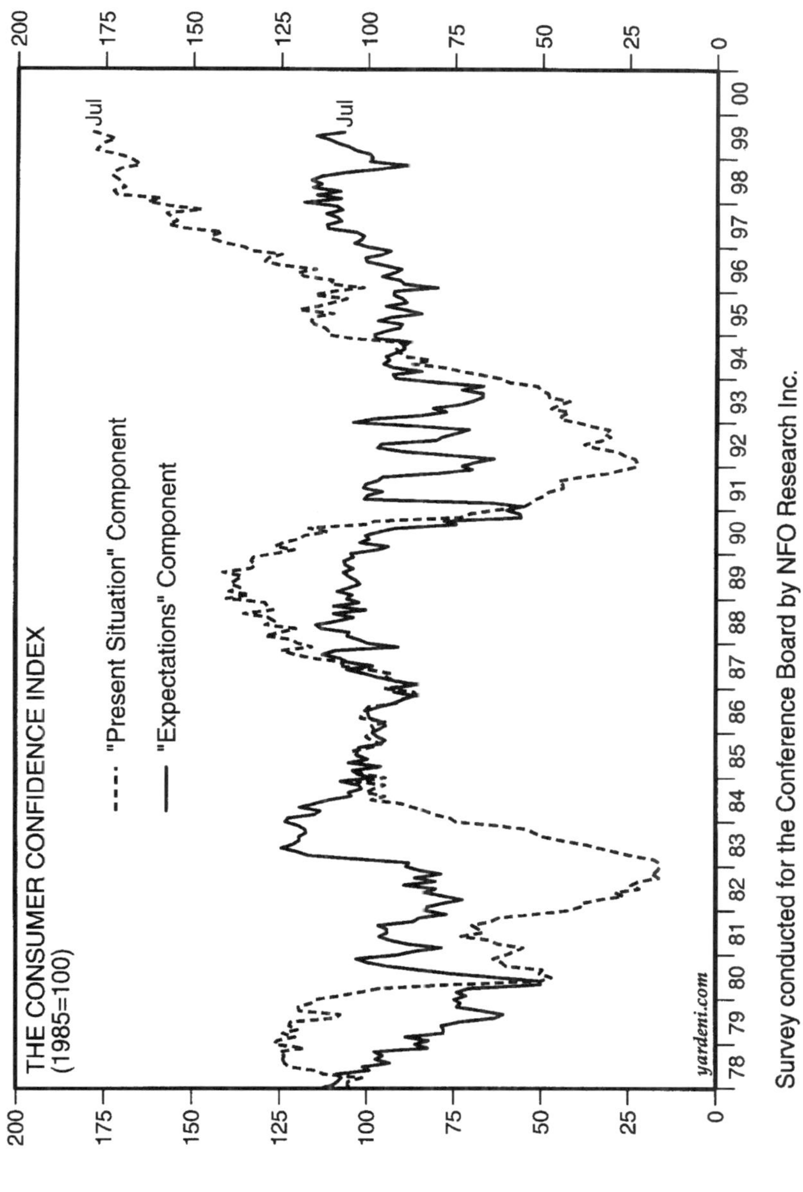
THE CONSUMER CONFIDENCE INDEX
(1985=100)
"Present Situation" Component
"Expectations" Component
Jul
Jul
yardeni.com
200
175
150
125
100
75
50
25
0
78
79
80
81
82
83
84
85
86
87
88
89
90
91
92
93
94
95
96
97
98
99
00
Survey conducted for the Conference Board by NFO Research Inc.

Figure D.14

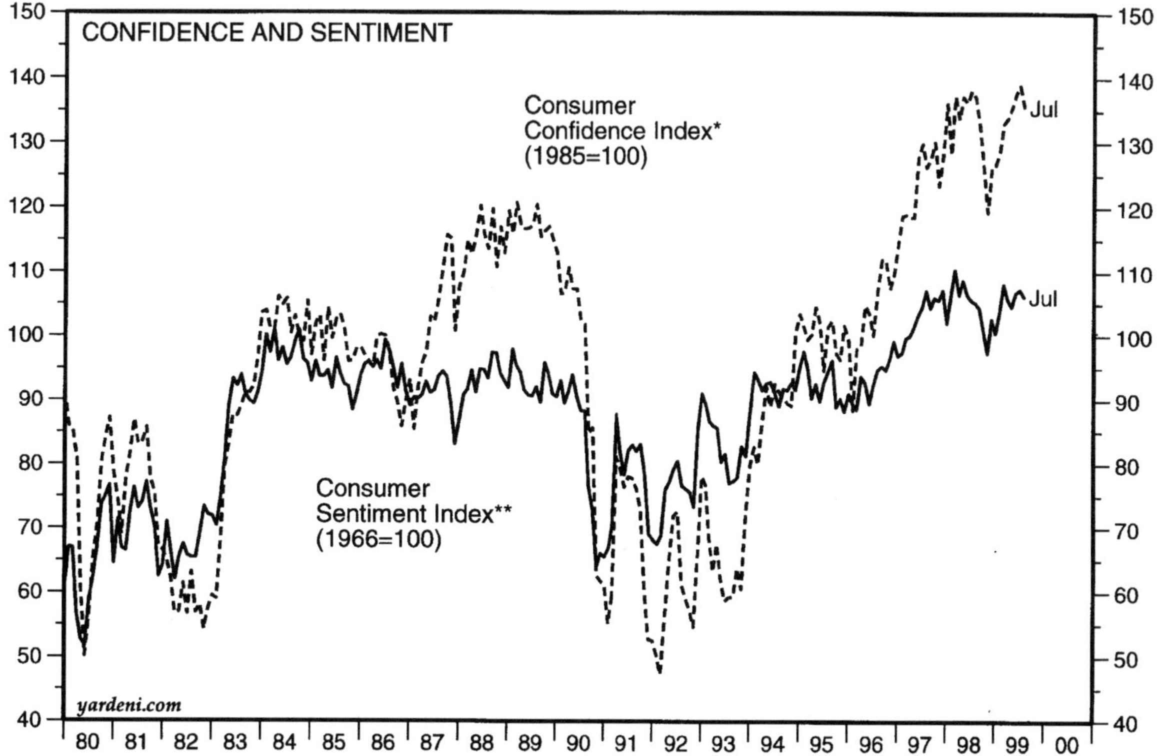

* Conference Board. ** Survey Research Center, University of Michigan.

Figure D.15

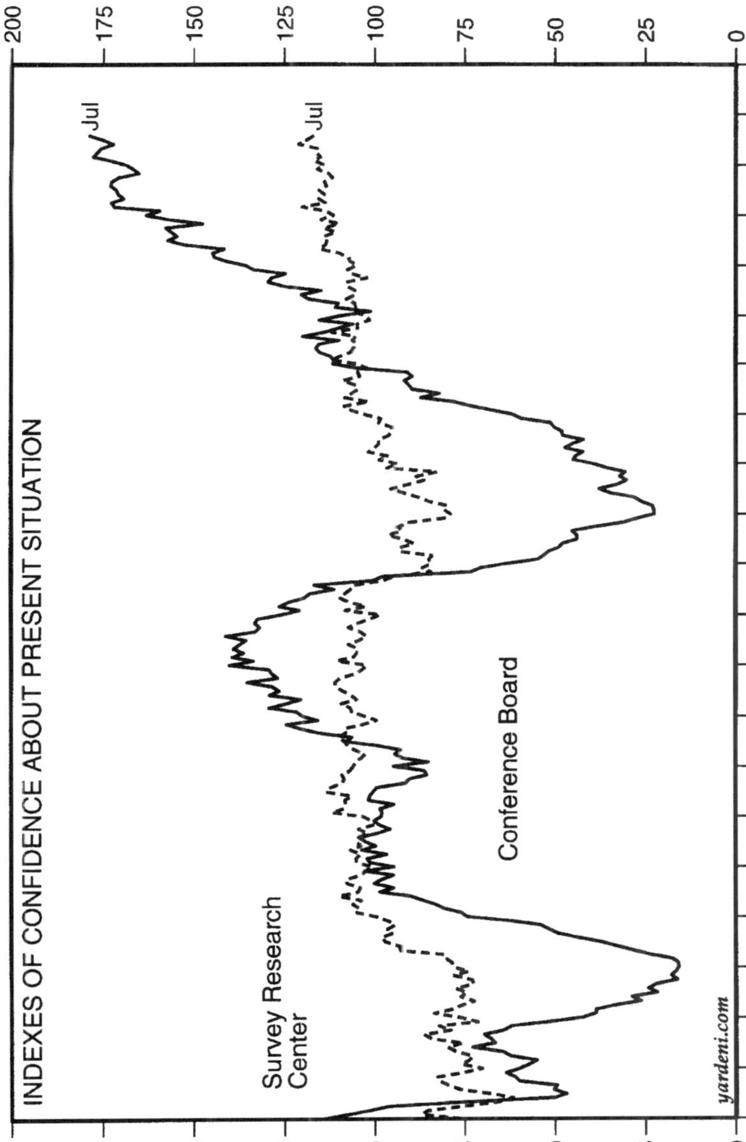
INDEXES OF CONFIDENCE ABOUT PRESENT SITUATION
Survey Research Center
Conference Board
Jul
Jul
yardeni.com
200
175
150
125
100
75
50
25
0
80 81 82 83 84 85 86 87 88 89 90 91 92 93 94 95 96 97 98 99 00

Figure D.16

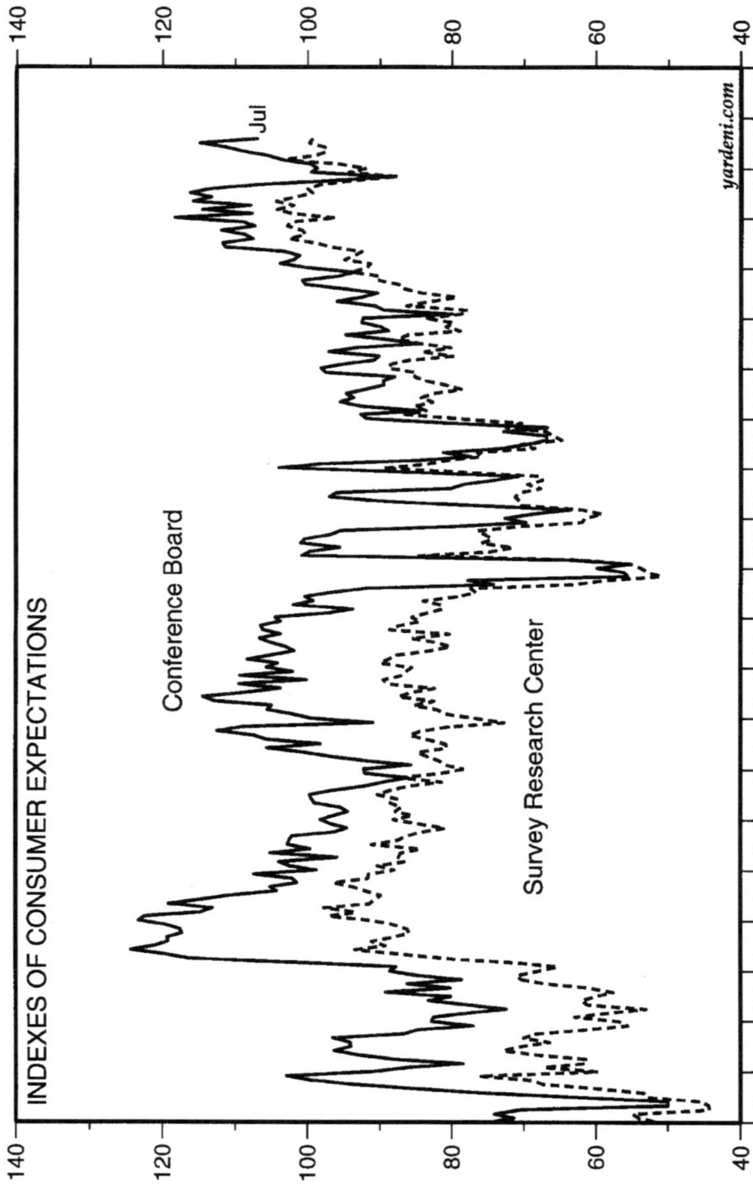
INDEXES OF CONSUMER EXPECTATIONS
Conference Board
Survey Research Center
Jul
yardeni.com
140
120
100
80
60
40
80
81
82
83
84
85
86
87
88
89
90
91
92
93
94
95
96
97
98
99
00

Figure D.17

Section E: International Telecommunications Industrial Structures: Selected Charts and Graphs for Strategists

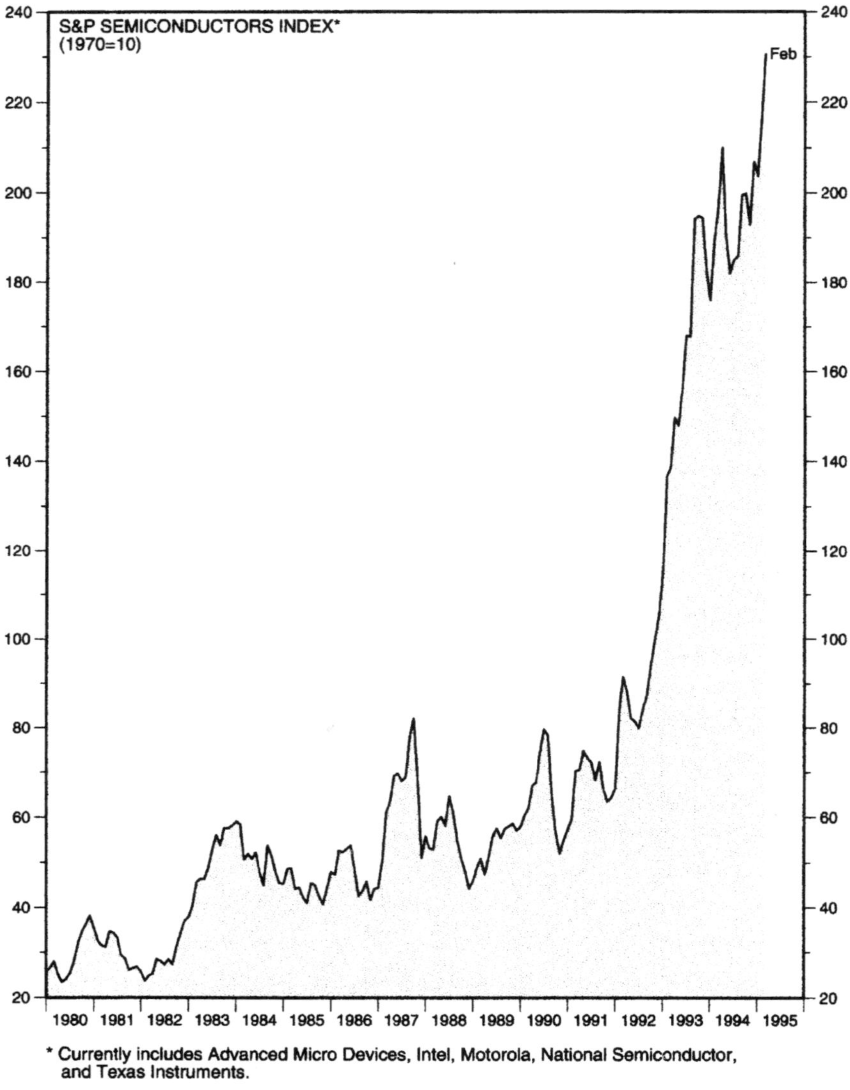
S&P SEMICONDUCTORS INDEX*
(1970=10)
Feb
240
220
200
180
160
140
120
100
80
60
40
20
1980 1981 1982 1983 1984 1985 1986 1987 1988 1989 1990 1991 1992 1993 1994 1995
* Currently includes Advanced Micro Devices, Intel, Motorola, National Semiconductor, and Texas Instruments.

Figure E.1

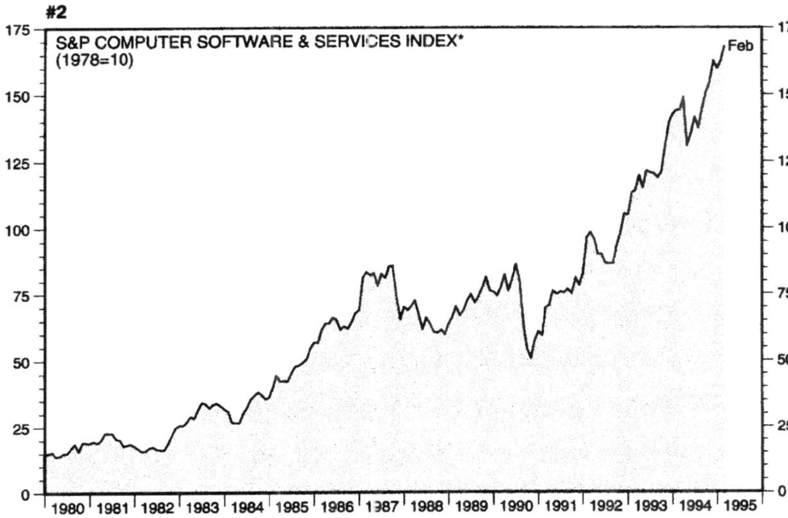

* Currently includes Autodesk, Automatic Data Processing, Ceridian, Computer Associates Int'l, Computer Sciences, Lotus Development, Novell, Oracle Systems and Shared Medical Systems.

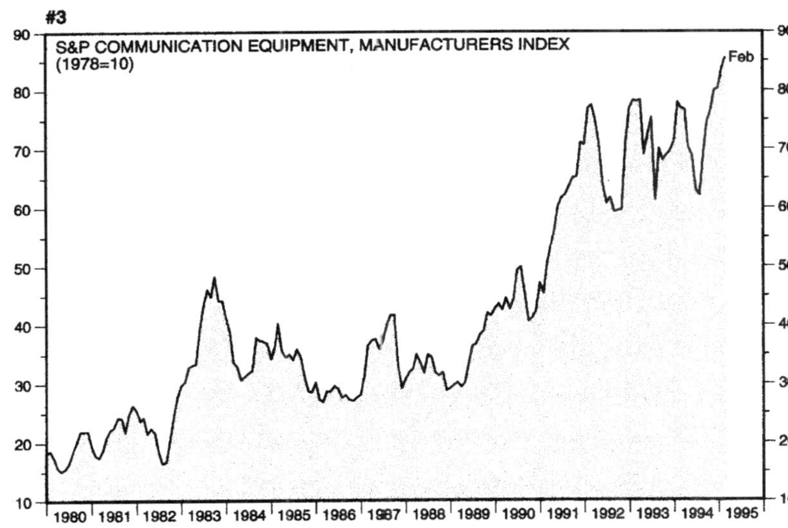

* Currently includes Andrew, DSC Communications, M/A Com, Northern Telecom, and Scientific-Atlanta.

Figure E.2

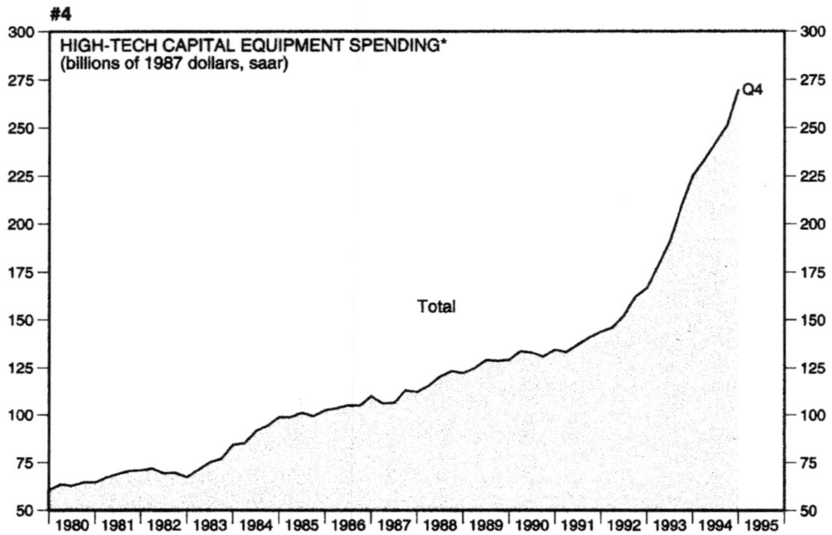

* Producers' durable equipment spending on office and store machinery, communication equipment, photographic equipment, and scientific and engineering instruments.

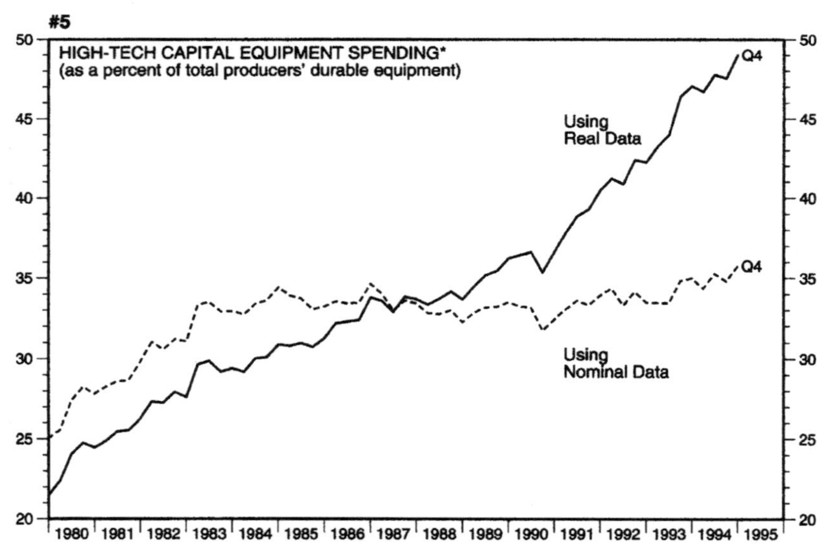

* Producers' durable equipment spending on office and store machinery, communication equipment, photographic equipment, and scientific and engineering instruments.

Figure E.3

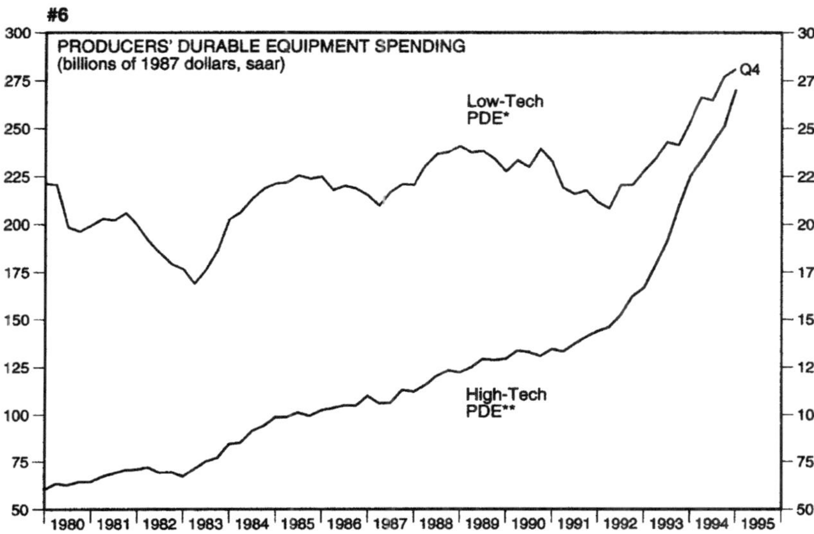

* Including heavy industrial equipment, transportation equipment, and other equipment.
** Producers' durable equipment spending on office and store machinery, communication equipment, photographic equipment, and scientific and engineering instruments.

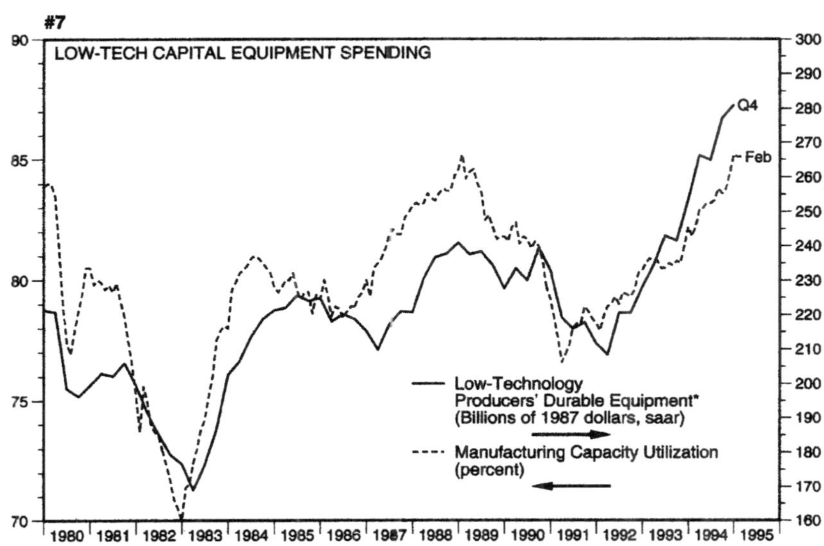

*Including heavy industrial equipment, transportation equipment, and other equipment.

Figure E.4

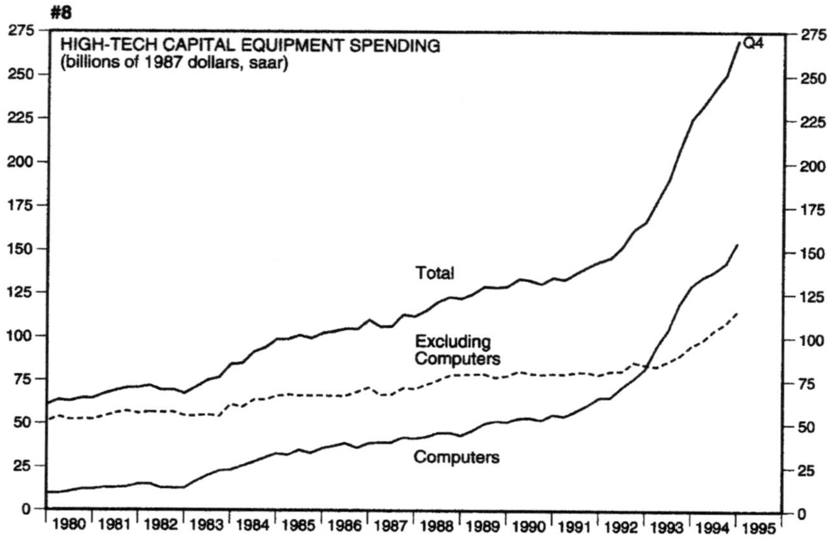
#8
HIGH-TECH CAPITAL EQUIPMENT SPENDING
(billions of 1987 dollars, saar)
Q4
Total
Excluding
Computers
Computers
275
250
225
200
175
150
125
100
75
50
25
0
1980 1981 1982 1983 1984 1985 1986 1987 1988 1989 1990 1991 1992 1993 1994 1995

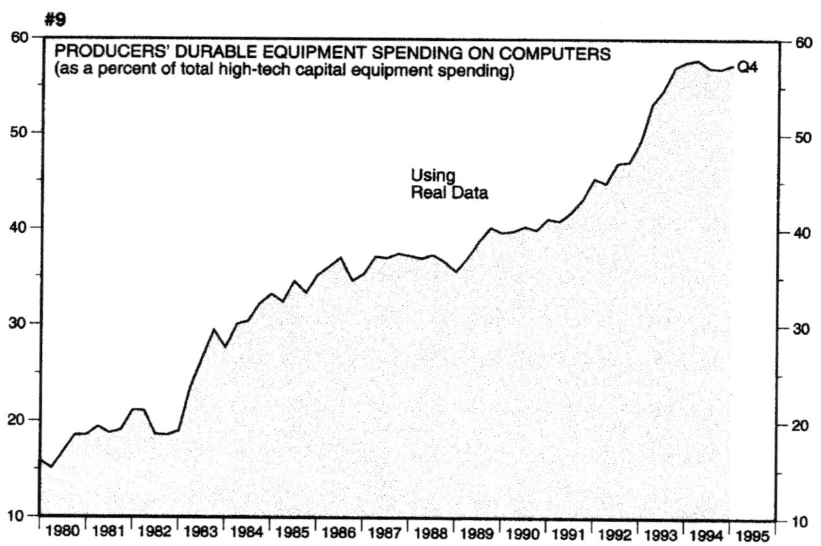
#9
PRODUCERS' DURABLE EQUIPMENT SPENDING ON COMPUTERS
(as a percent of total high-tech capital equipment spending)
Q4
Using
Real Data
60
50
40
30
20
10
1980 1981 1982 1983 1984 1985 1986 1987 1988 1989 1990 1991 1992 1993 1994 1995

Figure E.5

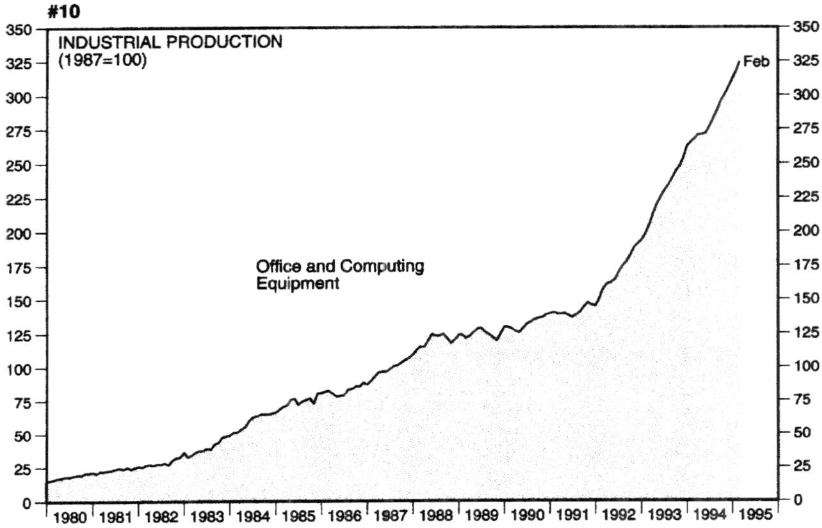
#10
INDUSTRIAL PRODUCTION
(1987=100)
Office and Computing
Equipment
Feb
350
325
300
275
250
225
200
175
150
125
100
75
50
25
0
1980 1981 1982 1983 1984 1985 1986 1987 1988 1989 1990 1991 1992 1993 1994 1995

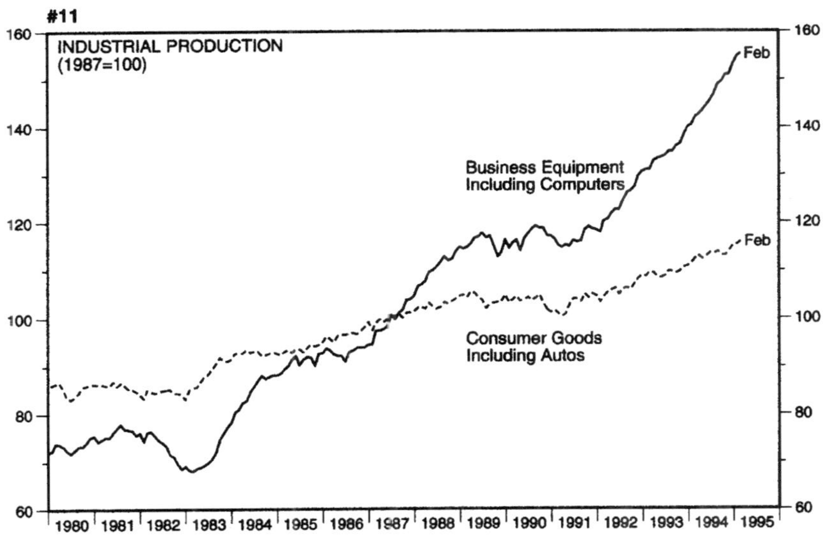
#11
INDUSTRIAL PRODUCTION
(1987=100)
Business Equipment
Including Computers
Consumer Goods
Including Autos
Feb
Feb
160
140
120
100
80
60
1980 1981 1982 1983 1984 1985 1986 1987 1988 1989 1990 1991 1992 1993 1994 1995

Figure E.6

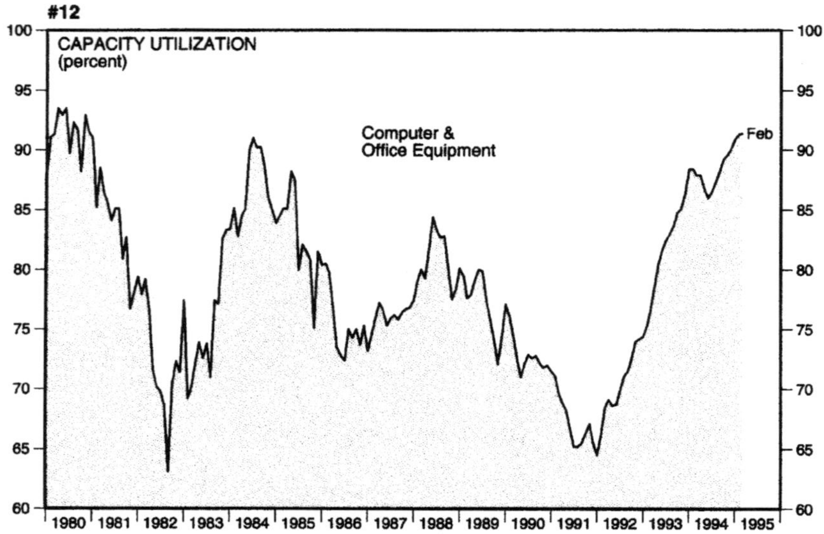
#12
CAPACITY UTILIZATION
(percent)
Computer &
Office Equipment
Feb
100
95
90
85
80
75
70
65
60
1980
1981
1982
1983
1984
1985
1986
1987
1988
1989
1990
1991
1992
1993
1994
1995

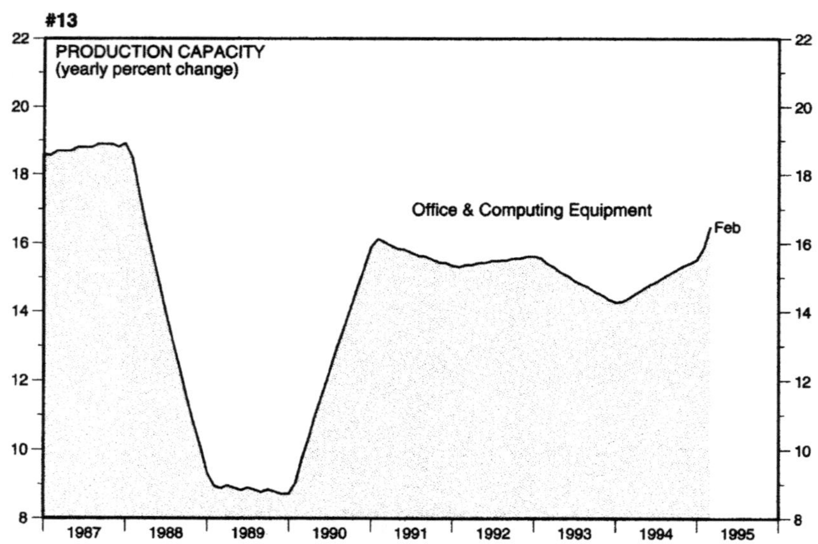
#13
PRODUCTION CAPACITY
(yearly percent change)
Office & Computing Equipment
Feb
22
20
18
16
14
12
10
8
1987
1988
1989
1990
1991
1992
1993
1994
1995

Figure E.7

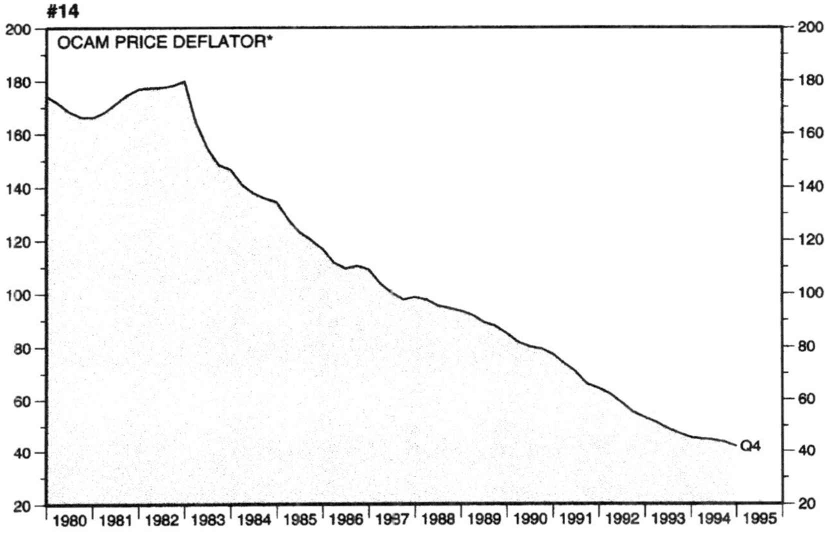

* OCAM = office, computing, and accounting machinery.

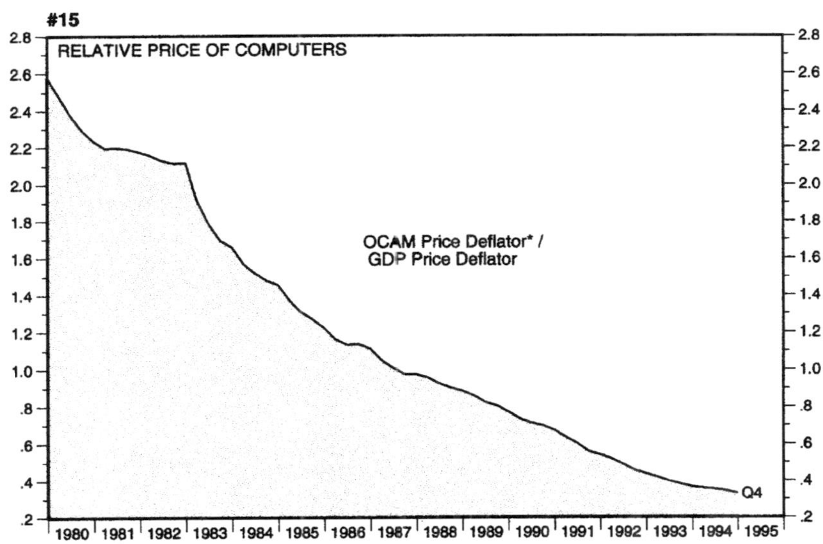

* OCAM = office, computing, and accounting machinery.

Figure E.8

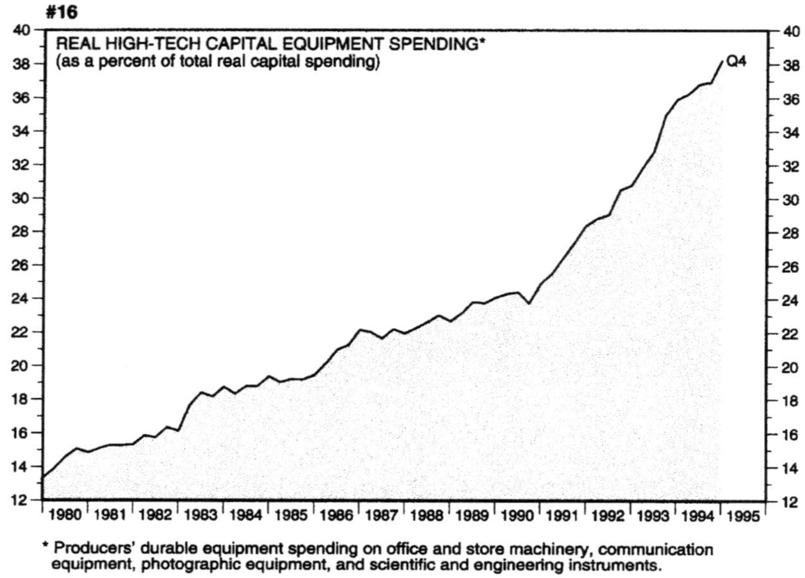
#16
REAL HIGH-TECH CAPITAL EQUIPMENT SPENDING*
(as a percent of total real capital spending)
Q4
40
38
36
34
32
30
28
26
24
22
20
18
16
14
12
1980
1981
1982
1983
1984
1985
1986
1987
1988
1989
1990
1991
1992
1993
1994
1995
* Producers' durable equipment spending on office and store machinery, communication equipment, photographic equipment, and scientific and engineering instruments.

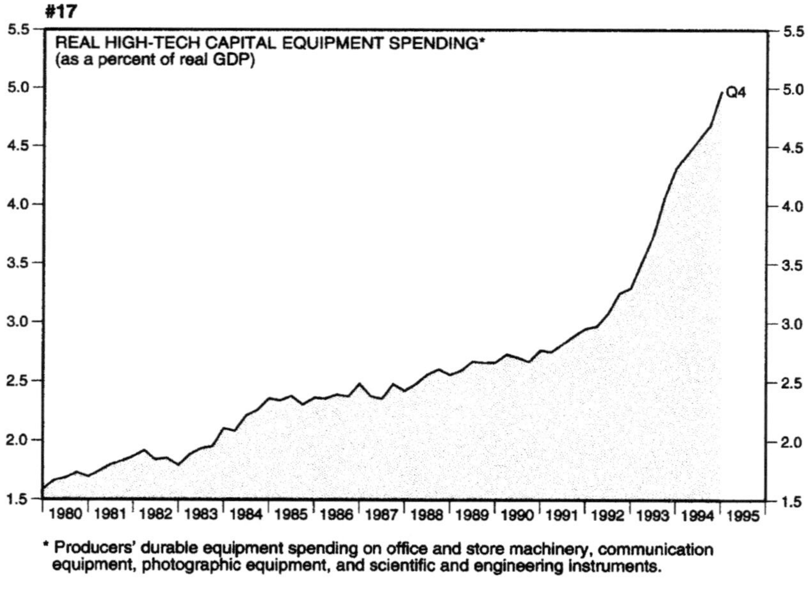
#17
REAL HIGH-TECH CAPITAL EQUIPMENT SPENDING*
(as a percent of real GDP)
Q4
5.5
5.0
4.5
4.0
3.5
3.0
2.5
2.0
1.5
1980
1981
1982
1983
1984
1985
1986
1987
1988
1989
1990
1991
1992
1993
1994
1995
* Producers' durable equipment spending on office and store machinery, communication equipment, photographic equipment, and scientific and engineering instruments.

Figure E.9

THE WORLD COMPETITIVENESS SCOREBOARD

Past results

Ranking as of April 13 1999

1	USA	100.00
2	Singapore	86.04
3	Finland	82.96
4	Luxembourg	81.20
5	Netherlands	81.06
6	Switzerland	80.11
7	Hong Kong	79.67
8	Denmark	77.53
9	Germany	76.72
10	Canada	76.47
11	Ireland	76.36
12	Australia	76.18
13	Norway	74.38
14	Sweden	74.29
15	U.K.	74.20
16	Japan	73.92
17	Iceland	72.73
18	Taiwan	72.08
19	Austria	71.34
20	New Zealand	71.24
21	France	70.76
22	Belgium	70.14
23	Spain	69.40
24	Israel	67.80
25	Chile	66.84
26	Hungary	63.46
27	Malaysia	62.58
28	Portugal	62.22
29	China	61.02
30	Italy	59.98
31	Greece	57.18
32	Philippines	55.05
33	Argentina	53.76
34	Thailand	53.27
35	Brazil	53.08
36	Mexico	52.57
37	Turkey	52.31
38	Korea	52.05
39	India	49.95
40	Slovenia	49.88
41	Czech Rep.	48.80
42	South Africa	48.36
43	Colombia	48.08
44	Poland	47.80

Figure E.10

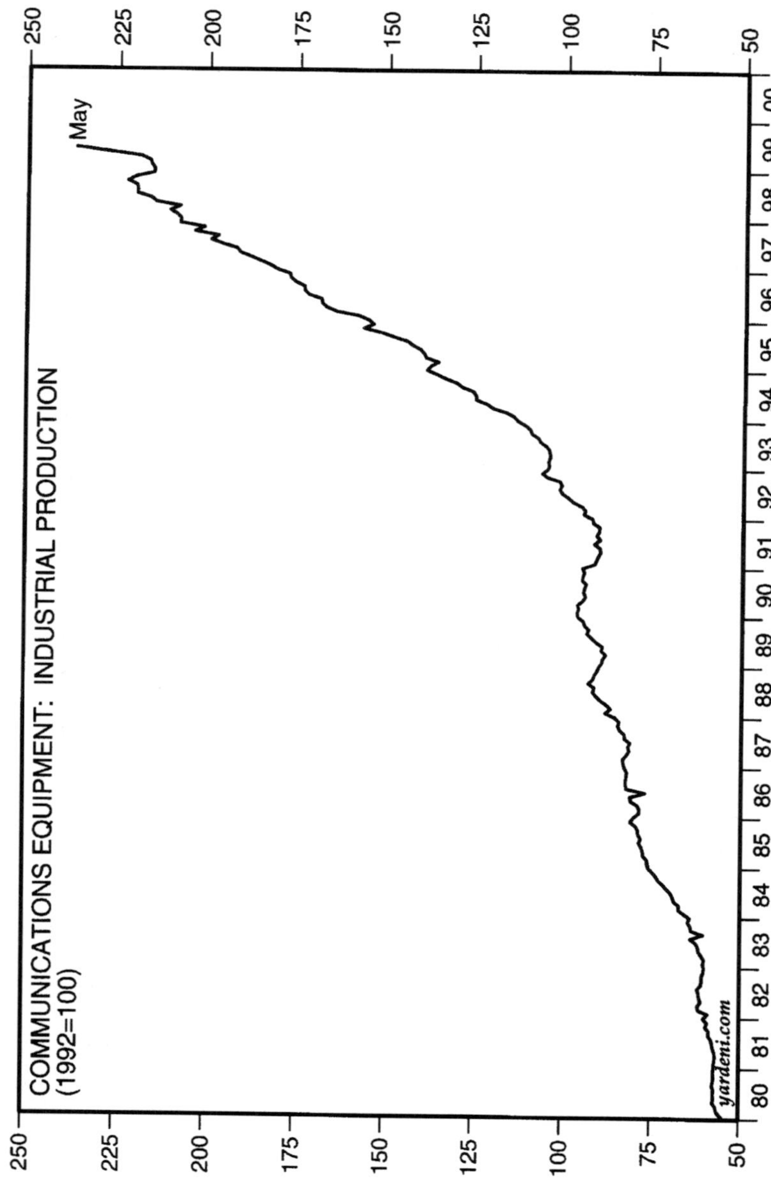

Source: Board of Governors of the Federal Reserve System.

Figure E.11

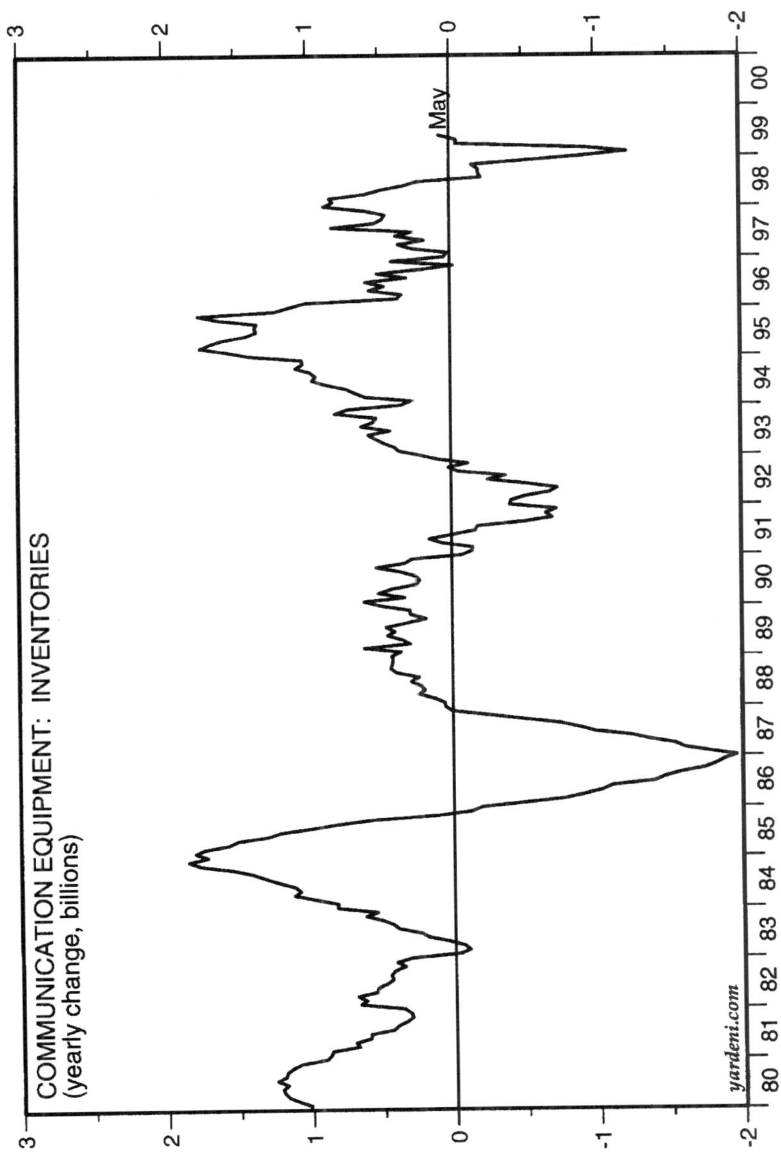

Source: US Department of Commerce, Bureau of the Census.

Figure E.12

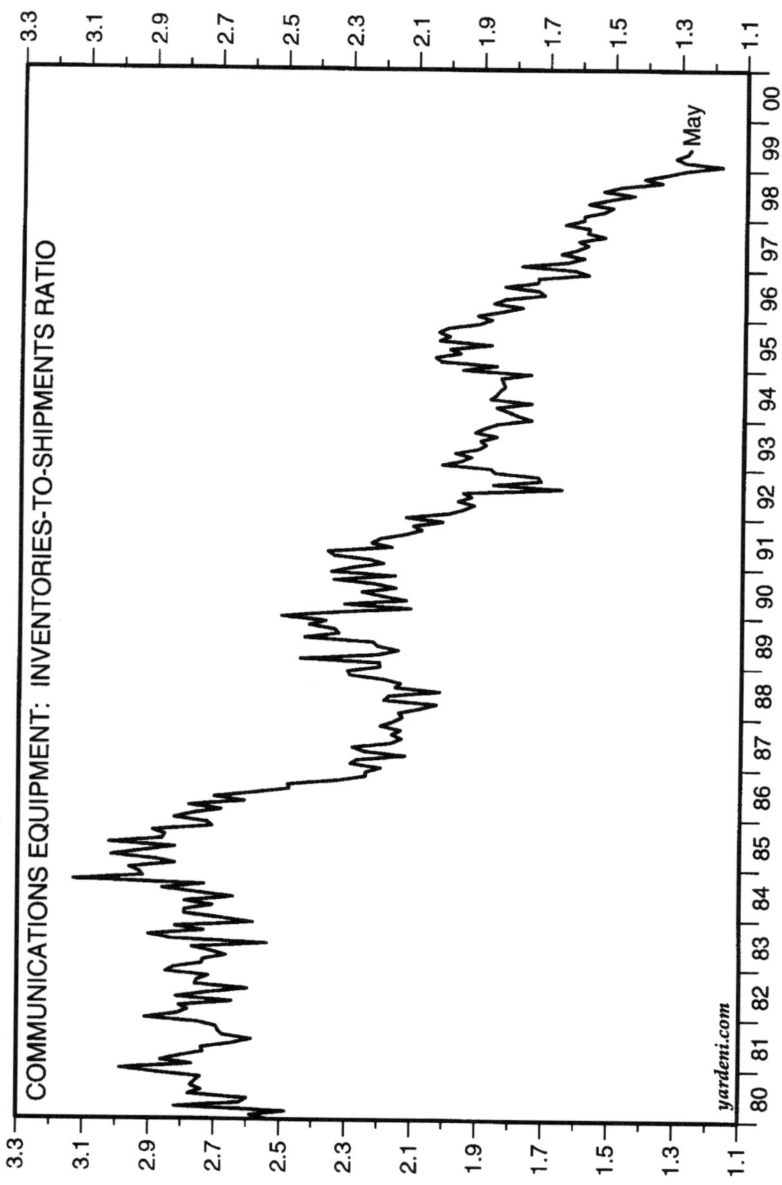

Source: US Department of Commerce, Bureau of the Census.

Figure E.13

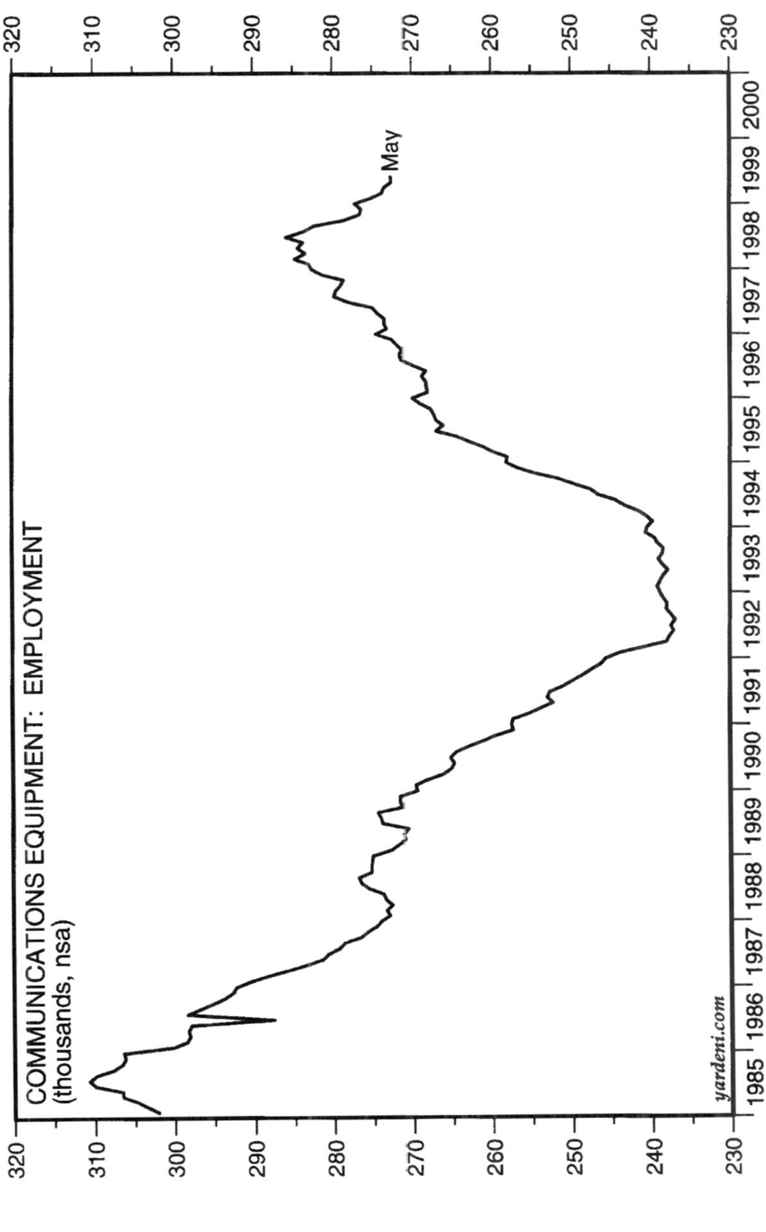

Source: US Department of Labor, Bureau of Labor Statistics.

Figure E.14

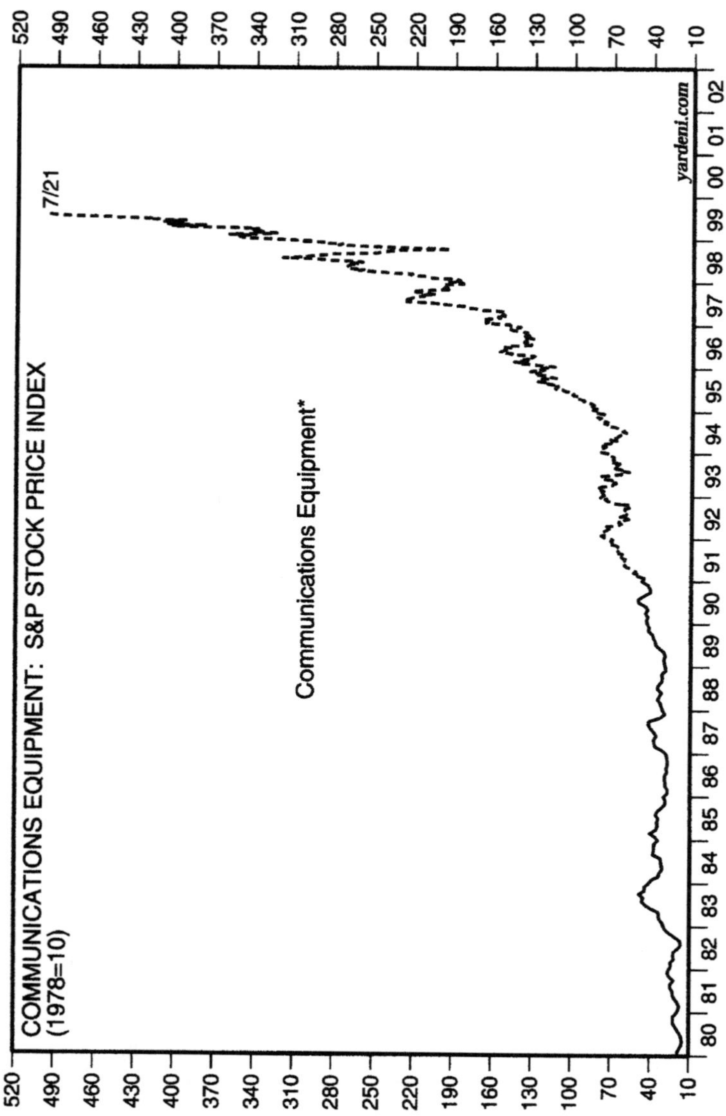

* Currently includes

Solid line is monthly, dashed line is weekly.

Figure E.15

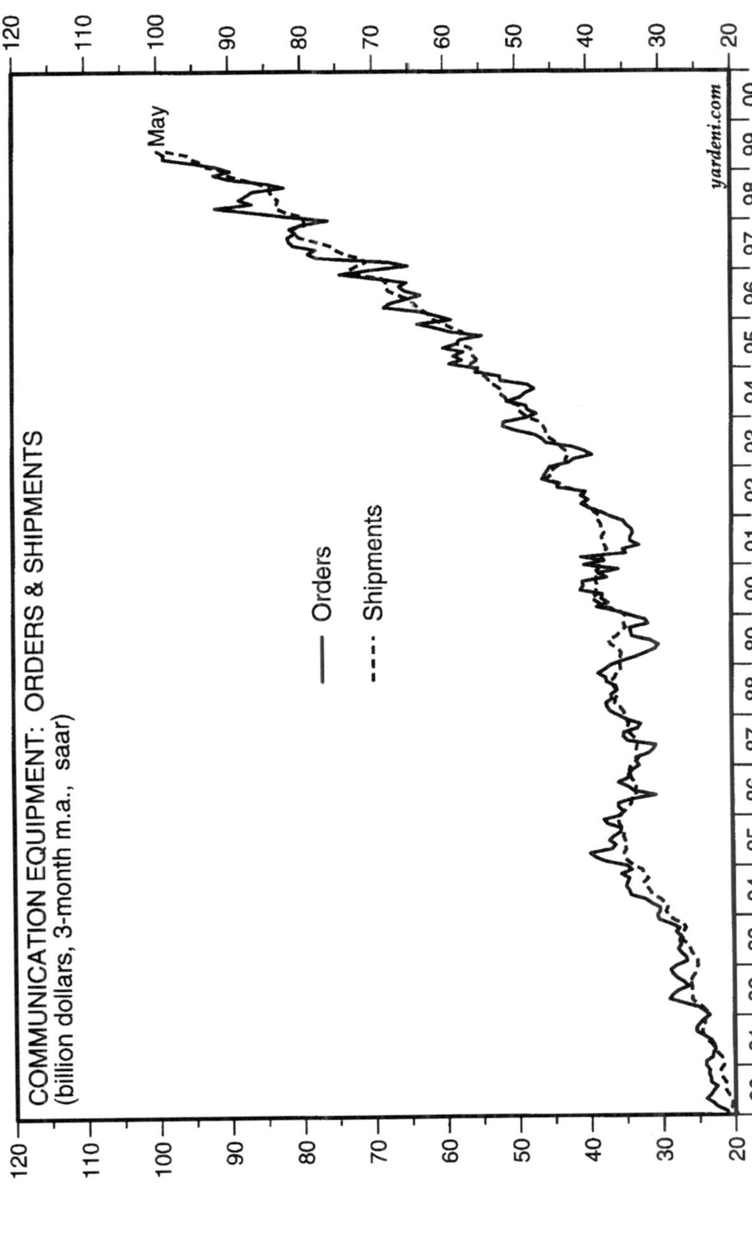

Source: US Department of Commerce, Bureau of the Census.

Figure E.16

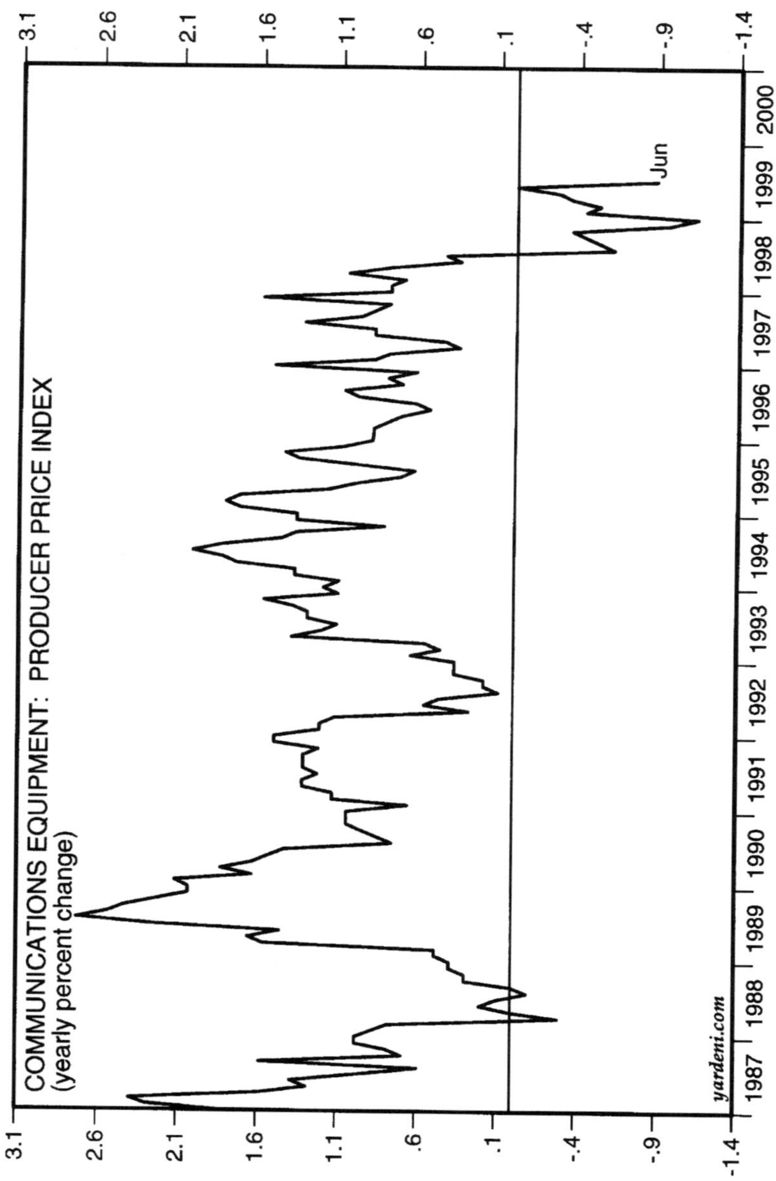

Source: US Department of Labor, Bureau of Labor Statistics.

Figure E.17

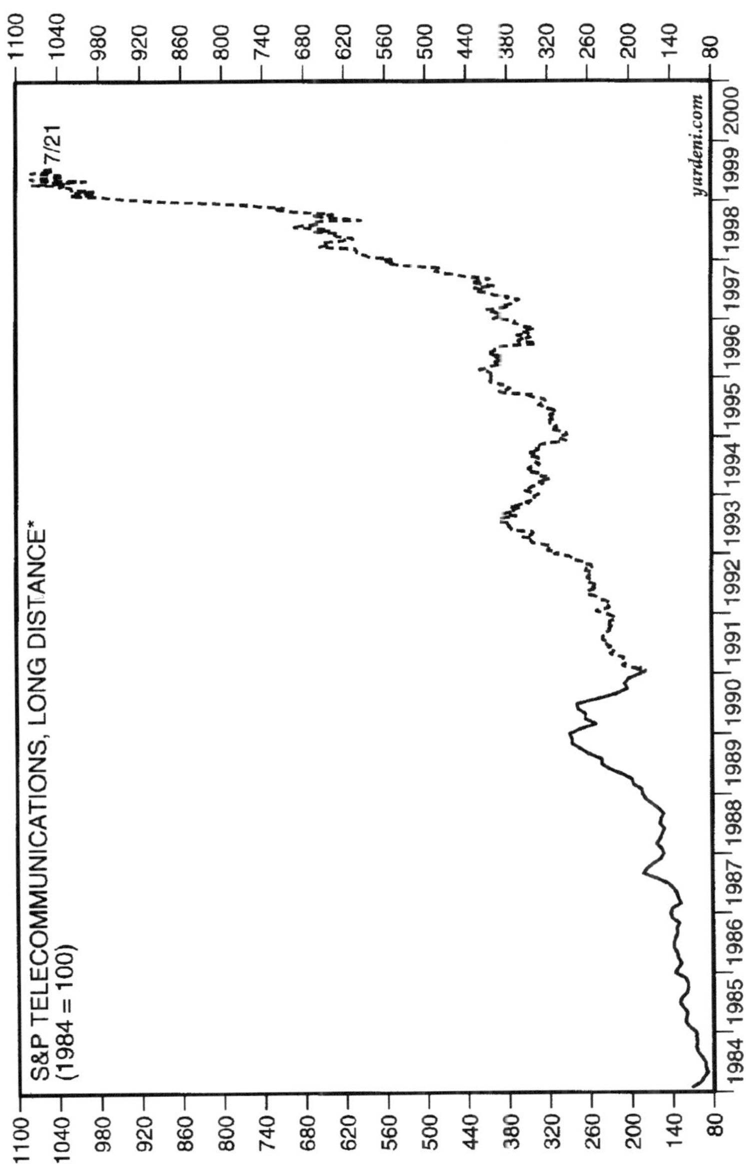

* Currently includes AT&T, MCI, and Sprint.
Solid line is monthly, dashed line is weekly data.

Figure E.18

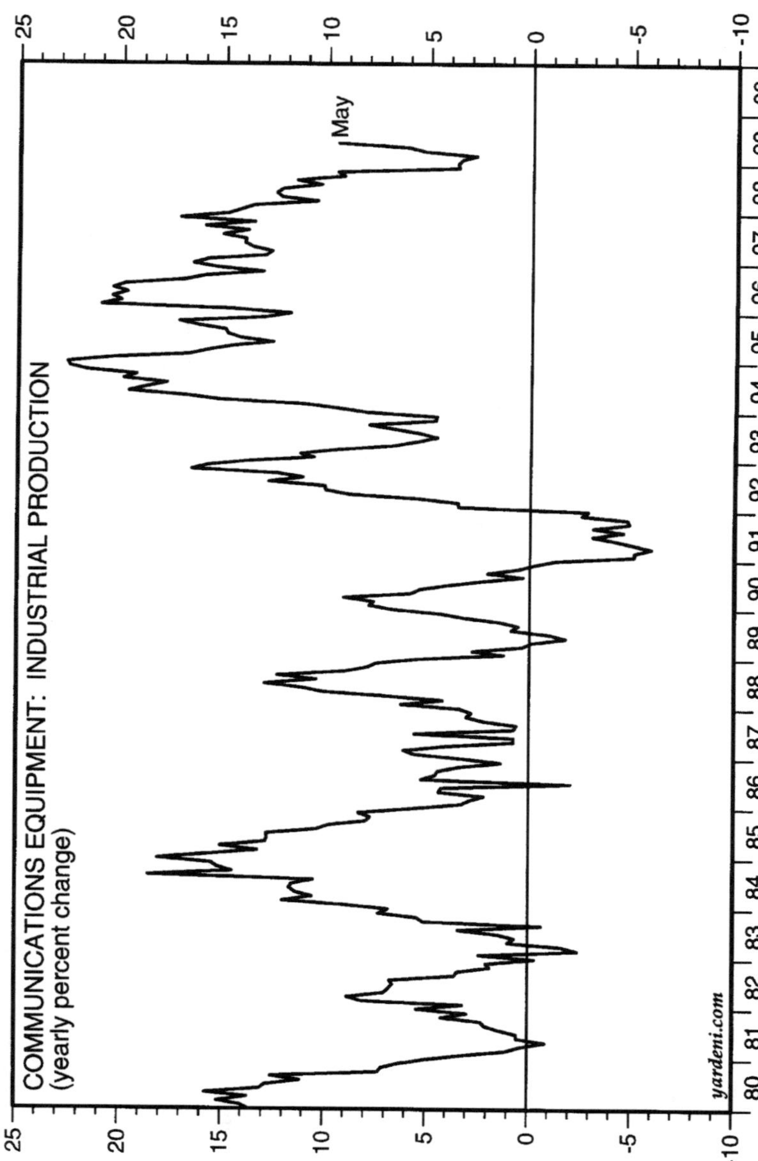

Source: Board of Governors of the Federal Reserve System.

Figure E.19

Section F: Web Site References for Strategists

Indicated below are Web site citations useful for telecommunications strategists. These references are divided according to the primary subfields of specialization in business management. Fields of specialization include finance, management, marketing, electronic commerce, economic analysis, and regulatory policy and law. Telecommunications strategic planners require conduits into each of these areas, both in terms of securing primary data as well as establishing methods for communicating with colleagues across disciplinary boundaries. Web sites identified below are highly recommended for strategic planners in finance, market research, and competitive analysis. Citations described below are strongly recommended by me.

Also inserted in this section are research institutes, news forums, associations, government directories and other sources of data of continuing interest to strategists. A majority of these Web sites are of seminal influence in the field of strategic management, but are subject to topical updates. It is strongly recommended that readers refer to www.umich.edu, www.brint.com, www.infoseek.go/titles, www.planware.org/strategy, www.burrelles.com, www.kiwiclub.edu for hyperlinked information relative to the following database categories:

- World Wide Web-based tools;
- Function-specific tools;
- News forums;
- Associations;
- Government directories;
- Research institutes;
- Specialized research on strategic management in telecommunications.

I. World Wide Web-based Tools

Unless otherwise noted, each of the following references is presumed to follow the standard URL (Uniform Resource Locator) code: http://www.name.com. For general articles and commercial data pertaining to subjects in telecommunications, strategic planning, competitive analysis, business management, marketing, finance, regulatory, or emerging legal issues:

tele.com
Internettelephony.com
Futuris.net
Dataquest.com
Ntq.com
Teledot.com
Telecoms-mag.com
Informationweek.com
Mmm.com
Ameritech.com
Att.com
Pacbell.com
Sbc.com
Ucop.slis.edu
Regulation.org
Cnet.com
News.com
Wired.com
Techwire.com
Zdnet.com
My.yahoo.com
Lib.depaul.edu
Vive.com/www/virtual/research/tel
Main.org/computer
Architel.com
Rpcp.mit.edu/diig/other sites
Millenniumtel.com/ref-gov
Otcog.com
Gfifax.com
Businessmonitor.co.uk
Ntsika.org
Fcc.gov.hotissue

Icanet.com
Iec.org
Computer.org
mmta.org
nab.org
tiaonline.org
tfi.com
imd.ch/way
mktg.inc.com
telecommagzine.com
burrelles.com
dryden.com/mgmt/topics
hbsp.harvard.edu/mgmt
planware.org/strategy
infoseek.go.com/titles
ieee.org

Additional Internet business and economics references directly pertaining to the communications industry can be located through Gopher at the following listing:
gopher://Niord.SHSU.edu:70/
11gopher_root%3A%5B_data.economics%5D.
Also refer to: http://www.wugate.wustl.edu.

The use of these tools will provide access to these important references (refer to the University of Michigan [www.u-mich.edu] Web site for URL citations):

Economics and Finance

Economics Resources from NetEcArchive
Economics Working Paper Archive
Bureau of Labor Statistics
Bell Atlantic Gopher
Business, Economics, Marketing
ECONsult International Consultancy and Forecasting
EconData
Economic Bulletin Board
Economic Research Service
Internet Business Journal
National Information Infrastructure Information

National Information Infrastructure Task Force
Resources for Economists on the Internet
World Bank Public Information Center

The following web pages can be accessed initially by tapping into www.dowjones.com and hyperlinking accordingly via the University of Michigan. Also refer to: www.sims.berkeley.edu for related hyperlinked information.

Macroeconomic Law and Regulation

ARL Federal Relations E-News
Cases, Statutes, & Topical Highlights, Cyberspace Law & Policy (UCLA On-Line Institute)
Clearinghouse on Computer Accommodation (COCA)
Computers and Society
Cyberlaw World Wide
Cyberlaw Center: Cybercrime Resources
Cyberspace Law Center: Law Directories and Other Sites
Cyberspace Law Center: Intellectual Property Resources
Cyberspace Law for Non-Lawyers
Digital Future Coalition
Economics of the Internet
Economics of Networks
Information Law Web
Intellectual Property Law Notes for Technology Companies
Legal Links in Electronic Commerce and Interactive Entertainment
Patent Law Web Server
Regulation of the Communications Functions of the Internet
State Public Utility Regulations
Technology Law Server
Telecom Law and Policy Resources
Technology Law Update
Telecommunications Policy Sites
WWW Multimedia Law
Citizen Internet Empowerment Coalition

Electronic Commerce

Committee on the Law of Commerce in Cyberspace
CommerceNet

Commercial Internet Exchange (CIX)
Information Innovation
Information Warfare
CNET.com
Dataquest Interactive
Year 2000ZDNet E-Business
OECD Electronic Commerce
Internet.com's Electronic Commerce Guide
IBM Institute for Advanced Commerce
Information Economies Project
NetBusiness
The Emerging Digital Economy
Competence Network
CYBER Review Newsletter
Center for Research in Electronic Commerce
A Framework for Global Electronic Commerce
Corporate Knowledge Management
Information Strategy Online
Governments and Electronic Commerce
NetAcademy on Business Media
Journal of Internet Banking and Commerce
Electronic Banking
Inventory of EU and Electronic Commerce Initiatives
OnLine Commerce and Models for the Future
Jupiter Communications
Transaction.Net
CyberSource
SEMPER
Business of the Internet
Internet Business Center
Frontiers of Electronic Commerce
Information Innovation Homepage
Fisher Center for Information Technology Management
Electronic Commerce Resource Guide
Electronic Commerce and the Internet
Association Francaise pour le Commerce et les Echanges Electroniques
Electronic Money in the United States
EFF's Digital Money
Clickshare
Electronic Money
Harvard Information Infrastructure Project

Internet Law and Policy Forum
Information Infrastructure Task Force
NII Virtual Library
Global Information Infrastructure: Agenda for Cooperation
Collaborations on Internet Security
Merit Network, Inc.
Alliance for Competitive Communication
Internet Telephony Consortium
Consumer Project on Technology

II. Function-Specific Tools

The following research tools contain periodic information on domestic and global markets and are listed by organizational function.

Strategic Planning

www.ioma.com
www.aom.pace.edu
www.imd.ch
http://future.sri.com
www.lumina.com/DA
www.hoovers.com
www.dnb.com
www.pathfinder.com
www.infoseek.com/business/economics
www.fourmilab.ch
www.enews.com
www.clark.net/pub
www.wueconb.wustl/edu/business
www.hbsp.edu/strategy
www.wbs.warwick.ac.uk./omindex
www.wst.com/library
www.apics.org/SIGs
www.skyenet.net/~leg/tqm
http://deming.eng.clemson.edu
http://deming.org/deminghtml/wedi
www.informs.org/jobs/jobs
www.itu.ch

www.wow-com.com
www.mpt.go.jp

Management

www.sbaon-line.gov www.imd.ch/
www.teleport.com
www.wp.com

Marketing

www.census.gov
www.marketingtools.com
www.advert.com
www.mci.com
www.esri.com
maps.esri.com
www.geo.ed
www.gis.umn.edu

Finance

www.actnet.org
www.moneypages.com
www.interaccess.com
www.investorweb.com
www.pcquote.com
www.hoovers.com
www.corpfinet.com
www.gnn.com
www.dbisna.com
www.sec.gov
www.princeton.edu
www.datasecurities.com
www.ventureconomics.com
www.federalreserve.com

III. News Forums

News forums are professional organizations who regularly report on the status of macroeconomic and business policy, as well as mergers, acquisitions and partnerships. Six primary news forums provide daily information on developments within the industry:

www.broadcast.com/business
www.prnewswire.com
www.businesswire.com
www.reuters.com
www.dowjones.com
www.wsj.com
www.cnbc.com
www.pctv.com
www.year2000.com
www.upside.com
www.economist.com
www.broadcast.com/business
www.djia.com
www.securitiesdata.com
www.industrystandard.com
www.bloomberg.com
www.businessweek.com
www.cbs.marketwatch.com
www.cnbc.com
www.cnnfn.com
www.forbes.com
www.msnbc.com/news/COM_Front.asp
www.moneynet.com
http://courses.dsu.edu/finance/www/resource.htm

IV. Associations

The following trade or citizen associations represent fertile avenues for the accumulation of data and economic analysis. You may obtain hypertext links via:

www.dowjones.com www.search.com.
Alliance for Competitive Communication
Benton Foundation's Communication Policy Project
Center for Civic Networking
Coalition for Networked Information Policy.Net
Counsel Connect

Cyberspace Law Institute
EDUCOM
MCI Public Policy
MCI Regulatory and Public Policy Department
MFJ Task Force
MIT Research Program on Communications Policy
Online Institute for Cyberspace Law and Policy
Society for Electronic Access
Video Information Providers for Non-Discriminatory Access
Benton Foundation
Canadian International Development Research Center (IDRC)
Centre for International Research on Communication and Information Technologies
Columbia University Institute for Tele-Information
Electronic Frontier Foundation
European Telecommunications Standards Institute
European Union Documents
International Maritime Satellite Organization
Intelsat
International Telecommunications Union
Leland Project
Massachusetts Institute of Technology Media Lab
National Computer Board of Singapore
Pacific Telecommunications Council
World Bank: Infodev Project

V. Government Directories

The following government agencies and their Web directories provide regular access to economic policy, regulation, demographic, legal and other data, as indicated. Additional federal agencies are expected to provide expanding web pages on communications services in the near future.

State and Local Government Regulatory Agencies: www.fcc.gov/state&local
Department of Justice: www.doj.gov
National Technology Transfer Center: www.lcweb.loc.gov
U.S. Patent and Trademark Office: www.uspto.gov
Commerce Information Locator Service: www.doc.gov
Federal Trade Commission: www.ftc.gov
U.S. Census Bureau: www.census.gov

Department of Commerce: www.doc.gov
Security and Exchange Commission: www.sec.gov
National Telecommunications Information Administration: www.ntia.gov
Federal Communications Commission: www.fcc.gov
International Trade Administration: www.ita.gov
Congressional Budget Office: www.cbo.gov

VI. Research Institutes

The following research institutes conduct research on regulation, deregulation and related matters. Web pages are available through the University of Michigan and www.regulation.org.

Progress and Freedom Foundation
Claremont Institute
Hoover Institute (Stanford University)
Center for the Study of American Business (Washington University)
Competitive Enterprise Institute
American Enterprise Institute
CATO Institute
Heritage Foundation
University of Michigan
Yale Documents Center
The Brookings Institution

VII. Academic and Trade Journals in Telecommunications Strategic Management

The following journals, covering various topics of relevance to strategic planners, can be hyperlinked through www.analysys.com.

ATM News Digest (daily journal for broadband networking)
BIT (The Spanish Telecommunication Engineers Association Monthly Magazine)
BT WATCH (weekly monitor of British Telecom produced by Apogeion Global Monitoring)
Business Communications Review
Business News Americas (Latin American daily news service, with telecoms section)

Butterworth-Heinemann (publisher of books and electronic products for students and professionals in the comms and telecoms industry)
Cable Foreman Magazine (bridging outside plant engineering and construction)
Cable World
Cabling Installation & Maintenance Magazine (fiber-optic communications technology and applications)
Call Centre focus (magazine covering call centre management)
Calling (Telia's on-line magazine, in Swedish)
CHA online (Cathey, Hutton and Associates: telecoms news service)
CLEC News (on-line telecommunications magazine)
Communications News magazine (monthly trade publication for network managers and IT professionals)
Communications Standards Review (publishes journals reporting on formal telecommunications standards in the ITU, ETSI and TIA)
Communications Week International
Computer-Mediated Communication Magazine (journal covering cmc issues from December Communications)
Connect-World (publications on telecoms development and liberalisation in Latin America, Caribbean and India)
Data Communications Magazine
Digital Horizons Telecommunications Newsletter (new product and technology announcements from vendors of telecom, cable, wireless and computer telephony)
Discount Long Distance Digest (an Internet Journal for the long distance industry)
e.biz (guide to electronic business, from Business Week Online)
e-Summit (monthly, online report on call center strategy published by MW Productions)
Editors' eNeWS (design house for electronic publishing in IT and communications)
E-Journal (The World-Wide Web Virtual Library: Electronic Journals)
ETHOS (recent telematics news provided by the European Telematics Horizontal Observatory Service)
exchange (weekly newsletter for the Australian telecommunications industry)
Fiber Optics Online (news, analysis, product information and virtual market place)
Global Communications (information centre for IT professionals)
GSM Info Online Search Page (information on networks, services, tariffs, roaming and coverage)
GSMag (magazine on GSM mobile telephony)
GTB (Global Telecoms Business)

HotWired (incorporating Wired)
ICB Toll Free News (regulatory and telecom industry reporting)
IEEE Computer Society (list of magazines published by IEEE)
infoHIGHWAY
Inside Cable (newsletter dedicated to the UK cable TV and telecommunications industry)
Interop Online Networld+Interop annual conference
Institute for International Research Websites for telecoms conferences (and associated newsgroups)
IT News (IT news, information on hi-tech products, software, books, etc)
Journal of Computer Mediated Communication at the Hebrew University of Jerusalem and its Annenberg School for Communication site
Lightwave Magazine
McKinsey Quarterly Telecoms Library
MIT Technology Review
Mobile Computing and Communications
Mobile Office Magazine
Mobile Week (journal of technology issues for users and operators of mobile networks)
Mobilis (features interviews, tutorials, reviews, concerning all aspects of PDAs, wireless communication, and mobile peripherals)
Netcomtalk from the Boston University College of Communication
Network Bibliography (keyword search)
NEWSdesk The Online IT News Network (on-line information service covering IT and telecommunications from Hewlett Packard)
NewsNow (up-to-the-minute news service with technology section)
NewsPage (news on communications, telecoms and datacomms inter alia)
Outside Plant Magazine (cabling, switches, poles, manholes, installation and maintenance for telephone and CATV voice and data)
Phillips Business Information Inc (satellite, cable and telecoms information)
Phillips TelecomWeb (news, research and analysis in the telecommunications industry)
PNE*wire* (on-line version of Public Network Europe magazine)
Point Topic Ltd (news and analysis on broadband access to the Internet)
Premises Networks Online (news, analysis, product information and virtual market place)
The Response Source (service that links journalists with IT sources)
Rural Telecoms in India (article by Abhinav Taneja)
Siemens Telcom Report (a Siemens publication covering global telecommunications trends, products, solutions and services)

Satellite Today (news service by Phillips TelecomWeb)
Satellite Week (journal of news and analysis of technology issues in the satellite industry)
Technology Research Institute (reports on telecom billing, data warehousing, provisioning and customer care)
Telecom Business Online (information source for carriers, convergence, competition and next-generation networks)
Telecom Sources (trade magazine with product information)
tele.com (publication for telecom service providers)
Telecom News Magazine (global news on telecoms, datacoms, multimedia, internet and events)
Telecom Publishing Group (produce a range of titles in telecommunications)
Telecom Today (telecommunications news and information)
Telecom Tribune (news, information and statistics on Japanese telecoms)
Telecoms OSS BSS and IN News (news service run by TelcOSS Executive International Inc)
Telecommunications Online
Telecommunications Policy
TRI: Telecommunications Reports International Inc (news and information services for the communications, multimedia and electric utility industries)
Telecoms Newsline monthly telecoms industry information source sponsored by Hewlett Packard
TeleGeography Home Page
Tele-M@rket (news and discussion forum for the call centre business)
TeleNews Asia (news on Asian telecommunications)
Telephony
TelephonyWorld.com (on-line resource for the computer telephony and communication industry)
Telirati (archive of telephony newsletters)
Test and Measurement.com (news, analysis, product information and virtual market place)
Total Telecom (daily news and information service for telecommunications professionals)
UCTIR: U.S.-China Telecommunications and Information Research (on-line newsletters on China's telecom and information industries)
U.K. Telecom Frequently Asked Questions
University of Iowa Libraries Gateway to the Internet (Communications Studies section of an Internet directory)
University of Miami Reading Room (comprehensive list of telecommunications journals)

Video Information Provider Newsletter
Virtual Institute of Information (V.I.I.) (On-line telecoms information from Communications and Mass Media research institute at Columbia University.)
Voice & Data (India communication magazine)
Warren Publishing Inc (daily telecom news)
Web-Vantage (technology guide for the internet development community)
Wireless Design Online (news and discussion forum for the wireless engineering market)
Wireless Integration Magazine
Wireless Today (news service by Phillips TelecomWeb)
Wireless Week

VIII. Specialized Research on Strategic Management in Telecommunications

Citations listed under this heading may be accessed through www.brint.com and www.sims.berkeley.edu.

The Internet Economy Indicators
Institute for Technology and Enterprise
Internet Economics
Netomics
Brazilian Journal of Economics
Internet 2
National Coordination Office for Computing, Information and Communications
Virtual Institute of Information
Local Switch Congestion
Internet and Telephony
Boardwatch Magazine
Network Economics
Intermediaries and Cybermediaries
Network Externalities
Internet Engineering and Planning Group
Interoperability and the Economics of Information Infrastructure
USA Global Interoperability for Broadband Networks
Interoperability and the National Information Infrastructure
NSF Networking and Communications Research
Telegeography
Internet Architecture Today

Policy Gateway
Berkman Center for Internet and Society
Internet Public Policy Network
Global Internet Project
Internet Law and Policy Forum
FindLaw
New Information Infrastructure: Strategies for U.S. Policy
Taxes and Electronic Commerce
Alliance for Public Technology
LawResearch
Telepathic

Legal Links in Electronic Commerce and Interactive Entertainment

These specific Web sites are recommended for state-of-the-art analysis and benchmarking in corporate strategic planning departments:

www.trac.org
www.trackingcommunications.com
www.ratexchange.com
http://smartone.svi.org
www.telstra.com/au/infocommunications
www.benton.com
http://china.si.umich.edu/telecom/telecom-info.html
www.ukans.edu/cwis/units/coms2/tech.html
www.its.bldrdoc.gov/its.html
www.onepassinfo.com/opteldef.html

These government and industrial organizations produce primary source data on strategic management problems in the telecommunications industry. Refer to for hyperlinked information to these sites and their assorted publications. Parenthetical commentary is provided by www.analysis.com and www.brint.com.

Arabsat (Arab Satellite Communications Organization)
Artis Software Corporation (OptSim: high-end optical communication systems simulator)
Batelco (Bahrain Telecommunications Company)
CIENA LightWorks (DWDM and intelligent optical networking)

Comviq (Swedish GSM operator)
Cyprus Telecommunications Authority
Etisalat (United Arab Emirates)
Global One (Jordan)
Graphnet Inc (data communications services)
GTO: General Telecommunications Organization, Sultanate of Oman
IBM Public Affairs (Living in the Information Society)
IDN: Inter Digital Networks (broadband and metropolitan network operator in UK)
Itlissalat Al Maghrib (Morocco)
IWE: Institute for Wireless Education (training courses in wireless technology)
Jinny Paging (Lebanon)
Kuwait: Ministry of Communications
McKinsey Quarterly e-commerce library
McKinsey Quarterly Telecoms Library
McKinsey & Company
MobiNil (Egypt)
MPT Lebanon (Lebanese Repulic Ministry of Posts and Telecommunications)
MTC (Kuwait)
New Telecom Policy 1999 (India's policy statement)
Newentrants.com (New Entrants Tracking Service, maintained by Analysys Ltd)
Noyes Fiber Systems (manufacturer of fiber optic test equipment)
ntl Group Ltd (UK operator)
OECD: Electronic Commerce.url
POLE-TEC INC (line test equipment)
Q-TEL (Qatar Telecom)
Ravi Visvesvaraya Prasad & Associates (telecom, datacom, networking and software cnsultants in India)
SATRA (South African Regulator)
STC: Saudi Telecom Company
Stratecast Partners (communications industry strategic intelligence)
Tele2 Mobil (Swedish GSM operator)
TeleCap Communications (investment information on new telecoms companies)
Telecom Business Online (information source for carriers, convergence, competition and next-generation networks)
Telecommunications Inc (Comprehensive Communications Consulting)
TeleYemen
Telsim (Turkey)
Thuraya Satellite Telecommunications Company

TSS: Total Software Solutions Inc (wireless and wireline billing systems)
Tunisie Telecom (Tunisia)
Turkcell (Turkish GSM operator)
web2FAX (tool for faxing over the internet, by Graphnet Inc)
Butterworth-Heinemann (publisher of books and electronic products for students and professionals in the comms and telecoms industry)
Comarco Inc (wireless communications products, engineering and management)
CommuniTech (headsets, speakerphones, audioconferencing and videoconferencing systems)
DSD: Distributed Software Development (billing and customer care solutions for paging operators world wide)
Dwyer Associates (courses in telecommunications policy and regulation)
e.biz (guide to electronic business, from Business Week Online)
Emil Kipriotis SA (manufacturer of communication terminals)
EnterSource (enterprise resource management in telecommunications, energy and utilities)
FirstLink Communications Inc (provider of wireless LAN/WAN/MAN, ISP, Voice IP, modems, and T1/E1 communications equipment)
Flagship TeleSystems (billing systems and customer care solutions for telecommunications)
FRIAP: French Internet Access Providers list (listing of French ISPs)
Global Cellular and Paging (wireless products, services and accessories)
Global Light Telecommunications Inc (Emerging International Carrier)
Grayson Wireless (wireless measurement systems)
Highpoint Telecommunications Inc (Emerging International Carrier)
iaxis corporation (European Carrier's Carrier)
RIT Distance Learning (online courses at the Rochester Institute of Technology)
SAFCO Technolgies Inc (data collection and analysis of wireless networks)
Salient 3 (designs, manufactures and markets equipment for use by communication network operators worldwide)
Studio Cazzani (telecom consultancy focusing on Italy)
The Helein Telecommunications Group (management, operational and technical consulting to the telecommunications industry)
TMNG: The Management Network Group Inc (management and operational consulting company for the telecommunications industry)
TONEX (telecommunications, internet and communications training and consulting service)
Trend Communications (hand-held test equipment for ISDN, Frame Relay and ATM, brand name "aurora")
Videoconference.com (forum for videoconference professionals)

World Telecom
3rd Millenium (two-way radio, telecoms, IT and broadcast equipment)
A Global Marketplace for SMEs (paper outlining protential benefits of e-commerce to SMEs)
ACEL: Atlantic Communications Enterprises Limited (turn-key supplier of telecommunications products and services)
ACIF: Australian Communications Industry Forum (standards, industry codes and industry support services)
Advanced Datacom Systems (network management, monitoring and performance management solutions)
AITEC: Australian Information Technology Engineering Centre (joint venture between four Australian universities)
Alan Warwick Consulting Ltd (U.K. consultant in IT and telecommunications procurement)
BCL (international and long-distance services in St Petersburg and NW Russia)
CBSI: Complete Business Solutions Inc (IT and telecom consulting services)
Cell-Tel International Inc (second market celluar and infastructure equipment)
Cyber Digital (digital data and voice telephone switches, PBXs, class VI and V switches)
Direct Visual Limited (videoconferencing supply and support)
El Homo Comunicatus (Spanish cable operator)
Fiber Optics Online (news, analysis, product information and virtual market place)
GE American Communications (Americom) (satellite service provider)
GFRC Shelters (modular concrete buildings for the telecom industry)
Gilat Interactive Distance Learning (IDL systems)
Gilat Satellite Communications Inc (international links and VSAT services)
Glocalnet (operates pan-European voice-over-IP network)
Guaranteed Electronic Markets (paper proposing new ways of organising e-commerce transactions)
ICT electronics (test and measurement solutions for telecommunications based on SDH, PDH, ATM and IP)
InnoWave Wireless Systems (fixed wireless access solutions.)
International Chamber of Commerce (Commission on Telecommunications and Information Technologies)
Lynk (remote LAN/WAN and IP telephony solutions)
MapInfo (spatial solutions for telecoms strategy issues)
Metrodata (high performance ATM and non-ATM access products)
Microtronix (telephony test equipment for testing to ITU-T, IEEE standards, and many country specific requirements)
NextCall Telecom plc (U.K. residential service provider)

OFTEL: Inquiry into Electronic Commerce (trade and industry select committee report)
Premises Networks Online (news, analysis, product information and virtual market place)
PTT: Providers of Telecommunications Training (computer based training courses for the telecoms industry)
Quallaby Corporation (analysis and reporting solutions for network service providers)
RSA Data Security (encryption and authentication technologies)
Taiwan Telecommunication Industry Company Ltd. (coin, card payphone system, digital pair gain system, facsimiles)
Telecommunications Engineering Technology (undergraduate degree at RIT: Rochester Institute of Technology)
TelephonyWorld.com (on-line resource for the computer telephony and communication industry)
Telesoft Design Ltd. (design and manufacture of medium-scale multi protocol converters, CTI OEM boards and protocol test solutions)
Telution (customer care and billing solution for competitive communications providers)
Test and Measurement.com (news, analysis, product information and virtual market place)
The Communications Group (forum for communications lawyers, consultants, regulators, policy-makers and other professionals)
Thomcast Group (multimedia equipment and AM, DAB, FM, MMDS and TV transmitters)
TLA Communications (international telecommunication access, with a variety of enhanced calling services)
UNCITRAL (United Nations Commission on International Trade Law)
VersaTel Telecom International (Benelux service provider)
Webmonkey (E-commerce tutorial)
WinSoft Corporation (custom software solutions for electronic test systems)
Wireless Design Online (news, analysis, product information and virtual market place)
BD Software (customer acquisition, loyalty, credit control, billing and management software solutions)
Bertsch International (Internet-based learning consultancy specializing in teaching the use of technology in the classroom)
BICC Communications (cables, cable systems and services)
BSW Telecoms (bespoke business and operations software solutions for telecoms network and service providers)
BTG: Banner Telecom Group plc (mobile telephone accessories)

Clear Communications (intelligent surveillance applications for network quality and performance monitoring)
Comtest Laboratories (Australian telecommunications and electrical safety compliance testing laboratory)
CrossKeys (software solutions to telecommunications service providers for network information and control)
CTIA (Cellular Telecommunications Information Association)
David's Sonet/SDH Page (SONET and SDH information, especially on testing)
EAP (Irish provider of data transport services over wireless point-to-multipoint broadband network)
Econet Wireless (GSM operator in Zimbabwe)
e-Summit (monthly, online report on call center strategy published by MW Productions)
Foxcom Wireless (fiberoptic transmission systems for wireless services)
Galaxy Power Inc (intelligent power management solutions)
GTB (Global Telecoms Business)
Intec Systems (interconnect billing systems developer)
Integration Technologies (cable telephony upgrades, cable modems, SONET and WDM networks)
In-Touch Associates (business strategy, marketing, sales and training through to network planning, engineering and operations)
INTUG (International Telecommunications User Group)
iPass Corporate Access (Internet connectivity for mobile employees)
Israel Ministry of Communications
JSI Capital Advisors (investment banking, industry information and education services to the independent communications provider industry)
K & M TEL INC (refurbished, new and used transmission, central office, switching, power, outside plant equipment, installation and deinstallation)
Motech Industries Incorporated (general electronic test and measurement equipment)
NETMANSYS (network management software products)
Norcom Inc (telecommunications and Internet service provider)
Nursat (licensed carrier of voice, data, video and Internet services in Kazakstan, with a satellite network)
OSI: Objective Systems Integrators (service and network management solutions to the worldwide communications industry)
Paralan Corporation (products for the SCSI enhancement market place)
Paul Budde Communication (telecommunications and information highways consultancy services and publications)
Payphone Connection (public communication solutions and equipment in Canada)

PCIA (Personal Communications Industry Association)
PrimeTEC (European based carrier's carrier)
Protek (network management specialists who develop Cortex, a network management tool)
Ratekona (Lithuanian broadcasting consultants)
Regulatel (association of regulators in Latin America)
Russell and Partners (telecoms recruitment consultancy)
Satyam Telecom (customer care and billing, CTI, mediation, clearing house, migration services, etc)
Senao International Co Ltd. (cellular phone accessories)
Starnet Engineering Ltd. (telehousing and technical support services)
Tait Electronics Ltd. (mobile radio equipment)
TDG: TeleCommons Development Group (international telecommunications and networking consulting firm specializing in solutions for rural areas)
Technology Investment Partners (European corporate finance and project management for growing technology businesses)
Telecom Consulting Group Inc (interconnection agreement negotiations and consulting on other telecom service and network design issues)
Telecom Solutions (low rate long distance services, toll-free numbers, international calling)
Telecommunications Management Program (Oklahoma State University)
The GSM Consultancy (consultancy services to the international wireless telecomms industry)
TSI: Telecom Services International (telecommunications engineers supplying integrated digital radio link systems for voice and data)
VoiceTrak (voice mail, call accounting and fax back software)
Wireless 2000 Engineering Advisory Group (professional engineering services to wireless communications industry)

The following trade and academic journals contain company-specific and industry-specific articles on telecommuincations strategic management (note that these publications can be accessed through www.analysys.com):

American Telesis Inc. (wholesaler of private lines or data lines to resellers and agents)
Arthur Andersen (global multidisciplinary professional services firm)
Band-X: the bandwidth exchange (independent virtual market for international wholesale telecom capacity)
Bloomberg Web site
Calculus Solutions Ltd. (customer care and billing product set for pan-European emerging and mature Telcos)

CalTech International Telecom Corporation (CLEC building VOIP network, operating in California, Illinois, Colorado and Arizona)
Cape Saffron Online (trader in international traffic)
CINTEL: Centro de Investigacion de las Telecomunicaciones (nonprofit organization offering consultancy on the strategic and technological aspects of telecommunications in Colombia and other Latin American countries)
CMMC: Customer Management Multimedia Callcentre (association for companies using Call Centre)
Comparable Performance Indicators (comparison of Quality of Service between UK telecoms companies)
DACOM (Korean operator)
Distance Education (World Wide Web Virtual Library)
E-TECH Inc (2-way cable modems, HDSL , ADSL and transmission products)
FCS: Federation of Communication Services (UK trade association for the mobile telecommunications industry)
Federal Gateway (consolidates federal sites and provides pre-searched information)
Fujitsu Broadband Services (high-capacity phone systems, voice messaging products, ATM equipment, SONET, IP telephony, call center products)
General Cable PLC (UK cable company covering Yorkshire, W London and Birmingham)
Global Telecom International (distributors of PhonClub: flat rate long distance service on IP)
Infozech (billing software services for telecom service providers)
Intelecon Research & Consultancy Ltd (focus on rural telecommunications, emerging markets and developing countries)
JSCSYS (Canadian consultancy and systems development for telecommunications service providers)
Kuyaké Technologies Inc (strategic telecommunications consulting services)
Loral Space & Communication
myfreeoffice.com (long distance, calling cards, and internet access)
NCI: Net Communications Inc. (global IP-based voice and data services)
Phillips Tarifica (consultancy specialising in tariffs)
Phillips TelecomWeb (news, research and analysis in the telecommunications industry)
Primal Systems (software solutions to aid customer acquisition and retention in the telecommunications industry)
Priority Communications Inc. (national telephone and voice processing solutions)
Profundus Consulting Ltd. (telecoms marketing consultancy with European focus)

RomTelecom (Romanian operator)
Shiron Satellite Communications (satellite communications products including satellite L-band modems, VSATs and InterSKY for two-way fast internet via satellite, based in Israel)
Shunra Software Ltd. (developed The Cloud, a tool which simulates Internet and IP WAN Intranet conditions)
Stormin Protection Products Inc (surge suppressor for communications equipment)
SureCom Solutions (digital business telephone and voice processing solutions)
TCIL: Telecommunications Consultants India Ltd. (telecommunication consultancy and engineering company)
Tele2 U.K. (high speed data connection over advanced digital wireless network)
Telecom Today (telecommunications news and information)
Telekom-Control (Austrian Regulator)
TelePost Office 2000 (virtual office provider)
WebTelecom (Internet phone numbers that support calls from any H323 standard phone)
XCP Inc (phone card printer vending machines)
@Home Network (Internet access over cable modem)
1st Family ('free' Internet access bundled with flat-rate long distance IP telephony)
ACTEL Enterprises Pty Ltd. (strategy consultants specializing in cellular, WLL and VAS)
Alcatel (site for Alcatel's space activities)
Analysys Satellite Communications Database
Andersen Management International A/S (management consultants specialized in telecommunications)
Arcome SA
Atlantic Fibre Optics (manufacturer of Fibre Optic test equipment)
BABT: British Approvals Board for Telecommunications (independent European approval body for telecoms equipment, as well as offering training and consultancy services)
Barry Reichelt Telecommunications Consulting Services Pty Ltd. (audits, tenders, Y2000, training)
BiblioNet (IT and telecoms resources)
BiblioNet: Standards Direct (ordering service for standards documents)
BTI Communications Group Ltd. (equipment, local, long distance, internet, computer-telephony and wireless)
Business Communications Ltd. (global consultancy specializing in telecommunications and media convergence)
Cable Community (representation group for Cable & Wireless customers)

Cadzow Communications Consulting Ltd. (consultants on TETRA, communications security and technology management)
CD3 Multimedia (graphic designers for multimedia)
Cell Tech Ltd. (transmission, RF on fibre, GSM /DCS turnkey solutions)
Centre for Telecommunication Networks (University of Durham, UK)
CLEC News (on-line telecommunications magazine)
CLEMESSY (telecom test systems)
Click GSM (mobile phone network in Egypt)
Comisión de Regulación de Telecomunicaciones (Colombian Telecommunications Regulatory Commission)
Connogue Limited (radiocommunications mobile and broadcasting, spectrum management, regulatory policy and strategy)
CUT: The Campaign for Unmetered Telecommunications (UK lobbying group)
Cutler & Company (specialists in convergence)
Datapoint (data, voice and video integration products)
Demon Internet
Domosys Corp (communication solutions for residential and commercial LANs)
DTS: Dynamic Technology Systems Inc (value-added service provider of telecommunications and information technology)
Economic TeleDevelopment Forum (forum on relationship between telecommunications infrastructure investment and enhanced community economic development)
EHPT: Ericsson Hewlett-Packard Telecommunications AB (telecommunications management solutions especially in fixed-mobile convergence)
E-Journal (The World-Wide Web Virtual Library: Electronic Journals)
El.pub (telecommunications) (telecoms section of EC sponsored R&D news and resources site)
Elektrobit Group (engineering consultancy for wireless communications)
EQUIPTRANS (French data transmission equipment manufacturer)
ESRI (GIS tools for geographical analysis in telecoms)
European Electronics Limited (transmission and reception solutions: terrestial, satellite, microwave, mobile, LAN and WAN)
European Interconnect Atlas (country-by-country description of the status of interconnection in Europe, created by Analysys Ltd. for the European Commission)
Evans Associates (telecommunications network planning, design, and implementation)
FreBon International Corporation (implements point-to-point and multipoint videoconferencing networks)

Genesys Telecommunications Laboratories Inc (provides open CTI framework and software for telecoms including call routing, reporting and multimedia applications)
Global Internetworking (provider of local and long distance high-speed bandwidth and leased line services)
Infrared Communication Systems Inc (infrared systems for data networking and telecommunication applications)
Insight Research Corporation (market research and competitive analysis exclusively for the telecommunications industry)
InterSatCom Inc (satellite Internet provider)
IPL Telecommunications (systems integrator and software engineering)
iSpring Interactive (multimedia consulting, design, and production)
ITU Regulatory Colloquium (news and discussion forum on telecoms regulation, free access to ITU Regulatory Colloquium reports)
LCPI
Linx Communications (one number follow-me services and virtual PBX solutions)
Matra Marconi Space
Min-X (marketplace for IP telephony minutes)
Mitel (manufacturer of telecommunications products and semiconductors)
Monte R Lee & Co (telecoms engineering company)
Mott MacDonald (telecommunications engineering arm of multi-disciplinary consultancy)
MSUA: Mobile Satellite Users Association(forum for users, and good industry links)
National Communication Service, Inc (full telecoms provision for multi-site companies)
O-m6 Telecommunications (manufactures fiber optic cable assemblies)
ONCTek LLC (network consulting services)
Optimus (GSM operator in Portugal)
Pacific Access (technology services and solutions)
Phillips Business Information Inc (satellite, cable and telecoms information)
Phillips Omnicom Training(telecommunications and datacommunications training, including distance learning programmes)
Phone Savings: ADM & Associates (full service telecommunication brokers)
PROD-EL: Prodotti Electtronici SpA (radio simulcast networks analog and digital, radio terminals and turnkey systems)
Route-2 (IP telephony rate broker)
Satellite Today (news service by Phillips TelecomWeb)
SomerData (digital communications data capture and recording systems)
Superintendencia de Telecomunicaciones (Regulator in Guatemala)

Tartanchase Computer Services Ltd (supplies data communications and computer networking equipment)
T-COM (T1, T3, and Sonet test equipment manufacturer)
Technical Communications Corporation (encryption systems for data, voice and fax network communications security)
Teldat (internetworking, routers, transmision equipments (XDSL, baseband), multimedia, etc)
Telecommunication and Informatics Unit (World Bank work program)
Telecommunications Industry at a Glance (ITU rankings of operators and country subscriber data)
Tele-M@rket (news and discussion forum for the call center business)
Teleplan Italy (Italian consultancy using billing and tariff analysis to minimize telephone expenses for clients)
Telsoft Inc (manufactures DTMF and telephone line analyzers)
The Cell Site (information resource on Digital Mobile Telephone Systems)
Transpower Technologies Inc (magnetic components for ethernet, telecom, datacom and power applications)
Ultra Active Technology Ltd. (videoconferencing supplier in Far East)
UUNET
Vega Group (systems and software engineering consultancy working in the space and telecommunications industry)
Vital Connections (service provider also offering autodialers, voicemail systems, message on hold systems)
Warren Publishing Inc (daily telecom news)

Glossary

Absolute advantage	Capacity of a country to produce a specific good with fewer resources (per unit of output) than other countries.
Acquisition	A purchase of one firm by another; in rare instances, a strategic alliance among multiple firms may result in acquisition of one or more companies.
Administrative law	That body of public law which governs bureaucratic management and regulatory policy.
Affiliation	Association between two or more firms through non-equity agreements, the principal objective of which concerns strengthening distribution channels for each affiliate.
Aggregation	The process of attracting customers so as to expand an existing consumer base.
Antitrust law	Federal statutes preventing, modifying or restricting proposed mergers or acquisitions which result in reduction of market competition. The Sherman, Clayton, FTC, and Robinson-Patman Acts (passed in 1890, 1914, 1915, and 1936, respectively), define contemporary antitrust law.
Baby Boom Generation	That generation born between 1946 and 1964; some demographers extend this definition to include births beginning in 1940 through 1965. This generation, comprising 80 million, is the largest American demographic group in history.

BOP	Business Opportunity Paradigm: a model, designated in this volume, and validated through recent empirical research, which suggests that strategic planning in many telecommunications enterprises should focus on the development of tertiary markets.
Brand	A symbol, image, term, name, design, or label which differentiates a firm from its competitors.
Brand awareness	A measure, statistic, or other evaluative tool which isolates a consumer's knowledge of a brand's image.
Business cycle	Variations in supply and demand in the macroeconomy which influence prospective consumption for all goods and services. Recent studies on the business cycle reveal that post-World War II prosperities are lengthening, while recessions are contracting in length.
Brand extension	The extension of a firm's reputation to the development of other products and markets.
Business processing reengineering	The restructuring of a firm's functions based on the introduction of advanced communications and information technologies.
Cherry picking	A tactic typically applied by a firm bringing a new product to market; the term is intended to convey that high-income consumers become the initial target of entrepreneurs seeking a market presence for the product.
Classic Strategic Planning	A simple six-step procedure for strategic planning, embracing mission definition, elaboration of company goals, an assessment of the firm's competitive environment, an internal audit of the firm's resources, identifiable action steps, and a plan to implement tactics designed to achieve strategic goals. The inability of classic planning methodology to contend with volatile competitive environments has inspired a host of alternative planning models in recent years.
Communications Act of 1934	The fundamental federal statute guiding communications policy through 1996.
Comparative advantage	The capacity of a country to produce a specific good at a lower opportunity cost than its trading partner.
Competitive advantage	An economic expression used to define the inherent benefits, gains, or other favorable circumstances that a firm can exploit relative to its competition; the expres-

	sion sustains particular efficacy in the field of industry analysis.
Competitive analysis	The procedure whereby the competitive environment of a firm is dissected.
Competitive environment	A reference in modern strategic management which emphasizes the intensification, volatility and uncertainty associated with non-incremental change in all sectors of the economy.
Competitive research	Also described as "surveillance research," in which a firm defines strategic threats likely to emerge from the competition.
Consumer decisionmaking	The process by which consumers assimilate, retain, store, and retrieve product characteristics prior to purchase.
Consumer primary	An organizational perspective which holds that consumers, during the impending age of customization, represent the exclusive focus of market research.
Continuous learning markets	The great long-term areas of consumer demand likely to be the strategic objective of major firms within the next decade; these markets are characterized by the need to certify education and training, as well as to link collaborative networks of businesspeople, communities and other groups of common interest.
Convergence	A word of dual meanings: may be referred to as the growing interaction and influence of telephony, computer, and broadcasting technologies designated to create new products and services; may also be applied to describe the recent spate of partnerships, alliances, affiliations and other business formations uniting former competitors in introducing new products to market.
Co-opetition	A business model which conveys the desire of two or more firms to cooperate, rather than directly compete, in producing or distributing a product or service.
Corporate consolidation	Reduction in the number of firms serving a market through mergers and acquisitions.
Current account	A term used to refer to a country's balance-of-payments position in international trade. American balance-of-payments (or net trade position) has been measurably enhanced in recent years through the export of commu-

nications products, equipment and services, despite the fact that the United States has not experienced a balanced-trade position since 1980.

Customization — The design of products and services geared to customer-directed specifications.

Demand — Consumption arising in response to pricing alternatives.

Demographics — The statistical characteristics or variations assigned to a group or population.

Deregulation — Reduction, relaxation or modification of regulatory powers as applied to agency of federal, state, or local governments; the American experiment with deregulation across a variety of industries began in 1978 and extends through the year 2000.

Determinants of demand — Those factors responsible for consumption, including income, taste, competitive alternatives, demographic variables, and future expectations of consumers, households and organizations.

Determinants of supply — Those variables responsible for production, including technology, cost, managerial innovation, number of sellers, and future expectations of firms regarding price and profitability.

Digitization — A term of multiple meanings, but typically used to refer to the generation, storage, distribution or assimilation of information in non-material form; distilled to its strategic significance, the term confers economic advantage to those companies able to leverage information technology or communications equipment in response to a customer base.

Disintermediation — The elimination of labor (reduction or elimination of intermediaries) involved in the distribution of goods and services to market, typically precipitated by advances in communications and information technology.

Duopoly — A market served by two producers.

Economic profit — The amount of profit a firm earns beyond the cost of production; this term is often distinguished from "accounting profit," which designates that economic costs can exceed explicit costs in certain instances.

Economies of scale	The decrease in cost per unit of a product or service, instigated by volume buying, division of labor and managerial efficiencies.
Economies of scope	The expertise that a firm can apply in utilizing complementary skills of its personnel to produce two or more goods or services.
Electronic Commerce	Also referred to as "e-commerce," the label applied to networked markets of consumers and producers; electronic commerce is growing exponentially relative to growth rates exhibited in both the domestic and global economies, implying it is supplanting traditional models of commerce and thus reshaping traditional strategic management models.
Externalities	Third-party consequences resulting from a transaction between buyers and sellers.
Five-Force Model	A method of competitive analysis, originated by Michael Porter at Harvard Business School, which integrates five analytical elements to discern optimal strategic plans for a firm.
Fixed costs	A firm's charges which remain constant regardless of increases or decreases in sales activity.
GATT	General Agreement on Tariffs and Trade, devised in 1948 as an international body of government officials whose principal objective has been the promotion of global economic growth through reduced governmental trade restrictions and lower tariffs.
GDP	Gross Domestic Product, a statistical measure gauging annual national wealth through the unduplicated count of all goods and services.
Generation X	That portion of the population born between 1965 and 1982.
Generation Y	That fraction of the population born between 1983 and 2002; sometimes referred to as the "millenial generation."
Geodemographics	The application of demographic variables to specify the relationship between consumer behavior and regional affiliation.
Geographic Information Systems (GIS)	Databases containing geodemographic information (see geodemographics).

Globalization	The increasing interaction of nations, firms and individuals in pursuing common economic objectives.
Government lag	Often applied to bureaucratic regulators who take an inordinate amount of time to supply licenses, respond to corporate requests for information, or who are otherwise slow to respond to business concerns.
Herfindahl-Hirshman Index	A method of economic analysis validating the presence of emerging monopolies through the accumulation of market share; increasingly, the index is decisive in launching anti-trust investigations in those firms or industries which exhibit a strategic predilection toward monopoly.
Horizontal integration	Absorption of competitors for purposes of enlarging market share; also defined as the acquisition of competitors at the same level in the distribution chain.
IBOP	International Business Opportunity Paradigm: a model implying that long-term communications demand will be fueled to a considerable degree by tertiary markets.
Income elasticity	The relationship between income and consumption, as measured in percentage changes.
Indifference curve	Utility derived from a choice between two alternative options for consumption; the curve measures the degree to which a consumer is able to purchase one product relative to the proportionate consumption of another.
Innovation	The capacity to improve or refine an existing product line, as distinguished from "creativity," which implies the power to promulgate birth of new products. Viewed from a strategic perspective, most firms can generally be categorized as either innovative or creative with respect to the products they bring to market.
Intellectual capital	The value of a firm's labor force, as measured in knowledge, intelligence, creativity or capacity to innovate; increasingly, the economic significance of labor is being recognized by strategists as mobility of the labor force defines future strategic advantage in the information technology and communications industries.
Intermediation	The creation or evolution of intermediaries necessary to produce, distribute, market or re-sell goods or services; after experiencing a century of growing economic inter-

	mediation, the American economy has recently exhibited definite signs of disintermediation in the production and distribution of goods and services to market.
International convergence	The interaction and evolving relationship between the computer, broadcasting, and telephony industries on a cross-national or cross-regional basis.
Internetworked enterprise	A firm able to leverage communications technologies so as to establish and strengthen relationships with suppliers, distributers and consumers.
Intrapreneurship	Creation of a new firm within an existing corporate structure, in which entrepreneurs are able to exploit the financial, marketing and managerial resources of the parent to facilitate production and distribution of a new product line.
Just-in-time (JIT) inventory/ purchasing	Refers to advances in information and manufacturing technology which facilitate congruence between production and demand; the telecommunications sectors have become critical in refining JIT management methods for all other industries.
Liberalization	In regulatory applications, the relaxation or existing government sanctions or constraints; in international trade, the relaxation of quotas, tariffs, or domestic content provisions.
Market	May be dichotomized as real versus theoretical; a real market consists of buyers and sellers interacting in the economy at any given time, while the theoretical (or maximum) market is one in which all potential buyers and sellers can simultaneously interact.
Market acceptance	A term applied to a product that has gained foothold in a market, or has attained a minimum of 10% market share. Also used to differentiate strategic management philosophy among high-technology firms: is it more advantageous to be "first-to-market" with new product, or "first-to-market acceptance" with an existing or veteran product line?
Market equilibrium	The quantity of a product or service that people purchase which is simultaneously equivalent to the quantity produced by sellers at any given time.

Market research	The surveying of demographic, economic, and other data in support of strategic plans and sales objectives.
Market structure	A label used to describe market characteristics of a given market sector. Market structures include monopoly, oligopoly, competitive, purely competitive, monopolistic competitive and duopoly forms.
Mass customization	The process of customizing a product line for mass consumption; the capacity of a firm to design or personalize a product or service according to the unique needs of every customer within a large and expanding customer base.
Merger	Integration of two firms which typically relinquish their individual identities in favor of a new title and complementary activities.
Monopolistic competition	A market served by many producers, but distinguished from perfect competition through brand, image, or reputation.
Monopoly	In theory, a market served by a single producer. An effective monopoly can operate holding as little as two-thirds market share.
NAFTA	Trade agreement negotiated and ratified by the United States, Canada, and Mexico during 1994. The agreement contains various provisions affecting telecommunications trade between these nations. NAFTA stands as a prototype for liberalized trade in multinational negotiations, and is likely to instigate eventual complete free trade throughout the western hemisphere.
Opportunity cost	The amount that is foregone by an individual or firm in pursuing one objective as against other available options.
Panel consensus	An organizational technique designed to produce a forecast such that members concur in its underlying reasoning.
Particle marketing	Market research aimed at the customization of products and services; also referred to as *one-to-one marketing,* the logic of this tactic holds that each customer is treated as a market unto him or herself.
Post-industrial society	Also referred to as post-industrialization, this expression has become a synonym for the *information society,*

and identifies: (1) the value of information to knowledge-based economies; (2) nations, including the United States, Western Europe, Canada, and Japan, whose highest percentage of the labor force works in service or knowledge-based industries.

Price elasticity Degree of responsiveness of buyers to changes in product price.

Price implosion A theoretical model projecting a market collapse as profits through "price wars" depress an industry; this disequilibrium results in eventual equilibrium, as failing firms are supplanted by survivors.

Primary demand Initial consumer demand satisfied by the introduction of a product or service.

Productivity Ordinarily referred to as "labor productivity, and measured in a firm's output relative to resources expended in capital and technology; may also be measured in labor output per hour. Since 1997, studies reveal that American labor productivity has been rising relative to the level of organizational expenditures invested in new information technologies and communications equipment.

Reengineering See business process reengineering.

Regulation Government rules imposed on the pricing, quality, distribution, or other characteristics of products and services.

Regulatory overview An expression indicating the breadth of influence of authority held by government regulators over their prescribed duties (also taken to mean regulators' own interpretation of their responsibilities as defined by statutory law).

Reintermediation The procedure by which individuals and firms reinstall themselves into the distribution chain by providing value (in pricing, quality or service) to consumers which did not formerly exist.

Relational marketing A term used to define and evaluate a firm's relationship to its best (most profitable) customers.

Secondary demand Ancillary demand, with resultant intermediate businesses, occurring after a new technology is introduced.

Service explosion An economic model intended to convey the intent of the framers of the 1996 deregulation bill; proliferation

	of providers whose competitive interaction results in price reduction and service improvement for consumers.
Statutory law	That body of law devised and passed by Congress (i.e., statutes are the province of elected officials as opposed to common or other areas of law). Of particular consequence for telecommunications strategists are statutes governing administrative and anti-trust law.
Strategic alliance	Relationship between two or more firms seeking entry to market not otherwise attainable through individual action; this relationship is not based on an equity interest, as is the case in strategic partnerships.
Strategic choice	Identification of the optimal strategy among a variety of available options.
Strategic management	Often used interchangeably with strategic planning, the expression is typically applied to the on-going development, monitoring and management of strategy within the organization.
Strategic partnership	A relationship between two or more firms, the purpose of which is to bring to market a product requiring long-term capital, labor, and R&D commitment.
Strategic variations	Isolation of the most desirable strategic options.
Strategy	Plans which encompass organizational decision-making, allocation of resources, specification of goals and objectives, and the firm's response to changes in the competitive landscape.
Systems analysis	As applied to economics, a tool used to identify all inputs, processes, outputs and feedback so as to predict the future demand for newly introduced products or services.
TCN	Technology Capital Network: an organization dedicated to connecting inventors and entrepreneurs with venture capitalists.
Telecommunications Act of 1996	Sometimes referred to as the Telecommunications Reform Act, this statute, passed by Congress in February of that year, stipulated, among other items, the elimination of sanctions preventing commercial interaction between the cable, telephony and other communications industries. The economic and strategic significance of the law lies in its long-term influence

over other sectors of the economy as the next century unfolds.

Tertiary analysis — A economic tool, based on systems theory, which forecasts emerging market opportunities precipitated by externalities associated with new product introduction.

Undifferentiated marketing — A marketing strategy aimed at establishing a general reputation and brand image in the market without regard to segmentation or niche definition.

Value-added — An economic term applied to the contribution of a firm's components to the production, sales, or service of a product.

Value chain — The collection of all activities that allow a firm to exploit its competitive advantage in the market.

Variable cost — Those costs incurred as the level of output adjusts to changes in and availability of labor, capital, and other resources.

Venture capital — Funds invested by individuals or firms in the creation of business start-ups. Venture capitalists seek annual returns in excess of 25% or more before committing funds, given the level of risk.

Vertical integration — The acquisition of firms whose complementary expertise results in reduced costs of production and greater strategic control.

Volume segmentation — The dichotomization of consumer and business markets on the basis of comparative usage; a telecommunications provider might differentiate, for instance, its consumer base according to minimal, average, or excessive users.

WTO — World Trade Organization: a multilateral negotiating body of particular significance to telecommunications global trade. Since 1997, this group has negotiated a reduction in the number of regulatory sanctions and quotas governing international trade in communications equipment and services.

About the Author

James K. Shaw, Ph.D., is an associate professor in the College of Professional Studies, and director of the applied economics program, at the University of San Francisco. He has engaged in management and economic consulting for both the public and private sectors since 1979. His recent research involves the forecasting of emerging markets for networked commerce.

Mr. Shaw has authored or coauthored three books and numerous articles on business-governement relations and telecomunications policy. He also teaches industry analysis at the Saint Mary's College Graduate School of Business in Moraga, California. Mr. Shaw is presently involved in designing distance-delivered coursework on the digital economy for communications professionals.

Index

Recent Titles in the Artech House Telecommunications Library

Vinton G. Cerf, Senior Series Editor

Access Networks: Technology and V5 Interfacing, Alex Gillespie

Achieving Global Information Networking, Eve L. Varma, Thierry Stephant, et al.

Advanced High-Frequency Radio Communications, Eric E. Johnson, Robert I. Desourdis, Jr., et al.

Asynchronous Transfer Mode Networks: Performance Issues, Second Edition, Raif O. Onvural

ATM Switches, Edwin R. Coover

ATM Switching Systems, Thomas M. Chen and Stephen S. Liu

Broadband Network Analysis and Design, Daniel Minoli

Broadband Networking: ATM, SDH, and SONET, Mike Sexton and Andy Reid

Broadband Telecommunications Technology, Second Edition, Byeong Lee, Minho Kang, and Jonghee Lee

Communication and Computing for Distributed Multimedia Systems, Guojun Lu

Communications Technology Guide for Business, Richard Downey, Seán Boland, and Phillip Walsh

Community Networks: Lessons from Blackburg, Virginia, Second Edition, Andrew M. Cohill and Andrea Kavanaugh, editors

Component-Based Network System Engineering, Mark Norris, Rob Davis, and Alan Pengelly

Computer Telephony Integration, Second Edition, Rob Walters

Desktop Encyclopedia of the Internet, Nathan J. Muller

Digital Modulation Techniques, Fuqin Xiong

Enterprise Networking: Fractional T1 to SONET, Frame Relay to BISDN, Daniel Minoli

Error-Control Block Codes for Communications Engineers, L. H. Charles Lee

FAX: Facsimile Technology and Systems, Third Edition, Kenneth R. McConnell, Dennis Bodson, and Stephen Urban

Guide to ATM Systems and Technology, Mohammad A. Rahman

A Guide to the TCP/IP Protocol Suite, Floyd Wilder

Information Superhighways Revisited: The Economics of Multimedia, Bruce Egan

Internet E-mail: Protocols, Standards, and Implementation, Lawrence Hughes

Introduction to Telecommunications Network Engineering, Tarmo Anttalainen

Introduction to Telephones and Telephone Systems, Third Edition, A. Michael Noll

IP Convergence: The Next Revolution in Telecommunications, Nathan J. Muller

The Law and Regulation of Telecommunications Carriers, Henk Brands and Evan T. Leo

Marketing Telecommunications Services: New Approaches for a Changing Environment, Karen G. Strouse

Multimedia Communications Networks: Technologies and Services, Mallikarjun Tatipamula and Bhumip Khashnabish, editors

Packet Video: Modeling and Signal Processing, Naohisa Ohta

Performance Evaluation of Communication Networks, Gary N. Higginbottom

Practical Guide for Implementing Secure Intranets and Extranets, Kaustubh M. Phaltankar

Practical Multiservice LANs: ATM and RF Broadband, Ernest O. Tunmann

Protocol Management in Computer Networking, Philippe Byrnes

Pulse Code Modulation Systems Design, William N. Waggener

Service Level Management for Enterprise Networks, Lundy Lewis

SNMP-Based ATM Network Management, Heng Pan

Strategic Management in Telecommunications, James K. Shaw

Successful Business Strategies Using Telecommunications Services, Martin F. Bartholomew

Telecommunications Department Management, Robert A. Gable

Telecommunications Deregulation, James Shaw

Understanding Modern Telecommunications and the Information Superhighway, John G. Nellist and Elliott M. Gilbert

Understanding Networking Technology: Concepts, Terms, and Trends, Second Edition, Mark Norris

Videoconferencing and Videotelephony: Technology and Standards, Second Edition, Richard Schaphorst

Visual Telephony, Edward A. Daly and Kathleen J. Hansell

Wide-Area Data Network Performance Engineering, Robert G. Cole and Ravi Ramaswamy

Winning Telco Customers Using Marketing Databases, Rob Mattison

World-Class Telecommunications Service Development, Ellen P. Ward

For further information on these and other Artech House titles, including previously considered out-of-print books now available through our In-Print-Forever® (IPF®) program, contact:

Artech House
685 Canton Street
Norwood, MA 02062
Phone: 781-769-9750
Fax: 781-769-6334
e-mail: artech@artechhouse.com

Artech House
46 Gillingham Street
London SW1V 1AH UK
Phone: +44 (0)20 7596-8750
Fax: +44 (0)20 7630-0166
e-mail: artech-uk@artechhouse.com

Find us on the World Wide Web at:
www.artechhouse.com